Smart Textiles

Scrivener Publishing
100 Cummings Center, Suite 541J
Beverly, MA 01915-6106

Publishers at Scrivener
Martin Scrivener (martin@scrivenerpublishing.com)
Phillip Carmical (pcarmical@scrivenerpublishing.com)

Smart Textiles

Wearable Nanotechnology

Edited by

Nazire D. Yilmaz

WILEY

This edition first published 2018 by John Wiley & Sons, Inc., 111 River Street, Hoboken, NJ 07030, USA
and Scrivener Publishing LLC, 100 Cummings Center, Suite 541J, Beverly, MA 01915, USA

For more information about Scrivener publications please visit www.scrivenerpublishing.com.

Wiley Global Headquarters
111 River Street, Hoboken, NJ 07030, USA

For details of our global editorial offices, customer services, and more information about Wiley products visit us at www.wiley.com.

Library of Congress Cataloging-in-Publication Data

ISBN 978-1-119-46022-0

Cover image: Pixabay.Com
Cover design by Russell Richardson

Set in size of 11pt and Minion Pro by Manila Typesetting Company, Makati, Philippines

10 9 8 7 6 5 4 3 2 1

Contents

Preface

Originally, the need for textiles and clothing was related to protecting the human body from exposure to the elements of nature. A more comprehensive definition of conventional textiles also includes home textiles utilized in furnishings and the ones that have found use in bedrooms and bathrooms. Following these basic needs, aesthetics have become one of the main drivers of our selection of clothing and textiles. Recently, more functionality has started to be required, so functional/technical textiles, which can serve more sophisticated needs, have emerged. The last generation of textiles, smart textiles, remain one step ahead of the others by sensing and reacting to environmental stimuli.

Nanotechnology has carried the level of smart textiles one step further. Textile materials receive smart functionalities without deteriorating their characteristics via application of nanosized components. Consequently, functions conventionally presented by nonflexible bulk electronic products are achieved by "clothes."

Smart wearables should be capable of recognizing the state of the wearer and/or his/her surroundings and responding to them. Based on the received stimulus, the smart system processes the input and consequently adjusts its state/functionality or present predetermined properties. Smart textiles should also cater to requirements concerning wearability. Through the incorporation of nanotechnology, the clothing itself becomes the sensor, while maintaining a reasonable cost, durability, fashionability, and comfort.

This book provides a comprehensive presentation of recent advancements in the area of smart nanotextiles, with an emphasis on the specific importance of materials and their production processes. Different materials, production routes, performance characteristics, application areas, and functionalization mechanisms are referred to. Not only are mainstream materials, processes, and functionalization mechanisms covered, but also alternatives that do not enjoy a wide state-of-the-art use but have the potential to bring smart nanotextile applications one step forward.

The basics of smart nanotextiles are covered in the first chapter. Nanofibers, nanosols, responsive polymers, nanowires, nanogenerators, and nanocomposites, which are smart textile components, are investigated in Chapters 2 through 7, respectively. Nanocoating is investigated in Chapter 8, and nanofiber production procedures are examined in Chapter 9. Characterization techniques, which have uppermost importance in ensuring proper functioning of the advanced features of smart nanotextiles, are covered in the last chapter.

Nazire Yilmaz
Denizli, Turkey
September 2018

Acknowledgments

I want to thank my mother, Henrietta, and father, Ulku, for giving me, their baby girl, their never-ending support and for turning their house into a home office for me.

My gratitude goes to my beloved husband, who contributed to this book with his love and prayers. Thanks also go to my kids for their patience during the preparation stage of this book.

I want to acknowledge the authorities who have made this book possible by shifting the burden of giving lectures away from me. How can I ever forget what they have done for me?

Finally, special thanks go to Martin Scrivener for his support and patience.

Section 1
INTRODUCTION

1

Introduction to Smart Nanotextiles

Nazire Deniz Yilmaz

Textile Technologist Consultant, Denizli, Turkey

Abstract
This chapter provides a comprehensive presentation of recent advancements in the area of smart nanotextiles giving specific importance to materials and their production processes. Different materials, production routes, performance characteristics, application areas, and functionalization mechanisms are referred to. Not only the mainstream materials, processes, and functionalization mechanisms, but also alternatives that do not enjoy wide state-of-the-art use, but have the potential to bring the smart nanotextile applications one step forward, have been covered. Basics of smart nanotextiles, introduction to smart nanotextile components such as nanofibers, nanosols, responsive polymers, nanowires, nanocomposites, nanogenerators, as well as fundamentals of production procedures have been explained. In addition to materials and production technologies, characterization techniques, which have uppermost importance in ensuring proper functioning of the advanced features of smart nanotextiles, have also been investigated.

Keywords: Smart textiles, nanofibers, nanosols, nanowires, responsive polymers, nanocomposites, nanogenerators, characterization, fiber production, nanocoating

1.1 Introduction

Originally, textiles/clothing relates to catering the needs for protecting the human body from cold, heat, and sun. A more comprehensive definition of conventional textiles also include home textiles utilized in furnishing and the ones that find use in the bedroom and the bathroom [1, 2]. Following these basic needs, aesthetics have become one of the main drivers for people to use clothing and textiles [3]. Recently, more functionality has started

Email: naziredyilmaz@gmail.com

Nazire D. Yilmaz (ed.) Smart Textiles, (3–38) © 2019 Scrivener Publishing LLC

to be required, so functional textiles/technical textiles, which can cater more sophisticated needs, have emerged. The last generation of textiles, smart textiles, is capable of one step ahead: sensing and reacting to environmental stimuli [2, 4, 5].

Smart textiles can be also named as "intelligent," "stimuli-sensitive," or "environmentally responsive" [6]. Smart textiles have been described as "fibers and filaments, yarns together with woven, knitted or non-woven structures, which can interact with the environment/user" [7, p. 11958]. Smart textiles have broadened the functionality and, consequently, application areas of conventional textiles [7], as they show promise for use in various applications including biomedicine, protection and safety, defense, aerospace, energy storage and harvesting, fashion, sports, recreation, and wireless communication [4, 8–10].

Smart textile components perform various functions such as sensing, data processing, communicating, accumulating energy, and actuating as shown in Figure 1.1 [11]. In these fields, textile structures present some advantages such as conformability to human body at rest and in motion, comfort in close contact to skin, and suitability as substrates for smart components [8].

"Smartness" refers to the ability to sense and react to external stimuli [6]. The stimulus of interest can be electrical, mechanical, chemical, thermal,

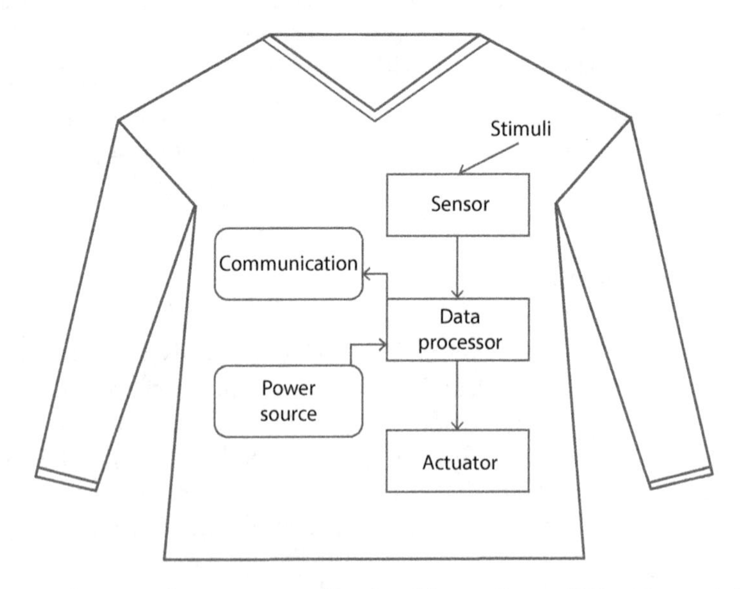

Figure 1.1. Smart textile components. (Reprinted from reference [11], with permission of Elsevier.)

magnetic, or light [4, 12]. Smart systems offer the capability of sensing and responding to environmental stimuli, preferably in a "reversible" manner, that is, they return to their original state once the stimulus is "off" [6].

Smart textiles can act in many ways for vast purposes including releasing medication in a predetermined way, monitoring health variables, following pregnancy parameters [13], aiding physical rehabilitation [14], regulating body temperature, promoting wound healing [15], facilitating tissue engineering applications [16], photocatalytic stain removing [17], preventing flame formation [18], absorbing microwaves [19], interfering with electromagnetic radiation [20], wireless communicating between persons, between person and device, and between devices (as in the case of IoT), and harvesting and storing energy [10]. In an everyday example, the smart textiles used for fashion, kids' toys, or entertainment can change color, illuminate, and display images and even animations [4, 10].

Smart textiles have attracted international research interest as reflected in the programs of the international funding bodies, for example, "Wear Sustain," a project funded by the European Commission. The Wear Sustain Project is directed by seven organizations, both public and private entities, across Europe, including universities, research centers, and short- and middle-scale enterprises (SMEs). This project has launched 2.4 million euros for funding teams to develop prototypes of next-generation smart textiles [21]. US-based National Science Foundation grants $218,000 to a career project titled Internet of Wearable E-Textiles for Telemedicine [22]. NSF of the USA has invested more than $30 million on projects studying smart wearables. The projects include belly bands tracking pregnancy variables, wearables alerting baby sleep apnea, and sutures that collect diagnostic data in real time wirelessly. NSF also supports the Nanosystems Engineering Research Center (NERC) for Advanced Systems for Integrated Sensors and Technologies (ASSIST) at North Carolina State University working on nanotechnological wearable sensors [23].

Different components are used for imparting smartness into textiles. These components include conductive fibers, conductive polymers, conductive inks/dyes, metallic alloys, optical fibers, environment-responsive hydrogels, phase change materials, and shape memory materials. These components are utilized in forming sensors as well as electrical conductors, and connection and data transmission elements [4]. Conductive materials added to fibers/yarns/fabrics include conductive polymers, carbon nanotubes, carbon nanofibers, or metallic nanoparticles [4, 24–26].

"Smartness" can be incorporated into textiles at different production/treatment steps including spinning weaving [27], knitting [28], braiding [29], nonwoven production [30], sewing [31], embroidering [3], coating/laminating [32], and printing [33] as shown in Figure 1.2.

Conventionally, conductive fibers and yarns are produced through adding conductive materials to fibers, or via incorporation of metallic wires/fibers such as stainless steel or other metal alloys [4, 25]. Another way to produce smart textiles is through incorporation of conductive yarns in fabrics, for example, by weaving. Drawbacks related with this method are the complexity, non-uniformity, as well as difficulty in maintaining comfortable textile properties [7].

Nanotechnology has carried the level of smart textiles one step further. Via application of nanosized components, textile materials receive smart functionalities without deteriorating textile characteristics [10, 34]. Consequently, functions conventionally presented by nonflexible rigid bulk electronic products are achieved by "clothes" [2].

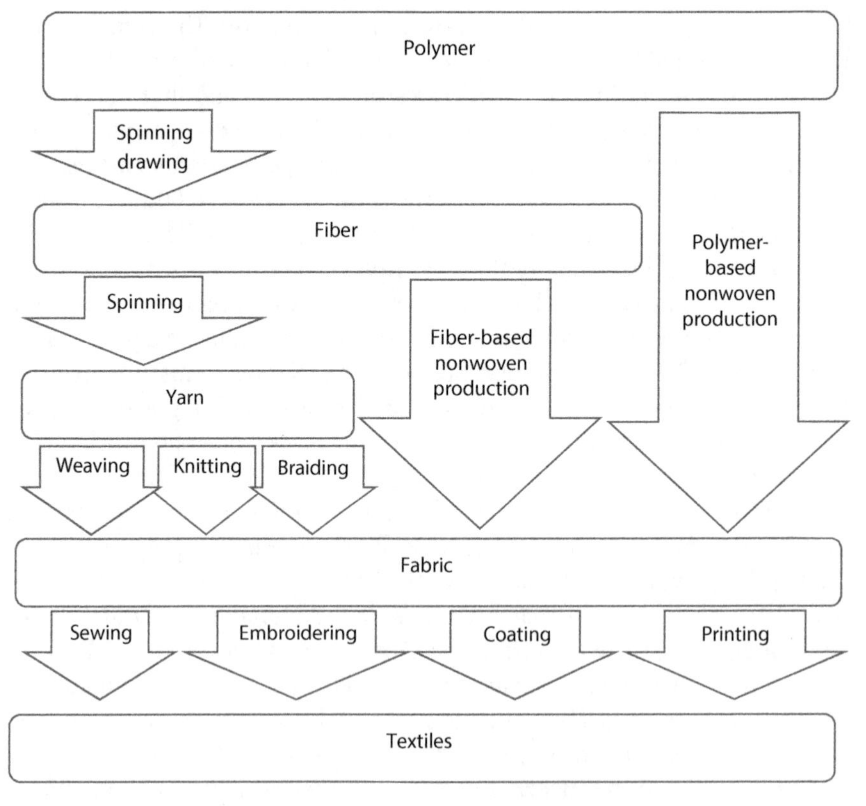

Figure 1.2. Production steps of textiles. (The image has been prepared by the author.)

Smart wearables should present capability of recognizing the state of the wearer and/or his/her surrounding. Based on the received stimulus, the smart system processes the input and consequently adjusts its state/functionality or present predetermined properties. Smart textiles should also cater needs regarding wearability [7]. Via incorporation of nanotechnology, the clothing itself becomes the sensor, while maintaining a reasonable cost, durability, fashionability, and comfort [35].

Based on their "smartness" level, smart textiles may be investigated under three categories [33]:

- Passive smart textiles
- Active smart textiles
- Very active smart textiles.

The first group can only detect environmental stimuli (sensor), whereas the second group senses and reacts to environmental stimuli (sensor plus actuator). On the other hand, the third group senses and reacts to environmental stimuli, and additionally adapts themselves based on the circumstances (sensor, actuator, and controlling unit) [2, 4].

1.1.1 Application Areas of Smart Nanotextiles

Potential application areas of smart textiles are innumerable. In terms of personal use, they can act for making us feel comfortable, warn and protect us against dangers, monitor biometric data, treat diseases and injuries, and improve athletic performance via use of sensor-embedded clothing. Furthermore, they can be used by military and other security staff for communication. Fashion and decoration are also irreplaceable applications for clothing, not excluding smart wearables. Related examples include color-changing, lighting-up, picture-video-displaying wearables [4, 33].

As textiles are in close contact with human body over a large surface area, sensors can be placed at different locations, which presents advantage for biomedical applications. This fact provides greater flexibility and closer self- and remote monitoring of health variables. Smart textile components responsive to pressure/strain can be used to measure heart rate, blood pressure, respiration, and other body motions. Accordingly, piezo-resistive fibers can be utilized as pressure/stress sensors [7, 13]. Smart textiles also show promise for sensing body temperature [2], movements of joints [14], blood pressure, cardiac variables [36], respiration [37], presence/concentration of saline, oxygen, and contamination or water. Thermocouples can be utilized in measuring temperature, whereas carbon electrodes are used for detecting concentrations of different biological fluids [38].

As expected, active smart functionality needs energy to act, which in turn necessitates generation or storage of power. Power generation may be attained via use of piezoelectric [5], photovoltaic [39], or triboelectric components [40], which can harvest energy from motion, light, or static electricity, respectively [10].

1.1.2 Incorporating Smartness into Textiles

Smart textile components include conductive polymers, conductive ink, conductive rubber, optical fibers, phase changing materials, thermochromic dyes, shape-memory substances, miniature electrical circuits, and so on. In terms of textile functionality, organic polymers pose advantages compared to stiff inorganic crystals. The former materials exhibit low weight, flexibility, resilience, cost efficiency, and easy processibility [33, 34].

As mentioned, these "smart" components can be included into the textile structure at different stages. At the fiber spinning stage, electrically conductive components may be added to the spinning dope. Smart components can be integrated into textiles in the course of fabric formation such as weaving or knitting. After fabric formation, the finishing stage provides practical solutions for adding active components on the fabric such as nanocoating procedures [3, 4, 33, 41].

Smart textiles present the capability of sensing, communicating, and interacting via use of sensors, connectors, and devices produced from environmental-responsive components [4]. Sensors may be considered as members of a nerve system that can detect signals. Based on the environmental stimulus, actuators react autonomously or as directed by a central control unit [7]. Conductive materials that exhibit property change based on environmental stimuli such as stretch, pressure, light, pH value, and so on can be used as sensors [7].

Smart activity can be achieved by incorporation of human interface components, power generation or capture, radio frequency (RF) functionality, or assisting techniques. By using these components, innumerable combinations can be obtained conventionally by introducing cables, electronics, and connectors. However, wearers prefer comfortable textiles rather than clothes resembling "Robocop" costumes. To achieve this, the smart functionality should be integrated into the textiles [3, 33]. This can be made possible by using nanotechnology.

1.1.3 Properties of Smart Nanotextiles

The components of smart nanotextiles should provide some characteristics including mechanical strength, conductivity, flexibility, washability, and biocompatibility. These features, indeed, are not easy to achieve concurrently. Textile properties, such as drape, stretch, resilience, and hand, are especially important once the final use is taken into consideration. In order to achieve these characteristics, the structures should not be coarse and the resultant fabric should not be heavy (not exceeding 300 gsm). Of course, these requirements cannot be met via use of conventional electrical appliances, metal wires, and so on. The challenge is to maintain connectivity and integrity through the interconnections among the components and devices during deformation throughout the intended use. An approach to solve this problem is to use sinus-wave or horseshoe-shaped designs of the conductive components to minimize the effect of deforming in the flexible textile substrates. Another potential solution is to encapsulate the conductive component in a stretchable polymeric substrate [7]. Nanotechnology presents advantage in terms of mechanical flexibility. Thinness provides flexibility based on the nanosizes of the elements. Accordingly, a smart textile structure that preserves the extensibility of a conventional textile fabric can be achieved. Durability against washing and aging is also very important. This can be attained via effective bonding of smart components with the textile substrate through nanocoating procedures [41].

Besides, thinness and flexibility, transparency is another plus for smart components to be used in wearables, due to minimized interference with the designed appearance. As expected, at a very high thinness level, even opaque materials, such as metals, exhibit transparent optical property. Ultrathinness results in decreased optical absorption and increased light transmission [42]. Indeed, this level of thinness can be obtained from nanoscale materials via nanotechnological applications.

1.1.4 Nanotechnology

Nanotechnology, which is an emerging interdisciplinary field, is considered to provide various impacts in different science and technology areas including, but not limited to, electronics, biomedicine, materials science, and aerospace [43]. Nanotechnology shows promise for use in higher and higher number of applications in different arenas such as textiles and clothing to impart enhanced properties and performance [32].

In the last two decades, we have witnessed that nanotechnology has found use in textiles for improving and/or imparting properties including smart functionalization [32]. Nanotechnology enables certain functions including antibacterial, antistatic, self-cleaning, UV-protective, oil and water repellency, stain proof, improved moisture regain, and comfort performance in textiles while maintaining breathability, durability, and the hand [43]. Nanotechnology applications on textiles have succeeded in attracting great interest by both research and commercial communities [32]. The studies related to nanotechnological practices, that is, application of nanomaterials, on textiles cover *in situ* synthesis, cross-linking, and immobilization on textile substrates [32].

1.1.5 Nanomaterials

Nanomaterials refer to materials at least one dimension of which is in the nanometer order, that is, generally lower than 100 nm [32]. These materials show promise for use in functional and high-performance textiles based on their high specific characteristics stemming from great surface area-to-volume ratios [43].

Although there is a perception that the nanoscale materials are novel materials, they have been used since the early decades of the 20th century. An example to this is carbon black, a nanomaterial that has been used in automobile tires since the 1930s. Indeed, the capabilities of nanosized materials have increased drastically since then [44].

The use of nanoscale materials in the textiles field is increasing rapidly, and they have found use in various applications catering industrial, apparel, and technical needs. The main aims of incorporating nanomaterials in textiles include imparting functionalities such as electrical conductivity, flame retardancy, antibacterial, superhydrophobic, superhydrophilic, self-cleaning, and electromagnetic shielding [34, 45].

Most of the nanomaterial applications necessitate definite particle dimensions with narrow variation. By controlling production parameters, different characteristics of nanomaterials can be manipulated. These characteristics include particle dimensions, chemical composition, crystallinity, and geometrical shape. And the production parameters are pH, temperature, chemical concentration, used chemical types, etc. [44]. Various shapes are observed in nanoparticles such as nanorods, nanospheres, nanowires, nanocubes, nanostars, and nanoprisms. Via manipulation of synthesis variables, it is possible to attain different nanoparticle shapes [34].

A critical matter related to use of nanostructures is difficulty in dispersion as nanoparticles tend to agglomerate due to van der Waals and electrostatic double-layer attractions. In order to form stable dispersions, some precautions should be taken such as using dispersing agents including surfactants and functionalization of nanostructures using organic compounds and monomers [34].

Another major problem related to nanomaterials is their durability on textile substrates. Due to lack of surface functional sites, nanomaterials do not show affinity to textile fibers. In order to address this problem, surface functionalization via physical or chemical techniques has been suggested. Another solution is embedding nanoparticles in polymer matrices on textiles substrates [34].

One of the novel abilities of nanoscale materials is "smartness," which shows promise for use in smart textile applications. Smart textiles include nanotechnological components such as nanofibers, nanowires, nanogenerators, nanocomposites, and nanostructured polymers. Smart nanotextiles are investigated for use in biomedical, aerospace, and defense applications, among others [43]. Development of smart nanotextiles requires knowledge on nanotechnological components, their properties, production techniques, and nanotechnical characterization methods.

This chapter provides a comprehensive presentation of recent advancements in the area of smart nanotextiles giving specific importance to materials and their production processes. Different materials, production routes, performance characteristics, application areas, and functionalization mechanisms are referred to. Not only the mainstream materials, processes, and functionalization mechanisms but also alternatives that do not enjoy wide state-of-the-art use, but have the potential to bring the smart nanotextile applications one step forward, have been covered. Basics of smart nanotextiles, introduction to smart nanotextile components such as nanofibers, nanosols, responsive polymers, nanowires, nanocomposites, nanogenerators, as well as fundamentals of production procedures have been explained. In addition to materials and production technologies, characterization techniques, which have uppermost importance in ensuring proper functioning of the advanced features of smart nanotextiles, have also been investigated.

1.2 Nanofibers

Among various forms that nanomaterials can take such as nanorods, nanospheres, and so on, the fiber form comes to the forefront due to its superior characteristics. The advantageous properties of this material form include flexibility, high specific surface area, and superior directional performance. These merits allow many uses from conventional clothing to reinforcement applications in aerospace vehicles. Nanofibers refer to solid state linear nanomaterials, which are flexible and have aspect ratios exceeding 1000:1. Nanomaterials are characterized by their dimensions at least one of which should be equal to or less than 100 nm. A million times increase in flexibility can be achieved via reduction of the fiber diameter from 10 μm to 10 nm, which also leads to increases in specific surface area, and in turn surface reactivity [46].

Numerous functionalizations can be attained by use of nanofibers produced from various polymers including polypyrrole, polyaniline [7, 47], polyacetylene [4], polyvinylidene fluoride, poly N-isopropylacrylamide (PNIPAAm), polyethylene glycol, and so on, and incorporation of different functional components such as carbon nanotube, graphene, azobenzene, and montmorillonite nanoclay [10, 34, 48, 49]. More of these polymers and functional components can be found in the following chapter [46]. Via use of these nanofibers, it is possible to achieve smart functionalities as follows.

1.2.1 Moisture Management

Moisture behavior of materials is determined not only by the chemical but also the topographical properties [50]. Nanofibers can be utilized for smart moisture management functions of textiles such as superhydrophobicity and switchable hydrophilicity–hydrophobicity. Superhydrophobicity can be obtained via mimicking the microstructure of various plant leaves, known as the "Lotus effect." This function is provided by two characteristics: a hybrid rough microstructure and a hydrophobic surface [51]. Nanofibrous membranes of polyurethane, polystyrene, and polyvinylidene fluoride have been studied for producing superhydrophobic structures. The nanofibrous structure emphasizes both hydrophilic and hydrophobic characteristics. The rough microstructure of superhydrophobic materials can be improved by incorporating beads, rods, microgrooves, or pores/dents in the nanofibrous structures during electrospinning procedures. By varying electrospinning, dope parameters fibers in bead-on-string form can be obtained [46, 52, 53].

Nanoscale bumps and dents can be formed by incorporating nanoparticles onto nanofibers and sonicating these nanoparticles away. In this way, superhydrophobic effect can be provided. By introducing fluorinated polymers with low surface energy on the nanofibrous membranes, hydrophobicity can be further improved. A study showed that hierarchical roughness positively affected amphiphobicity (hydrophobic and oleophobic at the same time). Another material popularly used for hydrophobicity is the hydrophobic SiO_2 nanoparticle, which allows enhanced surface roughness [9, 50].

Switchable moisture behavior of materials stimuli can also be provided by use of nanofibers. Here, switchable moisture behavior refers to reversible change of the material characteristic from hydrophilic to hydrophobic based on environmental stimuli such as pH, UV rays, and temperature [54]. In a related study, by use of electrospun poly(N-isopropylacrylamide)/polystyrene nanofibrous membranes, the wettability of which shows change from hydrophilic at room temperature to almost superhydrophobic at 65° [55]. In another example, nanofibers showing dual-responsive wettability were developed by Zhu *et al.* [56]. They produced electrospun core-shell polyaniline–polyacrylonitrile nanofibers presenting superhydrophobic property. The wettability of the nanofiber can be changed from superhydrophobic to superhydrophilic via change in pH or redox conditions.

1.2.2 Thermoregulation

The human body is resembling a heat generator that emits heat energy throughout the time. In order to maintain vital body functions, a relatively narrow temperature range is necessary: 36.8 ± 0.8 °C. Protection of the body from heat loss or from overheating is carried out via clothing [57]. As known, heat transfer takes place in three forms: conduction, convection, and radiation. Conduction is the form where heat transfer takes place in solid materials. Here, heat transfer is negatively affected by the air fraction of a specific material, that is, heat insulation. Electrospun nanofibrous materials possess high porosity; in other words, their air fraction is high. Thus, high thermal insulation is expected from them. However, there are other factors that hinder this property: very low thickness, low resistance to compression, and other mechanical shortcomings [46, 58].

Hence, direct application of nanofibrous membranes for thermal insulation purposes is not common. Rather than this, nanofibers have been used as carriers of phase change materials. Phase change materials have the ability to store heat energy at high temperatures and release that energy at low temperature via phase change. Thus, the temperature of the phase change materials does not show noteworthy change. By microencapsulation of phase change materials in nanofibrous networks, loss of these materials is prevented and prolonged service life is maintained [41, 46, 59].

1.2.3 Personal Protection

Utilization of nanofibers for personal protection applies to different fields including protection against fire, elevated temperatures, bacteria, liquid, gas, mechanical, and electromagnetic attacks, among others. Besides different effects, fire protection stands as a major field. Certain polymers are used for production of flame-retardant clothing including Nomex® and polybenzimidazole to be used by racers' costumes. Their use by the general community has been restricted due to their high cost. A more cost-effective alternative is introducing flame-retardant agents in nanofibrous networks to obtain composite structures [46]. Such an example can be given as flame-retardant polyamide 6 nanocomposite fibers produced by Wu *et al.* [49] via addition of intumescent non-halogenated flame-retardant (FR) agents and montmorillonite clay platelets.

In terms of protection against electromagnetic effects, there are two means for realization of electromagnetic interference shielding: reflection and absorption. The reflection effect necessitates inclusion of an electrical conducting component, whereas the latter corresponds to use of a magnetic one. In order to enhance the electromagnetic shielding effect, it is common to use conductive and magnetic components concurrently. As carbon is a conductive fiber, it has been studied with different magnetic substances. Zhu *et al.* [45] produced electrospun fibers from a polyvinyl alcohol–ferrous acetate solution. They calcinated the produced fibers at high temperature to obtain iron oxide (Fe_3O_4)–carbon nanofiber.

1.2.4 Biomedicine

Nanofibers offer numerous advantages for use in the biomedical area. Nanofiber structure presents an orientation path that mimics biosystems [46, 60]. In their natural environment, cells live in nano- and/or micro-detailed surroundings. So, nanofibers, which present dimensions lower than the cells, provide a suitable man-made medium to attach to and to proliferate on. In a series of studies, it was reported that the functions of cells, including cell adhesion, proliferation, alignment, and migration, are affected by the nanoscale surface topography [46]. More information related to nanofibers for smart textiles can be found in Chapter 2.

1.3 Nanosols

Nanosols are coating agents used for functionalizing textiles. Nanosols are colloidal solutions of metal oxide particles in nanoscale dimensions in water or organic solvents [61]. Nanosols include inorganic nanoparticles prepared via the sol–gel method [9, 50].

Nanosols present metastable property due to their high surface-to-volume ratio. Hence, 3D network structures can be formed of nanosols by aggregation of nanoparticles and successive solvent evaporation in course of coating [61]. Nanosols are formed through hydrolysis of a precursor material. The precursors can be inorganic metal salts or metal organic compounds such as acetylacetonate or metal alkoxides. Metal or semimetal alkoxides are commonly utilized, which turn into hydroxides via hydrolysis processes. At high concentrations, hydroxides are generally unstable; thus, they may be subjected to successive condensation reactions resulting in nanoscale particle formation. Some examples of nanosol precursors can be given as $Al(OC_4H_9)_3$, $Si(OC_2H_5)_4$, tetraethoxysilane (TEOS), and titanium (IV)isopropoxide $Ti(OC_3H_7)_4$ [50, 61].

Similar to other nanoscale materials, nanosols also enjoy great effectiveness based on high specific area in terms of their dimensions generally below 100 nm [61]. Accordingly, coatings prepared with nanosols exhibit thicknesses of several hundreds of nanometers [62]. Using nanosols, surface or bulk properties of different substrates such as textile materials can be altered [61]. Via application of nanosols, various functions can be imparted to textiles. These functions can be divided into four categories according to Mahltig: optical (coloration, UV, and X-ray protection), chemical (inflammability, self-cleaning), biological (antimicrobial, biocompatibility), and surface-functional (hydrophobic, hydrophilic, abrasion resistant) functions [9].

Nanosol coatings are usually prepared via the sol–gel method as mentioned before. Various solvents are used for nanosol preparation including water, isopropyl, or ethanol. It is possible to modify nanosols through simple methods resulting in a variety of functionalities. On the other hand, a shortcoming related to nanosols is limited stability caused by water if selected as the solvent [50, 62]. If proper post thermal treatment is not carried out, the applied nanosol coating will present an amorphous structure referred to as "xerogel." Nevertheless, water is generally chosen in order to avoid undesirable aspects of organic solvents related to flammability, safety, and cost effectiveness [9, 63].

An important aspect related with nanosol applications on textiles is the adhesion between nanosol coating and the textile substrate. This is especially problematic with synthetic-fiber textiles. In order to increase adhesion, various techniques are utilized including use of cross-linkers, applying thermal, plasma, and corona treatments to activate the mentioned surfaces [61].

1.3.1 Applications of Nanosols

Nanosols, including silica and titanium dioxide sols, offer bioactive, protective, and hydrophobic functions for textile applications via physical or chemical modification methods of nanosols. Another good aspect of nanosols is that the inorganic nanosols are inflammable materials. So they tend to have positive effect in fire protection based on barrier effect [61, 63].

Nanosols have found use in UV protection as well. ZnO or TiO_2 containing nanosols were reported to have good UV absorption capability. Furthermore, in case where zinc oxide or titania particle dimensions are smaller than 50 nm, a transparent and colorless coating effect can be achieved [62].

Hydrophobicity is a requirement for certain applications such as outdoor clothing or self-cleaning textiles. Hydrophobicity is improved by surface roughness where air pockets can be trapped. Low surface energy and roughness result in superhydrophobicity as explained above [50]. Superhydrophobic effect can be achieved by using nanosols of metal oxides to increase roughness of fiber surfaces [17]. In a relevant study, superhydrophobicity was imparted to cotton fabrics by applying tetraethoxysilane (TEOS)-based nanosols for increasing roughness, and 1H,1H,2H,2H–fluorooctyltriethoxysilane modification for lowering surface energy via padding method where a post thermal treatment was applied to increase durability against washing [50]. Fluorinated compounds are used for water-repellent, oil-repellent, and thus, self-cleaning effects [63]. Antimicrobial effect is another function commonly obtained from TiO_2, SiO_2, and ZnO nanosol coatings. This effect is also positively influenced by hydrophobicity. Versatile TiO_2 nanosols are also utilized for self-cleaning applications and to achieve antistatic property. A practical and cost-efficient means to obtain photocatalytic self-cleaning stain removing effect is called "ceramization" where nanosols like TiO_2 are applied via a dip-pad-dry-cure method [17].

Conventional methods related to metal oxide nanoparticle preparation do not offer feasible means due to the entailed energy- and time-consuming processes [61]. Thus, development of more practical means of nanosol applications on smart textiles will benefit attracting wider embracement in the commercial range. On the other hand, more research on increasing durability of nanosols on textiles is expected in the future. More detailed information related to nanosols can be found in the third chapter [9].

1.4 Responsive Polymers

Smart textiles are generally considered as textiles with miniaturized electronic devices integrated within [4]. This definition is not false, but it is incomplete. Apart from electrically conductive materials, some polymers show responses triggered by changes in the environmental conditions including pressure, temperature, light, magnetic field, and so on [12, 64–67]. These polymers are defined as environmentally sensitive, stimuli-responsive, intelligent, or smart polymers [6]. Even though the materials are nanostructured, the responses can be observed at the macroscopic level and can be reversible [65, 66].

The response-triggering stimuli can be generally categorized as physical (temperature, pressure, electrical field, magnetic field, and ultrasound), chemical (pH, solvent composition, ion type, and ionic strength), and biological (glucose, enzyme, and antibody) [68, 69]. Nevertheless, biological stimuli can be also considered as a sub-group of chemical stimuli. Physical stimuli lead to changes in molecular interactions to a certain extent. The advantage of physical stimulus-triggered systems is the possibility of local and remote activation. Nevertheless, the systems in the human body are very closely tied to (bio)chemical processes. This makes the systems responsive to (bio)chemical stimuli including pH, ions, and biomolecules very important. There are also systems responsive to multiple stimuli. These systems are referred to as dual- or multi-responsive polymer systems [24].

The response mechanisms vary from polymer to polymer. These include neutralization of charged groups upon pH change or addition of oppositely charged chemical species, or change in the hydrogen bond strength [65]. Switchable solubility is an important mechanism of smart functions. For most of the smart polymers, a critical point can be mentioned where the response, that is, the change in polymer's property, is observed [6, 66].

Among different environmentally responsive polymers, poly(*N*-isopropylacrylamide) (PNIPAAm) and its derivatives attract extensive interest. PNIPAAm solutions show reversible thermo-responsive solubility behavior. They present soluble characteristic below a specific temperature, called as lower critical solution temperature (LCST), and insoluble characteristic over this point [24]. Some monomers used in preparation of environmentally responsive polymers include hydroxyethyl methacrylate (HEMA), vinyl acetate, acrylic acid, and ethylene glycol [6].

By incorporation of nonsoluble components, solubility behavior of polymers can be manipulated. Nanoscale dimensions show promise for environmentally responsive polymers in terms of response rate in comparison to bulk materials based on higher surface area. Thus, polymer systems with different smart functionality, response mechanism, and rate can be attained through sound design of components and architecture. Systems responsive to different stimuli can be fabricated from different precursor materials [12].

When smart polymers are the subject, one thinks that these are highly engineered, advanced materials produced via sophisticated synthesis routes starting from chemical precursor species. However, there are also environmentally responsive polymers based on natural materials beside synthetic ones, as well as hybrid ones that contain synthetic and natural components together. The natural components that find use in environmentally responsive polymers include proteins (collagen, gelatin) and polysaccharides (chitosan and alginate) [6].

There are many ways in production of environmentally responsive polymers including physical cross-linking methods (heating/cooling, ionic interaction, complex coacervation, H-bonding, maturation, cryogelation) [70], chemical cross-linking methods (chemical grafting, radiation grafting) [6], and advanced techniques such as sliding cross-linking, double networks, and self-assembling from genetically engineered block copolymers [64]. In the fourth chapter [24], more insight into responsive polymers is given.

1.5 Nanowires

Nanowires present high aspect ratios with diameters at the nanoscale (5–100 nm), whereas the lengths are between 100 nm and several microns [71]. In comparison to nanoparticles in other shapes like nanospheres and nanorods, nanowires exhibit some advantages based on their high aspect ratios. These advantages include effective electrical and thermal conductivity as well as mechanical flexibility [8]. On the other hand, nanoscale diameters of nanowires offer advantages in obtaining transparency from the final product [71].

Nanowires, similar to other nanomaterials, offer some advantages over bulk counterparts. As an example, Si nanowires present high signal-to-noise ratios and ultra-high sensitivities compared to conventional materials that allow use for detecting single virus particles, analyte presence, and DNA sequencing [72]. The shape of nanowires provides a direct pathway for electrical transmission lowering resistance. This shape allows orientation easier compared to other shapes. Metal nanowires may be superior to carbon nanotubes, which show high resistance at junctures [8].

Nanowires are made of metals, metal oxides, conductive polymers, or semiconductor materials. Different metals can be used in nanowire preparation including silver, gold, and copper, among others. Conducting polymers can be named as polyaniline, polythiophene, poly(p-phenylenevinylene), and polypyrrole. Semiconducting materials include oxides (ZnO, CuO, SnO_2), sulfides (Cu_7S_4, CoS_2), and others (Si, ZnSe, CdTe,...). Semiconductive silicon nanowires have been used for preparing biosensors [8, 73]. Among oxide semiconductors, ZnO is heavily investigated, the conductivity of which can be controlled via addition of dopants from insulating to highly conductive levels. The conductivity and other properties can be fine-tuned via manipulating the chemical composition. Miniaturized electrical devices including resistors, transistors, diodes, logic gates, and similar devices have been produced via use of nanowires on rigid and flexible substrates [8, 71].

Based on flexibility, in terms of precursor materials as well as resultant properties, nanowires show promise for use in fabrication of nanoelectronics, optoelectronics, electrochemical, and electromechanical devices [8]. Nanowires have been investigated for use in molecular chemical and biological sensors [74, 75], nanodrug delivery systems [76], personal thermal management, photocatalysis, strain sensors, lithium batteries, photodetectors, supercapacitors, and nanogenerators [8, 10]. Nanowires can be treated in solutions and mounted on numerous substrates under moderate conditions. Thus, nanowires can be exploited for preparation of mini-devices that provide high-quality service over high surface areas, which just suits textile usage [77]. For more information on nanowires, Chapter 5 [8] can be reviewed.

1.6 Nanogenerators

Power-generating components supply electrical energy in smart textiles, which can be used for activation of smart functions and wearables such as MP3 players integrated in textiles, as well as charging other appliances including cell phones. Energy harvesting is an interesting field where smart nanotextiles can be used. Even though notable advancement has been attained regarding use of lithium rechargeable batteries, use of them in smart textiles poses challenges based on durability and comfort requirements [8]. Power generation may be achieved by collecting the energy dissipated by the body of the wearer as well as from the surrounding nature [5, 10, 40].

A number of functions can be obtained from smart textile devices. Yet the response mechanism needs energy to be activated. The selection of an appropriate energy source for smart textiles still remains an unsolved question. Conventional batteries need frequent replacement/recharge, so their use is not very practical. Additionally, they cannot cater the light weight, flexibility, safety, and energy density performance required for common textile use [40]. Many studies have been conducted to develop suitable power devices like batteries or supercapacitors that can be integrated into textiles. Based on the shortcomings of batteries, novel types of energy-harvesting devices have been developed. These devices have the capability to convert environmental energies into electricity. The mentioned environmental energy resources include sunlight, body thermal energy, and mechanical energies (human motion, heartbeat, wind, wave, tide, sound) [8, 10, 40, 78].

Various solar cells have been developed to generate electricity from solar energy, including novel ones that are flexible and can be integrated into textiles. The limitations of solar cells stem from high dependency on weather, location, and season that do not allow sustainable supply of power [10, 39].

As one can expect, thermoelectric nanogenerators can produce electricity from thermal energy in the presence of a temperature gradient. In these devices, solid-state p- and n-type semiconducting materials are utilized. Unfortunately, the output and efficiency of thermoelectric nanogenerators are not sufficient for use in smart textiles [10, 78].

Compared with other power sources, devices that produce electricity from mechanical motions exhibit advantages. Mechanical energy sources can be the wearer (human motion, heartbeat) or the environment (wind, wave). These types of nanogenerators can be studied in two classes: piezoelectric nanogenerators and triboelectric nanogenerators, which have been extensively investigated [10, 40, 79]. Li *et al.* [79] produced a triboelectric nanogenerator using poly(vinylidene fluoride) nanofibrous membrane coated with polydimethylsiloxane and polyacrylonitrile nanofibers coated with polyamide.

Triboelectric nanogenerators present a very interesting type of nanogenerators. These nanogenerators function based on triboelectrification, which is generally considered as an unwanted phenomenon. Energy generation takes place as a result of triboelectrification and electrostatic induction, where flexible and stretchable materials that have everyday common use can be utilized including polyamide, polytetrafluoroethylene, and silk [46, 80]. Triboelectric nanogenerators have first been announced by Prof. Wang of Georgia Institute of Technology and his research team in 2012 [81]. With improvement of triboelectric nanogenerators via selection of ideal materials and optimized designs, area power density of 500 W/m^2 and total conversion efficiency rates of 85% have been achieved [82, 83].

There are also electromagnetic generators that are conventionally utilized to produce electricity from mechanical energy. However, their use in textiles is not practical based on the necessity of a heavy rigid magnet and low efficiency for low frequency movements. On the contrary, nanogenerators allow use of different materials, design flexibility, and low-frequency performance. More research on nanogenerators can be found in Chapter 6 [10].

1.7 Nanocomposites

Nanocomposites are promising for use in various areas such as automotive, aerospace, defense, and biomedicine fields. Nanocomposites allow design and characteristic choices that are impossible with conventional composites. Based on their light weight and multifunctionality, nanocomposites cater the needs without compromising aesthetics and comfort of textiles. In smart textiles, nanocomposites take part in sensors, actuators, mediators, biosensors, thermoregulation, energy storing, and harvesting elements, among others. Nanocomposites are especially promising for sophisticated niche areas. Nanocomposites have already started to be used in a number of applications; nevertheless, there are still various potential areas where nanocomposites can be utilized in the future [26].

Nanocomposites can be classified in three groups in terms of their matrices: ceramic-matrix nanocomposites, metal-matrix nanocomposites, and polymer-matrix nanocomposites [84]. Their flexibility and conformability with textiles render polymer-matrix nanocomposites more suitable for smart textile use [26]. Polymer-based nanocomposites can be manufactured through different methods such as *in situ* polymerization [6, 85], melt homogenization [86], electrodeposition, solution dispersion [34, 85], sol–gel technique [87], template synthesis, and some advanced methods including atomic layer deposition and self-assembly [41, 84].

To improve properties of polymers, different reinforcement elements including particles, fibers, or platelets, of the micro- or nanoscale, are utilized. The reinforcement components contribute to the strength, thermal resistance, fire-retardancy, electrical conductivity properties, and so on. When nanostructured reinforcements are used, these special properties and more can be achieved without interfering with textile performance characteristics such as flexibility, stretchability, breathability, drape, softness, hand, and others. Moreover, via use of nanosized fillers, the desired properties can be achieved at concentrations much lower compared to conventional microfillers [84, 88]. The nanocomposite components that have been studied can be given as carbon nanotubes; metals, metal oxides, and inorganic nanoparticles [88]; conducting polymers, nanocellulose; and nanoclay [89].

Among nanoreinforcing elements, carbon nanotube addition in nanocomposites results in substantial improvements in mechanical properties [84, 88], antibacterial property, and conductivity [90].

Cellulose is one of the most abundant materials on earth [91]. Besides its abundancy, cellulose also presents biodegradability, biocompatibility, and renewability [60]. In its nanostructured form, cellulose has been utilized in sensors, biosensors, self-powering devices [92], and actuators [89, 93]. Bacterial cellulose is a kind of cellulose that is excreted by bacteria rather than plants. Unlike common plant cellulose, bacterial cellulose exhibits nanostructure in its pristine form [94]. In a related example, Lv *et al.* [92] developed a bacterial cellulose-based nanocomposite biosensor detecting glucose level. Nanocellulose has been utilized in pH and heavy metal sensors. In these applications, detector molecules are stabilized on cellulose [93, 95]. Nanocellulose-based nanocomposites are also utilized in strain sensors and actuators. Electroactive paper is an interesting example of nanocellulose-based nanocomposite exhibiting electrical field-induced motion [26, 96].

Conducting polymer is another group of polymers used in nanocomposites. Conducting polymers exhibit inherent conductivity. Incorporation of conducting polymers results in improved dielectric, catalytic, piezoresistive, magnetic susceptibility, or energy storage and harvesting properties [84, 88]. Polyaniline, polypyrrole, and poly(3,4-ethylenedioxythiophene), as polythiophene derivatives, are conducting polymers that show promise for use in smart textiles in terms of their mechanical strength and elasticity, and durability together with electrical properties. Mechanical properties, solubility, and dispersibility characteristics of conducting polymers are not generally as good as common polymers. The resistance of polymeric materials may show change over time. Furthermore, they may need long time to respond to environmental stimuli. Thus, it is feasible to use them with nanofillers in composite form to attain desired performance properties concurrently [93, 97].

Nanoparticles of metals, metal oxides, and nonmetal oxides are utilized in nanocomposites as reinforcement components [88, 98]. These nanocomposites show unique mechanical, thermal, and electrical characteristics. Metal nanoparticles have found use in nanocomposites developed for catalyst and biomedical applications. Nanoparticles of metal oxides are added to nanocomposites to obtain mechanical strength, electrical and thermal conductivity, barrier effect, antibacterial effect, UV protection, and self-cleaning property. Among metal oxide particles, TiO_2 and SiO_2 are commonly utilized [84, 88].

Nanoclay-reinforced nanocomposites have been extensively studied in terms of their special properties including thermal resistance, flame retardancy, stiffness, and strength [26, 84]. Nanowire-based nanocomposites have found use in energy storing and harvesting applications [8]. More information on nanocomposites can be found in Chapter 7 [26].

1.8 Nanocoating

Nanocoating, that is, coating a substrate with nanosized particles, is used to develop smart wearables with advanced functions and performance. Efficient nanocoating can provide advantages in terms of comfort, durability, and aesthetics. Via nanocoating, functions including UV protection, self-cleaning, antibacterial effect, and water repellence, among other smart properties, can be imparted to textiles, at the same time maintaining textile comfort characteristics required by the wearer such as breathability, drape, etc. [4].

There are different methods including conventional means like dip coating to apply nanocoating on smart textiles [32]. Various other coating methods have been developed on textiles such as transfer printing, immersion, padding, rolling, spraying, and simultaneous exhaust dyeing. Electrostatic spinning (electrospinning) can also be considered as a means of nanocoating where a web of nanoscale fibers is coated on a surface [34]. Rinsing and padding methods are widely applied for textile applications [34]. During the padding process, metal oxide nanoparticle systems are applied on textile substrates, followed by thermal processes [61]. Via coating, *in situ* synthesis, cross-linking, and immobilization of nanomaterials onto textile surface are carried out [34].

An inevitable necessity for nanomaterials in terms of coating is dispersibility in colloids. van der Waals and electrostatic double-layer attractions lead to agglomeration and microcluster formation of nanoparticles. Effective dispersion can be achieved via inclusion of dispersants like surfactants, grinding thoroughly, and functionalization by adding organic monomers and compounds [34, 84].

A major issue related with nanomaterials in terms of use on textiles is durability. Because of nonexistence of functional sites, nanoparticles do not present attraction to textile materials. In order to obtain durability of the designed characteristics, chemical or physical surface functionalization methods are applied. Another option is to embed nanomaterials in polymers applied on textiles and obtain nanocomposite coatings [26, 34].

Different nanotechnology-based coating methods are applied on textiles including sol–gel method, cross-linking method, and thin-film deposition methods (physical vapor deposition, vacuum evaporation, ion implantation, and sputter coating) [34].

More insight into nanocoating techniques can be found in Chapter 8 [41].

1.9 Nanofiber Formation

As mentioned before, nanofibers present very important advantages based on their high specific surface area. There are various methods to produce nanofibers. Electrostatic spinning, which is generally referred to as electrospinning, is the most commonly used nanofiber production method [46]. Other than electrospinning, there are a number of other techniques including force spinning, phase separation, melt blowing, bicomponent spinning, flash spinning, and so on. Via these methods, nanofibers are generally obtained in the form of nanowebs in random orientation where fiber diameters may range from several nanometers to hundreds of nanometers [99].

The electrospinning technique is advantageous in terms of its simplicity and applicability to various materials such as polymers, ceramics, and metals [46, 99]. In an electrospinning setting, a capillary tip is exposed to high voltage. This leads to formation of an electrical field between that capillary tip and a grounded collector. A pendent drop is formed at the tip by the solution dope flowing through it. The shape of the drop is first turned into a hemisphere than to a cone (referred to as the Taylor cone) due to the electrical force. As the electrical force exceeds the surface tension of the polymer drop, a jet of the solution ejects toward the collector. The jet presents an instable whipping motion due to the imbalance in the electrical charge. This motion leads the jet to stretch and to attenuate. The jet rapidly dries into nanofibers, which accumulate on the collector [34, 100].

If the viscosity is not sufficient, then drops instead of fibers are obtained. Parameters affecting electrospun fibers include solution characteristics, processing variables, and ambient conditions. The solution parameters include viscosity, surface tension, conductivity, concentration, and molecular weight. Processing variables consist of flow rate, voltage, collector structure, and syringe-to-collector distance, whereas the ambient conditions refer to temperature and humidity of the surrounding air medium [101].

Nanofibers, generally produced by electrospinning, suffer from lower mechanical properties. Post-treatments have been developed to enhance the mechanical properties, including cross-linking and welding. Using higher viscosity dopes to eliminate weaker points due to solvent-rich regions has also been reported to improve nanofiber strength. Sonication may be a means to overcome this disadvantage for decreased dynamic viscosity. Another option is to introduce suitable nanofillers to the solution [46].

In addition to fiber formation, the nanofibrous mat can be concurrently coated onto a substrate. The fiber diameter can range between 10 and 500 nm based on the conditions. Nanofibers of different polymers can be obtained such as polyvinyl alcohol, polyurethane, polyamide, polyacrylonitrile, and polylactic acid via electrospinning [34].

Electrospun nanofibrous mats show high porosity and low pore dimensions. These properties are required for a number of applications like filtration systems (for water treatment, climatization, etc.) [102, 103], separators in lithium ion rechargeable batteries [104], scaffolds in tissue engineering [105], wound dressing [106], and controlled drug delivery [107].

Novel nanofiber production techniques have been developed in order to overcome shortcomings of conventional electrospinning: low production rate; consequently, difficulty in utilization in large-scale production. Those novel procedures include modified versions of electrospinning and electroless spinning methods. Modifications of the electrospinning process can be listed as multi-needle electrospinning, needleless electrospinning, bubble electrospinning, electroblowing, microfluidic-manifold electrospinning, roller electrospinning, and melt electrospinning. Nanofiber production methods without use of electrostatic force include melt blowing, template melt extrusion, force spinning, flash spinning, bicomponent spinning, phase separation, and self-assembly [99].

Among modified electrospinning processes, multi-needle electrospinning was developed to increase the production rate compared to the conventional single-needle electrospinning system, which is constrained with fiber formation from only one needle. In multi-needle electrospinning, a number of needles are located in specific fashions in order to increase production rate without allowing interference among needles [108].

Needleless electrospinning systems utilize whole liquid surface rather than needles to increase production output. Multiple jets are ejected from the fiber forming liquid under strong electrical field [109].

In template melt extrusion, nanofibers are produced in the cylindrical holes of an impervious membrane via oxidative polymerization reactions. Nanofibers can be produced from polymers, ceramics, metals, and semiconductors. The length of the nanofibers is determined by the thickness of the membrane, which lies between 5 and 50 mm [110].

Self-assembly is a bottom-up method where molecules are utilized as building blocks to build nanofibers [6, 99].

Drawing process is another interesting means of nanofiber formation. In this technique, a millimetric droplet of a solution is placed on a surface and the solvent is allowed to vaporize. A micropipette is dipped into this droplet close to its periphery and is then withdrawn. During withdrawal from the drop, a nanofiber is pulled out. Each drop can produce several nanofibers by this means [111].

Force spinning is another novel nanofiber production method resembling electrospinning where centrifugal force replaces the electrical field. In this technique, fiber forming material is heated and rotated at a high speed at the nozzle where nanofibers are extruded. Force spinning is advantageous in terms of elimination of strong electrical field, and flexibility in the selection of conducting and nonconducting materials. Furthermore, no disadvantages entailing solvent removal and recovery are present in the force spinning method [112]. More information related to different nanofiber production approaches can be found in Chapter 9 [99].

1.10 Nanotechnology Characterization Methods

As mentioned above, incorporation of nanoscale or nanostructured materials can impart numerous characteristics to textiles including smart functionalities. Nevertheless, utilization of nanomaterials entails some difficulties due to their inherent properties. This necessitates close monitoring of properties such as average particle dimension, dimension range, particle diffusion, presence of elements, and so on [43].

These parameters can be determined by use of various advanced characterization methods including scanning electron microscopy (SEM), transmission electron microscopy (TEM), atomic force microscopy (AFM), X-ray diffraction (XRD), Raman spectroscopy, and X-ray photon spectroscopy (XPS) [43].

Nanotechnological characterization methods can be grouped as imaging methods, spectroscopic techniques, methods using X-rays, size distribution, and zeta potential analyses [113].

For a very long time, optical microscopes have been utilized to observe micron-level materials that cannot be perceived by the human eye in detail. However, at the nanoscale, the optical microscopes are not sufficient based on aberrations and the lower limit of light wavelengths. Here, other imaging techniques come into play such as scanning electron microscopy (SEM), transmission electron microscopy/high-resolution transmission electron microscopy (TEM/HRTEM), scanning tunneling microscopy (STM), atomic force microscopy (AFM), and alike, which can be used to observe sub-micron-sized structures by presenting images with a very high magnification [43].

In AFM, a sharp tip on a cantilever reads the surface topography of a studied material. SEM utilizes a focused electron beam, which strikes and backscatters [114]. Among nanotechnological characterization methods, the transmission electron microscopy technique uses an electron beam that is transmitted through an ultrathin film of the material. The electron beam interacts with the material as it passes through the specimen. The transmitted electrons are used to form an image that is magnified and focused [43]. In STM, by utilization of piezoelectric tube and tunneling current, the surface topographies of materials can be observed at the nanolevel [115].

FTIR (Fourier transform infrared), Raman, and UV–Vis spectrums can be listed as spectroscopy methods. These techniques are very important in characterization of nanoparticles. FTIR spectroscopy detects absorption and/or emission based on asymmetric bonds' movements; thus, it provides knowledge about the chemical structures. The Raman spectroscopy measures inelastic scattering of monochromatic light interacted with a sample. The changes in the incident and the re-emitted light give information about the chemical and physical structure of the studied material. UV–Vis spectroscopy (ultraviolet–visible spectroscopy) is a method of determining nanoparticle properties based on measuring 180–900 nm light rays [60, 113, 116–118].

Among X-ray methods, EDX (energy dispersive X-ray analysis) technology, which usually accompanies SEM, allows us to determine the composition and concentration of nanoparticles located at the observed material surface, if the material includes heavy metals such as gold palladium and silver nanoparticles [43, 119, 120]. XRD utilizes interaction of X-rays and the sample and gives information about the crystal structure, crystallite size, crystallinity, and phases of the sample [121]. XPS analysis is analogous to SEM imaging, where X-rays take the place of the electron beam [122].

1.11 Challenges and Future Studies

One of the major challenges of smart nanotextiles is implementation of laboratory-scale studies to the mass-production stage as in the case with other emerging technologies. This is especially valid for electrospinning process. Thus, different modifications of the electrospinning method as well as alternative nanomaterial production methods have been developed for optimization of this process to suit large-scale production. It should be noted that not only the production processes but also the characterization procedures should be compatible with the macro-scale applications. On the other hand, some nanomaterials of smart nanotextiles have already found use in commercial applications. These nanomaterials include carbon nanotubes, zinc oxide, titanium dioxide, silicon dioxide, Al_2O_3, and silver [46].

Another major matter related to smart nanotextiles is their impact on the environment and human health. Effects of nanomaterials on the environment and human health are not yet fully known. Thus, extensive investigations should be carried out to detect unwanted effects of this new class of materials, especially those that have close contact with the skin, and more importantly, textiles that are used *in vivo*. The effects of nanomaterials in the workplace during production and storage stages including those of byproducts and waste water, as well as the effects to the end user during service life and to the environment after disposal, should be closely considered. The effects on the end user relate with the desired/undesired release from the textiles to the human body. Effective bonding/encapsulation of nanomaterials can prevent unwanted release. Potential effects of nanomaterials on health may include DNA damage, passing through and impairing tissue membranes, skin reactions, digestive and respiratory system disorders, and nerve system impairment. To the best of our knowledge, short-term negative effects of nanomaterials on healthy skin have not been reported yet [34]. It should be noted that the most critical negative effect of nanomaterials, as in the case of other materials, is when respired, as it can directly mix with blood, than when digested or when contacted through skin [58].

In order for smart functionalities to take place, energy is, of course, needed. This energy is conventionally provided from an outside source including rechargeable batteries. On the other hand, in advanced examples, smart components are able to harvest and store energy, as in the case of nanogenerators. This revolutionary invention of nanogenerators renders smart textiles more comfortable, versatile, practical, and fashionable. In the future, these nanogenerators can harvest energy even for other devices such as mobile phones [10].

Biomedicine is an important area where there is a great market potential. Smart nanotextiles can be used for the elderly, the sick, and the sports people. IOT (Internet of things) is another future reality, in which the smart nanotextiles can play a role [10].

Major progress occurs when different disciplines meet. Smart nanotextiles are a very significant example of this where areas of textiles, electronics, polymer science, and nanotechnology intersect. We do not expect the textile of the future to be just cotton, viscose, or polyester fabrics. To the contrary, wearables "equipped" with smart functionality will form the future of textiles.

1.12 Conclusion

Textiles present a remarkable series of substrates in order to serve numerous smart functions in an array of fields. Smart functionalities exhibit reversible reactions triggered by an external stimulus, that is, properties of medium, physical, chemical, electrical, etc. Introduction of smart functionalities leads to conversion of textiles into environmental-responsive devices that possess the ability to sense, react to, and adapt to surrounding conditions. By incorporation of smart nanocomponents, it has become possible to obtain advanced properties to cater needs in vast areas of applications including biomedicine, defense, entertainment, and others without compromising the comfort and aesthetics of textiles.

When smart nanotextiles serve their intended duty, this should not be at the expense of textile performance, including flexibility, breathability, comfort, and aesthetics. This can be achieved by use of nanoscale components, which do not form a thick, heavy addition that alters the appearance and drape of clothing. On the other hand, smart nanocomponents should endure the conditions that textiles typically undergo: straining, laundering, and alike. This is where the smart nanocomponent should comply with the rest of the clothing, and a good bonding between the nanomaterial and the textile substrate should be obtained.

Use of smart nanotextiles is not without questions. Use of novel materials necessitates avoidance of adverse effects on human and environmental ecology. This can be attained by assuring successful nanomaterial production, treatment, and bonding procedures, as well as realizing effective nanotechnological characterization methods. Another major challenge is conveying the laboratory-scale successes into the commercial stage. The production rates of nanomaterials are far from catering the mass community needs at the moment. As these problems are going to be solved one by one, we will be seeing that the growing success of smart nanotextiles will exceed our wildest expectations.

References

1. Yilmaz, N. D., Cassill, N. L., and Powell, N. B., Turkish towel's place in the global market, *J. Text. Apparel Tech. Manag.*, 5, 4, 1–43, 2007.

2. Chika, Y.-B. and Adekunle, S. A., Smart fabrics—wearable technology, *Int. J. Eng. Technol. Manag. Res.*, 4, 10, 78–98, 2017.
3. Mecnika, V., Hoerr, M., Krievins, I., Jockenhoevel, S., and Gries, T., Technical embroidery for smart textiles: Review, *Mater. Sci. Text. Cloth. Technol.*, 9, 56–63, 2014.
4. Syduzzaman, M., Patwary, S. U., Farhana, K., and Ahmed, S., Textile science & engineering smart textiles and nano-technology: A general overview, *Text. Sci. Eng.*, 5, 1, 1–7, 2015.
5. Yang, E. *et al.*, Nanofibrous smart fabrics from twisted yarns of electrospun piezopolymer, *ACS Appl. Mater. Interfaces*, 9, 28, 24220–24229, 2017.
6. Yilmaz, N. D., Multi-component, semi-interpenetrating-polymer network and interpenetrating-polymer-network hydrogels: Smart materials for biomedical applications abstract, in *Functional Biopolymers*, Thakur, V. K. and Thakur, M. K. (Eds.) Springer, 2017.
7. Stoppa, M. and Chiolerio, A., Wearable electronics and smart textiles: A critical review, *Sensors*, 14, 11957–11992, 2014.
8. Song, J., Nanowires for smart textiles, in *Smart Textiles: Wearable Nanotechnology*, Yılmaz, N. D. (Ed.) Wiley-Scrivener, 2018.
9. Mahltig, B., Nanosols for smart textiles, in *Smart Textiles: Wearable Nanotechnology*, Yılmaz, N. D. (Ed.) Wiley-Scrivener, 2018.
10. Pu, X., Hu, W., and Wang, Z. L., Nanogenerators for smart textiles, in *Smart Textiles: Wearable Nanotechnology*, Yılmaz, N. D. (Ed.) Wiley-Scrivener, 2018.
11. Mattila, H., Yarn to Fabric: Intelligent Textiles, in *Textiles and Fashion*, Sinclair, R. (Ed.) Woodhead Publishing, pp. 355–376, 2015.
12. Ebara, M. *et al.*, *Smart Biomaterials*. Springer Japan, 2014.
13. Bougia, P., Karvounis, E., and Fotiadis, D. I., Smart medical textiles for monitoring pregnancy, in *Smart Textiles for Medicine and Healthcare*, Van Langenhove, L. (Ed.) Woodhead Publishing, pp. 183–205, 2007.
14. McCann, J., Smart medical textiles in rehabilitation, in *Smart Textiles for Medicine and Healthcare*, Van Langenhove, L. (Ed.) Woodhead Publishing, 2007.
15. Qin, Y., Smart wound-care materials, in *Smart Textiles for Medicine and Healthcare*, Van Langenhove, L. (Ed.) Woodhead Publishing, 2007.
16. Khan, F. and Tanaka, M., Designing smart biomaterials for tissue engineering, *Int. J. Mol. Sci.*, 19, 17, 1–14, 2018.
17. Ortelli, S., Costa, A. L., and Dondi, M., TiO_2 nanosols applied directly on textiles using different purification treatments, *Materials (Basel)*, 8, November, 7988–7996, 2015.
18. Guo, C., Zhou, L., and Lv, J., Effects of expandable graphite and modified ammonium polyphosphate on the flame-retardant and mechanical properties of wood flour-polypropylene composites, *Polym. Polym. Compos.*, 21, 7, 449–456, 2013.
19. Xu, F., Ma, L., Huo, Q., Gan, M., and Tang, J., Microwave absorbing properties and structural design of microwave absorbers based on polyaniline and

polyaniline/magnetite nanocomposite, *J. Magn. Magn. Mater.*, 374, 311–316, 2015.
20. Bayat, M., Yang, H., and Ko, F., Electrical and magnetic properties of Fe_3O_4/ carbon composite nanofibres, in *SAMPE*, 2010.
21. "EU project launches €2.4m competition to create ethical and sustainable wearable technologies and smart textiles," 2017.
22. *National Science Foundation*, 2018. https://www.nsf.gov/awardsearch/showAward?AWD_ID=1652538.
23. O'Brien, M. and Walton M., These smart threads could save lives, *National Science Foundation*, 2016. https://www.nsf.gov/news/special_reports/science_nation/biomedtextiles.jsp.
24. Niiyama, E., Fulati, A., and Ebara, M., Responsive polymers for smart textiles, in *Smart Textiles: Wearable Nanotechnology*, Yılmaz, N. D. (Ed.) Wiley-Scrivener, 2018.
25. Pause, B., Application of phase change and shape memory materials in medical textiles, in *Smart Textiles for Medicine and Healthcare*, Van Langenhove, L. (Ed.) Woodhead Publishing, 2007.
26. Yilmaz, N. D., Nanocomposites for smart textiles, in *Smart Textiles: Wearable Nanotechnology*, Yılmaz, N. D. (Ed.) Wiley-Scrivener, 2018.
27. Zysset, C., Cherenack, K., Kinkeldei, T., and Tröster, G., Weaving integrated circuits into textiles, in *International Conference on Wearable Computers*, 2010.
28. Peterson, J., Carlsson, J., and Bratt, M., Smart textiles for knitted products—prototype factory, in *AUTEX 2009 World Textile Conference*, 2009.
29. Aibibu, D., Hild, M., and Cherif, C., An overview of braiding structure in medical textile: Fiber-based implants and tissue engineering, in *Advances in Braiding Technology: Specialized Techniques and Applications*, Kyosev, Y. (Ed.) Woodhead Publishing, 2016.
30. Krucińska, I., Skrzetuska, E., Surma, B., and Gliścińska, E., Technologies involved in the manufacture of smart nonwoven fabrics, in *Non-woven Fabrics*, Jeon, H.-Y. (Ed.) Intechopen, 2016.
31. Šahta, I., Vališevskis, A., Baltiņa, I., and Ozola, S., Development of textile based sewn switches for smart textile, *Adv. Mater. Res.*, 1117, 235–238, 2015.
32. Gashti, M. P., Alimohammadi, F., Song, G., and Kiumarsi, A., Characterization of nanocomposite coatings on textiles: *A brief review on microscopic technology*, 1424–1437, 2012.
33. Botticini, F., *Make your own smart textile. Experimentation of EdM conductive ink on textiles and development of a specific writing instrument for ink deposition on fabric*. Politecnico Milanı, 2016.
34. Gashti, M. P., Pakdel, E., and Alimohammadi, F., Nanotechnology-based coating techniques for smart textiles, in *Active Coatings for Smart Textiles*, June. Elsevier Ltd, pp. 243–268, 2016.
35. Wilson, A., Ten sensors per fibre with Xelflex, *Smart Text. Nanotechnol.*, 1, 2017.

36. Amft, O. and Habetha, J., Smart medical textiles for monitoring patients with heart conditions, in *Smart Textiles for Medicine and Healthcare*, Van Langenhove, L. (Ed.) Woodhead Publishing, 2007.
37. Huang, C.-T., Tang, C.-F., and Shen, C.-L., A wearable textile for monitoring respiration, using a yarn-based sensor, in *Tenth IEEE International Symposium on Wearable Computers*, 2006.
38. Nair, A., Chowdhury, N., and Chowdhury, T., Smart clothes, *Int. J. Adv. Res. Eng. Technol.*, 7, 5, 18–27, 2016.
39. Chen, J. *et al.*, Micro-cable structured textile for simultaneously harvesting solar and mechanical energy, *Nat. Energy*, 1–8, 2016.
40. Lin, Z. *et al.*, Large-scale and washable smart textiles based on triboelectric nanogenerator arrays for self-powered sleeping monitoring, *Adv. Funct. Mater.*, 28, 1, 1704112, 2017.
41. Pakdel, E., Fang, J., Sun, L., and Wang, X., Nanocoatings for smart textiles, in *Smart Textiles: Wearable Nanotechnology*, Yılmaz, N. D. (Ed.) Wiley-Scrivener, 2018.
42. Ghosh, D. S., *Ultrathin Metal Transparent Electrodes for the Optoelectronics Industry*. Springer, 2013.
43. Joshia, M., Bhattacharyya, A., and Ali, S. W., Characterization techniques for nanotechnology applications in textiles, *Indian J. Fibre Text. Res.*, 33, 304–317, 2008.
44. Raab, C., Simkó, M., Fiedeler, U., Nentwich, M., and Gazsó, A., Production of nanoparticles and nanomaterials, *Nano Trust Dossiers*, 6, July, 1–4, 2017.
45. Zhu, Y., Zhang, J. C., Zhai, J., and Al, E., Multifunctional carbon nanofibers with conductive, magnetic and superhydrophobic properties, *Chemphyschem*, 7, 336–341, 2006.
46. Wan, L. Y., Nanofibers for smart textiles, in *Smart Textiles: Wearable Nanotechnology*, Yılmaz, N. D. (Ed.) Wiley-Scrivener, 2018.
47. Yang, M. *et al.*, Single flexible nanofiber to simultaneously realize electricity-magnetism bifunctionality, *Mater. Res.*, 19, 2, 2016.
48. Zhong, J. *et al.*, Fiber-based generator for wearable electronics and mobile medication, *ACS Nano*, 8, 6273, 2014.
49. Wu, H., Krifa, M., and Koo, J. H., Flame retardant polyamide 6/nanoclay/intumescent nanocomposite fibers through electrospinning, *Text. Res. J.*, 84, 1106–1118, 2014.
50. Periolatto, M., Ferrero, F., Montarsolo, A., and Mossotti, R., Hydrorepellent finishing of cotton fabrics by chemically modified TEOS based nanosol, *Cellulose*, 20, 1, 355–364, 2013.
51. Yamamoto, M. *et al.*, Theoretical explanation of the Lotus Effect: Superhydrophobic property changes by removal of nanostructures from the surface of a Lotus leaf, *Langmuir*, 31, 26, 7355–7363, 2015.
52. Zhan, N. *et al.*, A novel multinozzle electrospinning process for preparing superhydrophobic PS films with controllable bead-on-string/microfiber morphology, *J. Colloid Interface Sci.*, 345, 2, 491–495, 2010.

53. Jiang, L., Zhao, Y., and Zhai, J., A lotus-leaf-like superhydrophobic surface: A porous microsphere/nanofiber composite film prepared by electrohydrodynamics, *Angew Chem Int.*, 116, 4438–4441, 2004.
54. Wagner, N. and Theat, P., Light-induced wettability changes on polymer surfaces, *Polymer (Guildf)*, 5, 16, 3436–3453, 2014.
55. Muthiah, P., Boyle, T. J., and Sigmund, W., Thermally induced rapid wettability switching of electrospun blended polystyrene/poly(N-isopropylacrylamide) nanofiber mats, *Macromol. Mater. Eng.*, 298, 12, 2013.
56. Zhu, Y., Feng, L., Xia, F., and Jiang, L., Chemical dual-responsive wettability of superhydrophobic pani-pan coaxial nanofibers, *Macromol. Rapid Commun.*, 28, 10, 1135–1141, 2007.
57. Hanna, E. G. and Tait, P. W., Limitations to thermoregulation and acclimatization challenge human adaptation to global warming, *Int. J. Environ. Res. Public Health*, 12, 7, 8034–8074, 2015.
58. Yilmaz, N. D., Design of acoustic textiles: Environmental challenges and opportunities for future direction, in *Textiles for Acoustic Applications*, Nayak, R. and Padhye, R. (Eds.) Springer, 2016.
59. Chalco-Sandoval, W., Fabra, M. J., López-Rubio, A., and Lagaron, J. M., Development of an encapsulated phase change material via emulsion and coaxial electrospinning, *J. Appl. Polym. Sci.*, 133, 36, 43903, 2016.
60. Yilmaz, N. D., Konak, S., Yilmaz, K., Kartal, A. A., and Kayahan, E., Characterization, modification and use of biomass: Okra fibers, *Bioinspired, Biomim. Nanobiomaterials*, 5, 3, 85–95, 2016.
61. Berendjchi, A., Khajavi, R., and Yazdanshenas, M. E., Application of nanosols in textile industry, *Int. J. Green Nanotechnol.*, 1, 1–7, 2013.
62. Mahltig, B. and Textor, G., *Nanosols and Textiles*. World Scientific Publishing, 2007.
63. Stegmaier, T., Recent advances in textile manufacturing technology, in *The Global Textile and Clothing Industry*, Shishoo, R. (Ed.) Woodhead Publishing, 2012.
64. Kopeček, J., Hydrogel biomaterials: A smart future?, *Biomaterials*, 28, 34, 5185–5192, 2007.
65. Kumar, A., Smart polymeric biomaterials: Where chemistry & biology can merge.
66. Peppas, N. A., Hydrogels, in *Biomaterials Science: An Introduction to Materials in Medicine*, Ratner, B. D. (Ed.) California: Academic Press, USA, pp. 100–107, 2004.
67. Paleos, G. A., *What are hydrogels?*, https://www.scribd.com/document/366622248/What-Are-Hydrogels, 2012.
68. Lee, S. C., Kwon, I. K., and Park, K., Hydrogels for delivery of bioactive agents: A historical perspective, *Adv. Drug Deliv. Rev.*, 65, 1, 17–20, 2013.
69. Yilmaz, N. D., Khan, G. M. A., and Yilmaz, K., Biofiber reinforced acrylated epoxidized soybean oil (AESO) composites, in *Handbook of Composites from Renewable Materials, Physico-Chemical and Mechanical*

Characterization, Thakur, V. K. and Thakur, M. K. (Eds.) Wiley-Scrivener, pp. 211–251, 2017.
70. Sandeep, C., Harikumar, S. L., and Kanupriya, Hydrogels: A smart drug delivery system, *Int. J. Res. Pharm. Chem.*, 2, 3, 603–614, 2012.
71. Ranzoni, A. and Cooper, M. A., The growing influence of nanotechnology in our lives, in *Micro and Nanotechnology in Vaccine Development*, Skwarczynski, M. and Toth, I. (Eds.) Elsevier Inc., pp. 1–10, 2017.
72. Singh, A. K., The past, present, and the future of nanotechnology, in *Engineered Nanoparticles Structure, Properties and Mechanisms of Toxicity*, Singh, A. K. (Ed.) Elsevier, pp. 515–525, 2016.
73. Clark, D. P. and Pazdernik, N. J., Nanobiotechnology, in *Biotechnology*, 2nd ed., Clark, D. P. and Pazdernik, N. (Eds.) Elsevier, 2016.
74. Helman, A., Borgström, M. T., van Weert, M., Verheijen, M. A., and Bakkers, E. P. A. M., Synthesis and electronic devices of III–V nanowires, in *Encyclopedia of Materials: Science and Technology*, 2nd ed., Buschow, K. H. J., Cahn, R. W., Flemings, M. C., Ilschner, B., Kramer, E. J., Mahajan, S., and Veyssière, P. (Eds.) Elsevier Ltd, pp. 1–6, 2008.
75. Tisch, U. and Haick, H., Arrays of nanomaterial-based sensors for breath testing, in *Volatile Biomarkers*, Amann, A. and Smith, D. (Eds.) Elsevier B V, 2013.
76. Sharma, H. S., Muresanu, D. F., Sharma, A., Patnaik, R., and Lafuente, J. V., Nanoparticles influence pathophysiology of spinal cord injury and repair, in *Progress in Brain Research*, Howard, C. J. (Ed.) Elsevier, pp. 154–180, 2009.
77. Duan, X., Nanowire thin films for flexible macroelectronics, in *Encyclopedia of Materials: Science and Technology*, 2nd ed., Buschow, K. H. J., Cahn, R. W., Flemings, M. C., Ilschner, B., Kramer, E. J., Mahajan, S., and Veyssière, P. (Eds.) Elsevier, pp. 1–10, 2010.
78. Hyland, M., Hunter, H., Liu, J., Veety, E., and Vashaee, D., Wearable thermoelectric generators for human body heat harvesting, *Appl. Energy*, 182, 518–524, 2016.
79. Li, Z., Shen, J., Abdalla, I., Yu, J., and Ding, B., Nanofibrous membrane constructed wearable triboelectric nanogenerator for high performance biomechanical energy harvesting, *Nano Energy*, 36, 341–348, 2017.
80. Wang, Y., Yang, Y., and Wang, Z. L., Triboelectric nanogenerators as flexible power sources, *Npj Flex. Electron.*, 1, 2017.
81. Fan, F. R., Tian, Z. Q., and Wang, Z. L., Flexible triboelectric generator, *Nano Energy*, 1, 328–334, 2012.
82. Zhu, G. *et al.*, A shape-adaptive thin-film-based approach for 50% high-efficiency energy generation through micro-grating sliding electrification, *Adv. Mater.*, 26, 23, 3788–3796, 2014.
83. Xie, Y. *et al.*, Grating-structured freestanding triboelectric-layer nanogenerator for harvesting mechanical energy at 85% total conversion efficiency, *Adv. Mater.*, 26, 38, 6599–6607, 2014.

84. Camargo, P. H. C., Satyanarayana, K. G., and Wypych, F., Nanocomposites: Synthesis, structure, properties and new application opportunities, *Mater. Res.*, 12, 1, 1–39, 2009.
85. Nguyen-Tri, P., Nguyen, T. A., Carriere, P., and Xuan, C. N., Nanocomposite coatings: Preparation, characterization, properties, and applications, *Int. J. Corros.*, 2018, 1–19, 2018.
86. Hári, J. and Pukánszky, B., Nanocomposites: Preparation, structure, and properties, in *Applied Plastics Engineering Handbook*, Kutz, M. (Ed.) William Andrew, pp. 109–142, 2011.
87. Lateef, A. and Nazir, R., Metal nanocomposites: Synthesis, characterization and their applications, in *Science and Applications of Tailored Nanostructures*, Di Sia, P. (Ed.) One Central Press, pp. 239–256, 2018.
88. Bratovčić, A., Odobašić, A., Ćatić, S., and Šestan, I., Application of polymer nanocomposite materials in food packaging, *Croat. J. Food Sci. Technol.*, 7, 2, 86–94, 2015.
89. Edwards, J. V., Prevost, N., French, A., Concha, M., DeLucca, A., and Wu, Q., Nanocellulose-based biosensors: Design, preparation, and activity of peptide-linked cotton cellulose nanocrystals having fluorimetric and colorimetric elastase detection sensitivity, *Engineering*, 5, 20–28, 2013.
90. Chowdhury, S., Olima, M., Liu, Y., Saha, M., Bergman, J., and Robison, T., Poly dimethylsiloxane/carbon nanofiber nanocomposites: Fabrication and characterization of electrical and thermal properties, *Int. J. Smart Nanomater.*, 7, 4, 236–247, 2016.
91. Yilmaz, N. D., Koyundereli Cilgi, G., and Yilmaz, K., Natural polysaccharides as pharmaceutical excipients, in *Handbook of Polymers for Pharmaceutical Technologies, Volume 3, Biodegradable Polymers*, Thakur, V. K. and Thakur, M. K. (Eds.) Wiley-Scrivener, pp. 483–516, 2015.
92. Lv, P. *et al.*, A highly flexible self-powered biosensor for glucose detection by epitaxial deposition of gold nanoparticles on conductive bacterial cellulose, *Chem. Eng. J.*, 351, 177–188, 2018.
93. Abdi, M. M., Abdullah, L. C., Tahir, P. M., and Zaini, L. H., Cellulosic nanomaterials for sensing applications, in *Handbook of Green Materials 3 Self- and Direct Assembling of Bionanomaterials*, Oksman, K., Mathew, A. P., Bismarck, A., Rojas, O., and Sain, M. (Eds.) World Scientific Publishing, pp. 197–212, 2014.
94. Yilmaz, N. D., Biomedical applications of microbial cellulose nanocomposites, in *Biodegradable Polymeric Nanocomposites: Advances in Biomedical Applications*, Wiley-Scrivener, pp. 231–249, 2015.
95. Kim, J.-H. *et al.*, Review of nanocellulose for sustainable future materials, *Int. J. Precis. Eng. Manuf. Technol.*, 2, 2, 197–213, 2015.
96. Kim, J. and Seo, Y. B., Electro-active paper actuators, *Smart Mater. Struct.*, 11, 355–360, 2002.
97. Grancaric, A. M. *et al.*, Conductive polymers for smart textile applications, *J. Ind. Text.*, 48, 3, 612–642, 2017.

98. Kurahatti, R. V., Surendranathan, A. O., Kori, S., Singh, N., Kumar, A. V. R., and Srivastava, S., Defence applications of polymer nanocomposites, *Def. Sci. J.*, 60, 5, 551–563, 2010.
99. Nayak, R., Production methods of nanofibers for smart textiles, in *Smart Textiles: Wearable Nanotechnology*, Yılmaz, N. D. (Ed.) Wiley-Scrivener, 2018.
100. Nayak, R., Padhye, R., Kyratzis, I. L., Truong, Y. B., and Arnold, L., Recent advances in nanofibre fabrication techniques, *Text. Res. J.*, 82, 2, 129–147, 2011.
101. Li, Z. and Wang, C., *One-Dimensional Nanostructures*. Berlin, Heidelberg, Springer-Verlag, 2013.
102. Shi, J., Guobao, W., Chen, H., Zhong, W., Qiu, X., and Xing, M. M. Q., Schiff based injectable hydrogel for *in situ* pH-triggered delivery of doxorubicin for breast tumor treatment, *Polym. Chem.*, 5, 6180–6189, 2014.
103. Li, J. *et al.*, Filtration of fine particles in atmospheric aerosol with electrospinning nanofibers and its size distribution, *Sci. China Technol. Sci.*, 231, 7, 597–616, 2017.
104. Chen, S., Ye, W., and Hou, H., Electrospun fibrous membranes as separators of lithium-ion batteries, in *Electrospun Nanofibers for Energy and Environmental Applications. Nanostructure Science and Technology*, Ding, B., and Yu, J. (Eds.) Berlin, Heidelberg, Springer, 2014.
105. Ngadiman, N. H. A., Noordin, M. Y., Idris, A., and Kurniawan, D., A review of evolution of electrospun tissue engineering scaffold: From two dimensions to three dimensions, *Proc. Inst. Mech. Eng. Part H J. Eng. Med.*, 231, 7, 597–616, 2017.
106. Liu, M., Duan, X.-P., Li, Y.-M., Yang, D.-P., and Long, Y.-Z., Electrospun nanofibers for wound healing, *Mater. Sci. Eng. C*, 76, 1413–1423, 2017.
107. Weng, L. and Xie, J., Smart electrospun nanofibers for controlled drug release: Recent advances and new perspectives, *Curr. Pharm. Des.*, 21, 15, 1944–1959, 2015.
108. Tian, L., Zhao, C., Li, J., and Pan, Z., Multi-needle, electrospun, nanofiber filaments: Effects of the needle arrangement on the nanofiber alignment degree and electrostatic field distribution, *Text. Res. J.*, 85, 6, 621–631, 2015.
109. Yu, M. *et al.*, Recent advances in needleless electrospinning of ultrathin fibers: From academia to industrial production, *Macromol. Mater. Eng.*, 302, 7, 1700002, 2017.
110. Li, H., Ke, Y., and Hu, Y., Polymer nanofibers prepared by template melt extrusion, *J. Appl. Polym. Sci.*, 99, 3, 1018–1023, 2006.
111. Xing, X., Wang, Y., and Li, B., Nanofiber drawing and nanodevice assembly in poly(trimethylene terephthalate), *Opt. Express*, 16, 14, 2018.
112. Hammami, M. A., Krifa, M., and Harzallah, O., Centrifugal force spinning of PA6 nanofibers—processability and morphology of solution-spun fibers, *J. Text. Inst.*, 105, 6, 637–647, 2014.

113. Pillai, M. M., Senthilkumar, R., Selvakumar, R., and Bhattacharyya, A., Characterization methods of nanotechnology based smart textiles, in *Smart Textiles: Wearable Nanotechnology*, N. D. Yılmaz, (Ed.) Wiley-Scrivener, 2018.
114. Smith, B., The differences between atomic force microscopy and scanning electron microscopy, *Azo Mater.*, 2015.
115. Schmid, M., The scanning tunneling microscope, Technical University of Wien, 2011. https://www.iap.tuwien.ac.at/www/surface/stm_gallery/stm_schematic.
116. Wu, J.-B., Lin, M.-L., Cong, X., Liua, H.-N., and Tan, P.-H., Raman spectroscopy of graphene-based materials and its applications in related devices, *Chem. Soc. Rev.*, 47, 5, 1822–1873, 2018.
117. Ultraviolet–visible spectrometer, *Royal Society of Chemistry*, 2018. http://www.rsc.org/publishing/journals/prospect/ontology.asp?id=CMO:0001803&MSID=B821416F.
118. *Raman spectroscopy basics*, Princeton Instruments. http://web.pdx.edu/~larosaa/Applied_Optics_464-564/Projects_Optics/Raman_Spectrocopy/Raman_Spectroscopy_Basics_PRINCETON-INSTRUMENTS.pdf.
119. Piazza, G. J., Nuñez, A., and Foglia, T. A., Epoxidation of fatty acids, fatty methyl esters, and alkenes by immobilized oat seed peroxygenase, *J. Mol. Catal. B Enzym.*, 21, 3, 143–151, 2003.
120. Fay, F., Linossier, I., Langlois, V., Haras, D., and Vallee-Rehel, K., SEM and EDX analysis: Two powerful techniques for the study of antifouling paints, *Prog. Org. Coatings*, 54, 3, 216–223, 2005.
121. *The X-ray diffraction small research facility: What is XRD?*, The University of Sheffield, 2018. https://www.sheffield.ac.uk/materials/centresandfacilities/x-ray-diffraction/whatxrd.
122. XPS/ESCA, *Physical Electronics*, 2018. https://www.phi.com/surface-analysis-techniques/xps-esca.html.

Section 2

MATERIALS FOR SMART NANOTEXTILES

2

Nanofibers for Smart Textiles

Lynn Yuqin Wan

Fiber and Polymer Scientist, Lululemon Athletica
Honorary Research Associate, Department of Material Engineering, University of British Columbia Vancouver, BC, Canada

Abstract

Due to the nano-effect, nanofibers have found numerous applications in many areas such as filtration, energy generation and storage, and biomedical and textile industries. Comparing with conventional textile materials, nanofiber membranes have super high porosity, surface area-to-volume ratio, and sensibility. Therefore, nanofiber membranes have demonstrated excellent protective properties against environmental threats to human beings as a new type of smart textile material. In this chapter, the definition and the advantages of nanofibers are first introduced, followed by comparisons of different fabrication technologies. As the most widely adopted fabrication method, electrospinning technology is emphasized with more detail. The types and properties of smart nanofibers as well as their applications in textiles are then elaborated. As a conclusion, the challenges confronting nanofibers as a special type of textile material are summarized and the future development trends are suggested.

Keywords: Smart textiles, nanofiber, electrospinning, moisture management, waterproof, thermoregulation, protection, wearables, medical care

2.1 Introduction

Recognizing the potential nano-effect that will be created when fibers are reduced to the nanoscale, there has been an explosive growth in research efforts around the world. There are several alternative methods for generating fibers in the nanometer scale, but none of them matches the popularity of the electrospinning technology largely due to the great simplicity and

Email: wanyuqin@gmail.com

Nazire D. Yilmaz (ed.) Smart Textiles, (41–90) © 2019 Scrivener Publishing LLC

versatility of the electrospinning process. The ability to create multifunctional nanoscale fibers from a broad range of polymeric materials in a relatively simple manner using electrospinning technique, coupled with the rapid growth of nanotechnology in the recent years, has greatly accelerated the growth of nanofiber technology. Ascribed to the characteristics of ultra-high surface area, surface reactivity, and sensitivity to stimuli, nanofibers find great potential applications in a wide range of areas such as filtration, energy generation and storage, biomedicine, tissue engineering scaffolds, chemistry, sensor, and smart textiles. In this chapter, we will first review nanofibers and nanofiber fabrication technologies, then narrow down to electrospun nanofibers and their applications for smart textiles including moisture management, waterproof, thermal regulation, personal protection, wearables, sensors, and medical care. We will conclude this chapter with current challenges of nanofiber materials and forecast of the development trend.

2.2 Nanofibers and Their Advantages

Fibrous materials are of great practical importance as a fundamental building block of the physical world. Comparing with bulk materials, fibers have high aspect ratio and specific surface area, good flexibility and anisotropic physical properties such as mechanical properties, and thermal and electrical conductivity levels, which turn fibers into a great material for many unique applications ranging from clothing to reinforcements for aerospace structures. Nanofibers are fibers with diameters falling into nanoscale lengths, characterized by their superior flexibility and high aspect ratio that is greater than 1000:1. According to the National Science Foundation (NSF), nanomaterials are matters that have at least one dimension equal to or less than 100 nm [1]. By reducing the fiber diameter from 10 μm to 10 nm, a million times increase in flexibility is expected. Specifically, the role of fiber size has been recognized in significant increases in surface area, surface reactivity, and sensitivity to stimuli.

One of the most significant characteristics of nanofibers is their extraordinary surface area per unit mass. For fibers having diameters from 5 to 500 nm, as shown in Figure 2.1, the surface area per unit mass is around 10,000 to 1,000,000 m^2/kg. In nanofibers that are 3 nm in diameter, about half of the molecules are on the surface. Since the majority of the atoms are located on the surface of a nanofiber, the intrinsic properties of nanofibers are different from conventional fibers where the majority of atoms are buried inside. The high surface area of nanofibers provides remarkable

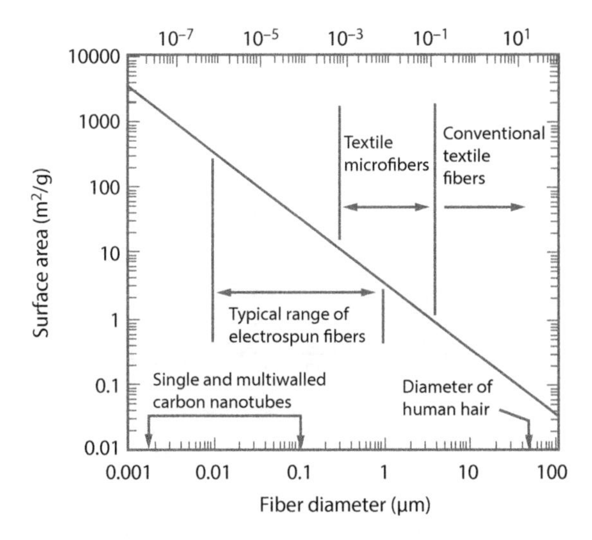

Figure 2.1 Effect of fiber diameter on surface area. Reprinted with permission from [3]; Copyright 2001 Elsevier.

capacity for attachment or release of functional groups, absorbed molecules, ions, catalytic moieties, and nanoparticles [2].

The size of conductive fiber has an important effect on system response time to outside stimuli such as electricity, temperature, light, and humidity. According to Nabet [4], by reducing the size of an electrical conductive wire, we can expect to simultaneously achieve better rectification properties as well as better electron transport in a nanowire [5]. This could be attributed to intrinsic fiber conductivity effect or the geometric surface and packing density effect or both, as a result of the reduction in fiber diameter. Similar principles are applicable to the size effect on other types of stimuli sensitivity of nanofibers.

Nanofibers can be categorized in different ways. They can be continuous filaments or discrete staple fibers. They can be classified as naturally resourced or synthetic based on their material source. They may be organic or inorganic depending on their chemical nature. They can be distinguished as blend, core shell, side by side, or island-in-sea with regards their structure. One of the most popular ways to differentiate nanofibers is according to their functional component/potential application as the initial burst of enthusiasm for nanofiber technology is the diverse advanced applications. Table 2.1 summarizes the properties of common nanofiber materials and their applications.

Table 2.1 The materials for nanofiber fabrication, their properties, and applications. Note: The table has been arranged by the author.

Properties	Materials		Applications
	Functional component	Polymer	
Mechanical reinforcement	CNT, graphene	Carbon precursor (cellulose, lignin, PAN, etc.), PI	Structural reinforcement [6–9]
Electrical conductive	CNT, graphene, Ag, Cu, Ni, Ag	PEDOT, PANi, PPy, PAN	Battery [10], supercapacitor [11, 12], sensor [13–16], wearables [5]
Thermal insulative		PAN, PU	Thermal insulator [17, 18]
Magnetic responsive	Fe_3O_4, Fe_2O_3		EMI shielding [19–21]
Semiconductive	MnO_2, Si		Supercapacitor [22, 23], battery [24–26]
Optical	Quantum dot (CdS, CdSe, CdTe, CuS)		Sensor [27], solar cell [28–30], fluorescence material [31]
Photocatalytic	TiO_2, ZnO, SnO_2, ZrO_2, TiO_2/SnO_2, TiO_2(B), NiO/ZnO, TiO_2/ZnO		Photocatalysts [32–35], optoelectronic [36], hydrogen evolution [37], solar cell [38–42], sensor [43, 44]
Piezoelectric	ZnO, PZT	PVDF, PVDF-TrFe	Pressure sensor [45, 46], nanogenerator [47–50]
Photosensitive	Azobenzene	PDBTT, P3HT, NO_2SP	Photocurrent therapy [51, 52], photochromic material [53], drug release [54]
Thermal sensitive		PANIPAAm	Drug release [55], thermoresponsive wettable surface [56, 57]

Phase changing	Paraffin wax, soy wax, fatty acid (lauric acid, myristic acid, palmitic acid, stearic acid)	PEG, PEG-g-PAN	Temperature regulation [58–60]
Antibacterial/ biological	TiO_2, ZnO, ZrO_2, Ag, peppermint oil, chitosan, OCT, CHX, chitin, sodium alginate	Chitosan, PLGA, PLA, PLLA, HA, PCL, PHBV, gelatin, collagen, etc.	Tissue engineering scaffold [39, 61, 62], wound dressing [39, 63, 64], antifouling [65], drug release [66]
Responsive wettability	TiO_2, ZnO	PANi, PNIPAAm	Oil/water separation [67, 68], control release [69]
Superhydrophilic		PVA, PAA, HA, PU, PEO	Water filtration [70], oil/water separation [71, 72], scaffold [73], antifogging [74]
Superhydrophobic	SiO_2, TEOS	PTFE, P(VDF-TrFE), PP, PVDF, PU, PS	Anti-fog coating [75], self-cleaning surface [75, 76], oil/water separation [77], water-repellent surface [57, 78, 79]
Superoleophobic		PVDF-HFP, PFOTS, PDMS	Oil/water separation [18, 80–82], self-cleaning surface [83]
Fire retardant	MMT nanoclay		Flame retardant [84, 85]

CNT: carbon nanotube, PAN: polyacrylonitrile, PI: polyimide, PEDOT: poly(3,4-ethylenedioxythiophene), PANi: polyaniline, PPy: polypyrrole, PU: polyurethane, PVDF: poly(vinylidene fluoride), P(VDF-TrFE): poly(vinylidene fluoride-trifluoroethylene), PDBTT: poly(N,N-bis(2-octyldodecyl)-3,6-di(thiophen-2-yl)-2,5-dihydropyrrolo[3,4-c]pyrrole-1,4-dione-alt-thieno[3,2-b]thiophene), P3HT: poly(3-hexylthiophene), NO_2SP: 1',3',3'-trimethyl-6-nitrospiro(2H-1-benzopyran-2,2'-indoline), PNIPAAm: poly(N-isopropylacrylamide), PEG: polyethylene glycol, OCT: octenidine·2HCl, CHX: chlorhexidine digluconate, PLGA: poly(lactic-co-glycolic acid), PLA: polylactic acid, PLLA: poly(l-lactide), HA: hyaluronic acid, PCL: polycaprolactone, PHBV: poly(3-hydroxybutyrate-co-3-hydroxyvalerate), PVA: polyvinyl alcohol, PAA: polyacrylic acid, PEO: polyethylene oxide, PTFE: polytetrafluoroethylene, PP: polypropylene, PS: polystyrene, PVDF-HFP: polyvinylidene fluoride-co-hexafluoropropylene, PFOTS: 1h,1h,2h,2h-perfluorooctyltrichlorosilane, PDMS: polydimethylsiloxane, MMT: montmorillonite.

2.3 Nanofiber Fabrication Technologies and Electrospinning

There are a number of techniques for fabrication of nanofibers. These techniques include composite (conjugated) spinning, chemical vapor deposition (CVD), drawing, template synthesis, self-assembly, melt blowing, electrospinning, and more. Each technique has its advantages and disadvantages. For instance, composite spinning derives from conventional fiber spinning technology, which means it can produce continuous nanofibers at a high throughput. However, this method is only suitable for a small group of polymers, the diameter of the fibers is usually close to a few micrometers, and the polymeric "sea" has to be dissolved to separate the "island" fibrils. Melt blowing is another technique that can massively produce nanofibers. But it also has disadvantages: restricted applicable polymers, microscale dimension, and nonwoven mat product form. The majority of the rest of the technologies such as CVD, drawing, template synthesis, and self-assembly are mainly employed in lab research studies for their well-known disadvantages of low productivity, discrete fiber form, and restricted product types. Therefore, the practical generation of fibers down to the nanometer scale was not realized until the rediscovery and popularization of the electrospinning technology by Professor Darrell Reneker [86, 87] in the 1990s.

According to Doshi *et al.* [86], electrospinning can be considered as a variation of the electrospray process (see Figure 2.2). Different from conventional fiber spinning techniques (melt, dry, or wet spinning) that rely on mechanical forces to produce fibers by extruding the polymer melt or

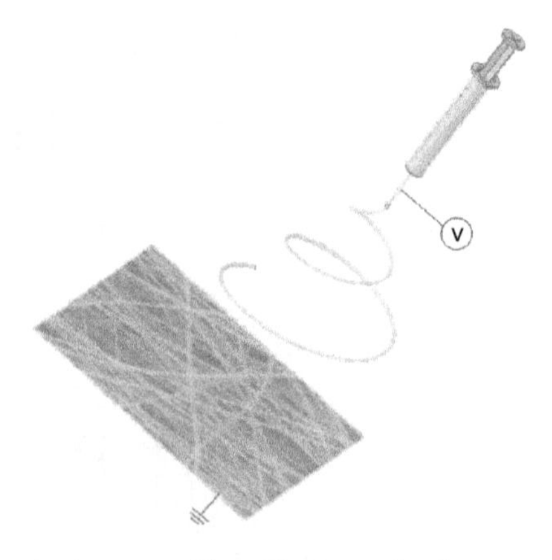

Figure 2.2 An electrospinning setup. Note: The image has been prepared by the author.

solution through a spinneret and subsequently drawing the resulting filaments as they solidify and coagulate, electrospinning produces ultra-fine fibers at a commercially viable level using electrostatic forces to attenuate polymer solution/melt jet. In a typical electrospinning process, high-voltage static electricity is applied to a capillary tip generating an electric field between the capillary tip and a grounded collector; polymer dope flows through the capillary and forms a pendent droplet, which is transformed to a hemispherical shape and then into a conical shape (known as Taylor cone) by the electrical force, at the tip of the capillary. When the electric force surpasses the surface tension of the polymer droplet, a jet is ejected flying towards the grounded collector. The jet undergoes whipping motion (bending instability) in a cone-shaped volume induced by electrical charge imbalance resulting in stretching and attenuating of the jet. The jet solidifies as a continuous ultra-fine filament and deposits on the grounded surface. The fibers can be collected as randomly oriented or aligned in certain directions [88–90].

The advantages of electrospinning technology can be summarized as applicability to variety of polymers, easy adjustability of product properties, and versatility in terms of composition, function, and product form. Hundreds of polymers have been successfully electrospun into nanofibers including natural and synthetic polymers, small-molecular and large-molecular polymers, organic and inorganic chemicals, etc. Numerous parameters can affect the transformation of polymer solution/melt into nanofibers in electrospinning. These parameters include a) dynamic properties of the spinning dope such as viscosity, conductivity, surface tension, and elasticity; b) governing variables such as applied voltage at the spinneret, the tip-to-ground distance, the hydrostatic pressure in the capillary tube, and the parameters of the capillary; and c) ambient conditions such as ambient temperature, air flow, and humidity in the electrospinning chamber. Through adjusting these parameters, the morphology and diameter of electrospun nanofibers, as well as the thickness and porosity of the fiber assemblies, can be easily controlled. Electrospun nanofibers having smooth, beaded, or porous morphologies have been fabricated simply by controlling the polymer solution viscoelasticity or by selection of the solvent system. Bicomponent or multicomponent nanofibers can be produced from a polymer blend or through conjugated electrospinning. Mixing functional fillers in the polymer dopes yields functional nanofibers with magnetic, electric conductive, thermal conductive, bioactive, stimulus sensitive, and catalytic properties. Therefore, electrospun nanofibers have found numerous applications in a wide range of fields such as filtration [91–93] and separation, catalysts [32–35], electrodes for batteries, fuel cells, and supercapacitors [10–12], drug delivery [94–97], tissue engineering [98–103], sensors [43, 104–110], and, of course, smart textiles by introducing the above-mentioned functionalities to fabrics [111, 112].

2.4 Smart Nanofibers and Their Applications in Textiles

As the society evolves, people are caring more about their life quality. Clothing as one of the necessities of life is drawing more and more attention. Besides the aesthetic requirement, functionality and adaptiveness become the most important selling points that most brands are competing on. From the point of view of the manufacturer, functional and adaptive textiles help them gain loyal customers and earn greater interests for the added value.

The functionalities of a garment are more subjective while the adaptiveness is a more objective concept. Subjective functions focus on satisfying a wearer's aesthetic and comfort requirements. Adaptive functions relate more to transitional conditions. An adaptive garment is required to satisfy a wearer's aesthetic and comfort requirements regardless of his/her activities and the varying environment. In the field of textiles, comfort is described mainly in "terms" of softness, smoothness, warmth/coolness, and dryness. Softness and smoothness are believed to be determined by the stiffness and roughness of a fabric, which can be achieved by selection of a proper material. Generally, a textile material possessing low initial Young's modulus and low friction of coefficient offers the feel of softness and smoothness. On the other hand, warmth/coolness and dryness are more challenging to be realized as they are always associated with activity and the environment. An adaptive garment is often required to be able to keep body temperature constant and be independent of the wearer's activity and the ambient environment, provide comfort feel even when sweating, prevent permeation of rain and wind, be easy caring, etc. These functions are usually summarized as thermal regulation, moisture regulation, waterproof, windproof, and self-cleaning. Special functions, such as biological and chemical protection, are also requested for specific textiles [113].

Nowadays, the boundaries of smart textiles have been pushed wider as intelligent wearables have become a highlighted technology. The future of wearables is believed to reside in smart fabrics that can adapt to the user's needs or ambient conditions in a variety of contexts, from sports and fitness, protection, and health monitoring to chronic disease management.

2.4.1 Moisture Management and Waterproof

Permeability of textiles to air, water vapor, and water plays critical roles with regard to the comfort of the wearer. The air permeability of a fabric is a measure of the rate of air flow passing perpendicularly through a known area under a prescribed air pressure differential between the two surfaces of a material the fabric [114]. It is generally expressed in SI units as $cm^3/s/cm^2$ and in inch-pound units as $ft^3/min/ft^2$. Air permeability is an important factor in the performance of such textile materials as gas filter, fabrics for air bags, clothing, parachutes, tentage, etc. In clothing, air permeability is tightly related to the moisture regulation capability of a fabric. It can also be an indicator of the breathability of weather-resistant and waterproof fabrics. Water vapor permeability or moisture vapor transmission rate (MVTR), also known as breathability, measures how quickly water vapor/moisture passes through a fabric or other substance. It is usually measured in $g/m^2/day$, or the mass of moisture that passes through a square meter of fabric in 24 hours. Breathability is an important parameter for the wearing comfortability of water-resistant and waterproof fabric. While air-permeable waterproof fabrics tend to have relatively high moisture vapor transmission, it is not necessary to be air permeable but breathable. When it comes to the water permeability of a fabric, people are practically trying to find out how well a fabric can prevent the penetration of water/rain drops. Three terms can come across regarding the water-blocking capability: water-resistant, water-repellent, and waterproof. Water-resistant fabrics are only able to resist the penetration of water to some degree but not entirely. Water-repellent fabrics cannot be easily penetrated by water, especially as a result of being treated for such a purpose with a surface coating, while waterproof fabrics are impervious to water. Waterproof membranes developed for garments so far exhibit good performance in terms of waterproofness. In practice, multilayered membrane structures are commonly employed. For instance, the Gore-Tex XCR is fabricated with a superhydrophobic PTFE membrane microporous structure and a hydrophilic PU layer. The PTFE membrane provides a superior waterproof performance as the micropore structure is introduced to improve its breathability, and the PU layer seals the pores from pollutants.

Gas and liquid permeate through fabric in two ways: diffusion through the polymer matrix or open structures or pores. Obviously, the diffusion rate is much higher by going through air than through polymer matrix, i.e., the larger the pore volume of a fabric, the higher the permeability the fabric will have. For planar orientations, the porosity of an electrospun nanofiber membrane is typically around 90% [115], which guarantees a high moisture permeability of electrospun nanofiber membranes. Kadam *et al.* [116] correlated the porosity of electrospun PAN nanofiber to air permeability and found a strong positive correlation between the air permeability and porosity. Comparing PAN nanofiber mats with different porosities, Borhani *et al.* [117] also proved that the nanofiber mat with higher porosity has higher air permeability, as well as heat and moisture transfer (see Figure 2.3).

Wettability is a critical factor that affects the moisture and sweat management performance of fabric. Hydrophobic surface usually shows water-repellent property. When a fabric is made of a hydrophobic fiber, a thin layer of hydrophilic coat will be applied to make the fabric wettable, and thus capillary wicking of moisture and sweat can take effect (see Figure 2.4). Coating of PVDF nanofiber membrane with 10% PVA aqueous solution reduced the contact angle of the membrane from 91±2.4° to 0° [118].

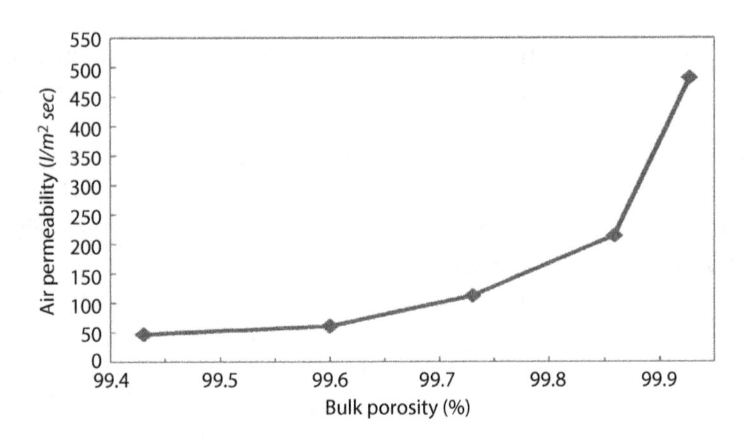

Figure 2.3 Air permeability versus bulk porosity of nanofiber mats. Reprinted with permission from [117]; Copyright 2011 Springer Nature.

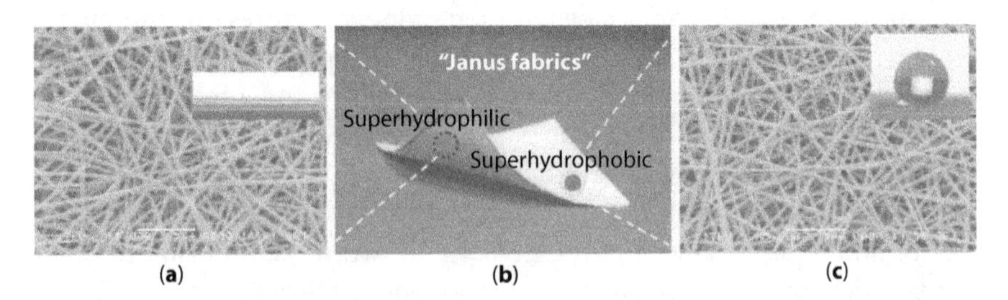

Figure 2.4 Electrospun Janus fabric (a and c) SEM images and contact angles of the electrospun PAN-tetraethyl orthosilicate nanofibrous mats (a) thermally hydrolyzed and (c) pristine, respectively. (b) Dual-layer electrospun Janus fabric demonstrating asymmetric wettability. Reprinted with permission from [79]; Copyright 2010 American Chemical Society.

High air permeability fabrics and waterproof fabrics are the two extremes of functional textiles. High air-permeable fabrics have good moisture management capability but poor water resistance. On the other hand, waterproof fabrics are always blamed for their low breathability. The emergence of nanofiber membranes brings the possibility to have water-repellent textile materials with high breathability, i.e., good moisture management. The radius of smoky rain droplet and clean cloud rain droplet is around 9 and 14 μm, respectively [119]. Most of the heavy rain droplets have radii much larger than 150 μm [120]. The pore diameters of electrospun nanofiber membranes range from 0.1 to 0.8 μm [3], and the pore size can be significantly modified by controlling the overall porosity [115] and the spinning process parameters [121]. Therefore, it is easy to see that the pore size of a nanofiber membrane can be designed to be air and water vapor permeable but impervious to rain droplets. Electrospun PU nanofiber membrane-laminated fabric systems exhibited better water vapor transport and air permeability compared to PU and porous PTFE-coated fabrics, and better water resistance than PU fabric [122]. By replacing a bilayer PTFE/PU membrane with an electrospun PU nanofiber membrane, Hong *et al.* [123] successfully fabricated a highly breathable waterproof fabric. This fabric showed a considerably higher degree of breathability but relatively lower waterproof capability when compared with Gore-Tex XCR.

Water droplets on smooth and flat hydrophobic solid surfaces usually form contact angles larger than 90° but smaller than 120° even on surfaces with very low surface free energy [124]. The principle connections between surface roughness and water repellency were investigated by Cassie and Baxter [125]. Scanning electron microscope (SEM) studies on microstructural surfaces of plants showed that the water repellency of plant leaves is mainly caused by epicuticular wax crystalloids, which cover the cuticular surface in a regular microrelief (slight irregularities elevation of a surface) of about 1–5 μm in height [126–128]. Therefore, mimic of superhydrophobic surfaces often copies the mentioned two characteristics: hybrid rough microstructure and hydrophobic surface. Electrospinning has been proved a facile technology for fabrication of superhydrophobic nano/microfibrous membranes. Study on electrospun PNIPAAm fibers at room temperature and elevated temperature proved that nanofibrous structure increases hydrophilicity of a hydrophilic material and the hydrophobicity of the hydrophobic material [57]. Hydrophobic PU [129–132], PS [133–136], and PVDF [137, 138] are the most commonly utilized polymers for fabrication of hierarchical rough microstructures. The rough microstructures are usually enhanced by introducing beads [139, 140], rods [141], microgrooves [142], or pores/dents [135, 143]. Bead-on-string PS fibers can be simply prepared by optimizing the spinning dopes [133, 136]. The water contact angle can be as high as 160.4° [133]. The water contact angle of bead-on-string PVDF nanofiber membrane was reported at 148.5° [138]. By attaching nanoparticles onto nanofibers and then sonicating the particles away, Birajdar and Lee [144] demonstrated that both nanoscale bumps and dents can make a nanofiber membrane superhydrophobic. Incorporation of low surface energy fluorinated polymers to nanofiber membranes can further enhance the water resistance, air permeability, and water vapor transmittance [145–148]. Fluorinated PU (FPU) modified PU membrane was found to exhibit amphiphobicity showing a water contact angle of 156° and an oil contact angle of 145°. Fractal dimensional analysis of the membrane confirmed a positive correlation between hierarchical roughness and amphiphobicity [147]. Hydrophobic SiO_2 nanoparticle is also a popular material for improved surface roughness and hence better hydrophobicity [149, 150]. It is noteworthy that waterproof is commonly used to describe a superhydrophobic membrane, though the membrane may be just water repellent or water resistant, especially when it comes to nanofiber membranes.

Regarding the smart waterproof function, Chen and Besenbacher [151] conjugated light-sensitive azobenzene with PCL nanofibers and created a light-responsive membrane. Upon UV irradiation, azobenzene undergoes a reversible trans-to-cis isomerization and, thus, a reversible change in the wettability of the nanofiber membrane from hydrophobic to hydrophilic. Using thermoresponsive polymer PNIPAAm for spinning, nanofiber membranes that switch wettability from hydrophilic to hydrophobic when temperature raised from 25°C to 50°C were also reported [56, 57].

2.4.2 Thermoregulation

The human body is like a heater emitting heat all the time. In order to maintain a suitable environment for vital organs and central nervous system, the body temperature is regulated within a fairly narrow range (36–40°C) considered as the functional working range, 36.5–37.5°C at rest. Extended protection of the human organism against heat loss/overheat in specified climatic conditions is mainly reliant on clothing.

Heat transfer from materials is based on convection, conduction, and radiation. Heat transfer through a solid object is via conduction. Thermal insulation grows as a directly proportional function of the air fraction in the material. The greater the volumetric porosity, the smaller the thermal conductivity of the material [152]. Electrospun structures are highly porous with much of the volume filled by air. Therefore, heat conduction through the solid portion of the matrix (the fibers) is negligible. Heat transfer through convection is dominant where fluid or air is allowed to move where there is a temperature gradient. For the electrospun nonwoven structure, the high surface area of the fine fibers and the fiber entanglements impede the free flow of air within the structure and thus limit convection [17, 153]. However, slip flow may allow heated air to travel across more easily, which should be thoroughly studied. The radiation component in heat transfer through fibrous insulation is more important. Thermal radiation can account for 40–50% of the total heat transfer in low-density fibrous insulation at moderate temperatures [17]. Based on theoretical analysis, it is believed that the thermal insulating efficiency of fiber-based insulation increases as the fiber size is reduced. Therefore, many investigations have been conducted on electrospun nanofibers for the application of thermal insulation.

One of the earliest studies is by Gibson and Lee [17] using electrospun PAN nanofibers. This nanomaterial showed excellent reduction in overall heat transfer compared to standard low-density fibrous insulating materials (at bulk densities below 50 kg/m^3). Thermal conductivity testing confirmed that decreasing fiber diameter tends to increase the thermal resistance of fibrous insulation materials, which is in accordance with the research studies conducted by Wang and Wang [154] and Sabetzadeh *et al.* [155]. However, Gibson and Lee [17] found that thick high loft electrospun insulation materials did not show significant thermal performance improvement over commercially available insulation materials, as pure nanofiber mats have poor insulation properties at low densities, though the insulation properties of compressed nanofiber mat were found to be approaching those of the aerogel/fiber composites at high bulk densities. They suggested that better thermal insulation properties may be achieved by incorporating a proportion of nanofibers into existing insulation structures or modifying the infrared radiation properties of nanofiber. Correspondingly, Kim and Park [156] found that the thermal resistance of electrospun PU web increased by 30–40% compared with untreated nanowebs, by depositing thermal reflective aluminum (Al) on the surface of electrospun PU nanofibers.

Due to the lack of supportive structural fibers, electrospun nanofiber mats have low thickness and weak resistance to compression. Thus, direct application of electrospun nanofiber mats for thermal insulation in textile is rare. Most research efforts have focused on using electrospun nanofibers as the holding matrix of phase changing materials for the application of thermoregulation. Phase change materials (PCMs) can store heat energy in a high-temperature environment and release the heat energy in a low-temperature environment through phase change while the temperature of the PCMs does not change much. However, the thermal conductivity levels of most PCMs are too low to provide a required heat exchange rate between the PCM and the substrate [157]. Microencapsulation of PCMs was proposed to improve the thermal performance of conventional PCMs by providing more heat transfer area per unit volume [158]. Microencapsulation also brings the advantage of preventing PCMs from contaminating the matrix materials. Accompanying the emergency of electrospinning technology, electrospun nanofiber has become a favorite carrier matrix for PCMs for its advantages based on high surface area, high porosity, and easy encapsulation process. PEG-g-PAN copolymers with various molecular weights of PEG were synthesized for fabrication of solid–solid PCM nanowebs via two-step free radical polymer grafting and coaxial electrospinning the copolymer with PAN [159]. The nanowebs demonstrated repeatable solid–solid phase change with the heat storage capacities varying between 35 and 75 J/g.

2.4.3 Personal Protection

Electrospinning boomed as a nanofiber fabrication technology after the U.S. Army Natick Soldier Research and Development Center proposed their nanofiber research for soldier protection, which evoked people's enormous enthusiasm in exploring the potential applications of electrospun nanofibers. The proposal brought up the concept of nanofiber-based uniform that can provide ballistic protection, extreme weather protection, and chemical/biological protection. For commercial textiles, the research of nanofiber is more focused on body protection from environmental threats as illustrated in Figure 2.5.

Fire protection is one of the major functions of protective clothing. Currently, high-performance flame-retardant fibers, such as Nomex® and polybenzimidazole (PBI), have been used for firefighter and racer suits demanding excellent fire resistance. However, application of these fibers for mass-market consumer products is restricted by their high cost. The potential of composite nanofibers with low-cost flame-retardant components has been explored considering the excellent thermal insulating property of porous nanofiber mats. Wu *et al.* [84] prepared flame-retardant polyamide (PA) 6 (nylon 6) nanocomposite nanofibers by introducing montmorillonite clay (MMT) platelets and intumescent non-halogenated flame-retardant (FR) additives to nylon 6 nanofibers. Their microscale combustion calorimeter testing results showed that FR particles played a major role in reducing flammability of the material in both solution- and melt-compounded

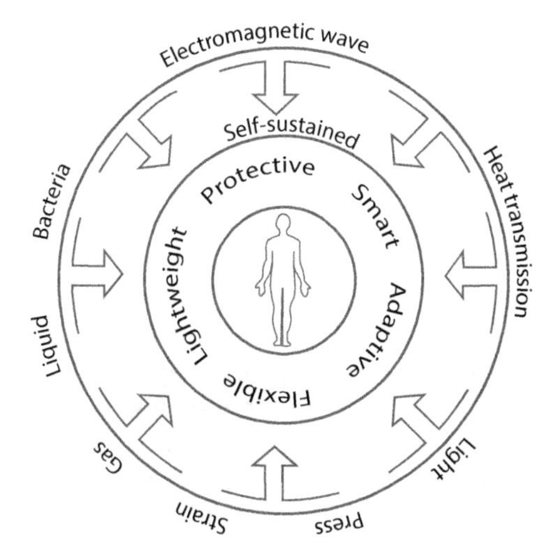

Figure 2.5 Textiles for personal protection. Note: The image has been prepared by the author.

samples and exhibited higher effectiveness as it is in melt-compounded samples. Through a similar method, Yin *et al.* [85] produced multicomponent FR-carbon nanotube (CNT)-PA6 nanocomposite nanofibers. The combustion properties of the fibers including both heat release rate and total heat release were significantly improved while thermal stability appeared to be compromised. With proper FR additive concentrations, synergism between multiwalled carbon nanotubes (MWCNTs) and nanoclay was observed.

Metal oxides are photocatalytic materials that absorb light, degrade chemicals, and kill bacteria through catalysis. ZnO is one of the most commonly used catalysts. By simply mixing ZnO nanoparticles with a polymer solution and electrospinning the mixture into fibers, UV-protective fiber mat was obtained [160]. This ZnO-embedded nanofiber mat was layered with a fabric. The layered fabric systems obtained UV-protective property from the ZnO nanofiber mat. The protection effect increased with increasing zinc oxide concentrations of the nanocomposite fiber mat and the area density of the mat. Aiming to solve the aggregating problem, Vitchuli *et al.* [161] combined electrospray process with electrospinning. ZnO nanoparticles were dispersed on the surface of Nylon 6 nanofibers. The prepared ZnO/Nylon 6 nanofiber had excellent antibacterial efficiency (99.99%) against both Gram-negative *Escherichia coli* and Gram-positive *Bacillus cereus* bacteria, and good detoxifying efficiency (95%) against paraoxon, a simulant of highly toxic chemicals. Depositing this nanofiber mat onto nylon/cotton woven fabrics did not impair the water vapor and air permeability of the fabrics significantly. Therefore, ZnO/Nylon 6 nanofiber mat can be a promising material for protective textiles. Besides ZnO, other antibacterial and chemical-warfare agents such as silver [162], activated charcoal [163], zirconium metal–organic frameworks (UiO-66-NH_2) [164], silica-encapsulated peppermint oil (PO), octenidine·2HCl (OCT), and chlorhexidine digluconate [165] were also investigated for electrospinning technology-based antibacterial and chemical decontamination materials.

The electromagnetic interference (EMI) shielding effect of electrospun nanofibers can be gained in two ways: embodiment of an electrical conductive material (reflection mechanism) or a magnetic material (absorption mechanism). Im *et al.* [21] combined conductive polyaniline (PANI) with conductive MWCNTs using polyethylene oxide (PEO) as the fiber matrix. To improve the electromagnetic interference shielding efficiency, carbon nanotubes (CNTs) were fluorinated for better dispersion and adhesion in the polyaniline electrospun fibers. The electrical conductivity of the fibers, improved with the fluorinated CNTs, reached the value of 4.8×10^3 siemens/m with an EMI shielding effectiveness (SE) of 42 dB. For better EMI SE, materials having electrical conductive and magnetic properties are usually jointly used. Kim *et al.* [166] prepared EMI shielding nanowebs by metallizing electrospun PVA nanofibers with various metals (Cu, Ni, Ag). They found that the EMI SE was dominated by the absorption mechanism as the metal thickness strongly affected the SE value of each metal-deposited nanofiber mat. They believed that the unique nanofibrous porous structure also played an important role in enhancing the absorption effect. In another research, they prepared Ag-coated PVA/Fe_2O_3 composite nanofibers, and with the resultant fiber mat, they obtained an EMI SE as high as ~45.2 dB [167]. Since carbon, as a conductive fiber matrix, is more attractive for EMI shielding application, magnetic nanoparticles-incorporated carbon nanofibers have been explored by research groups. By electrospinning of ferrous acetate ($FeAc_2$)/PVA solution and subsequent calcination of the resultant fibers at high temperature, hybrid Fe_3O_4/carbon nanofibers were produced by Zhu *et al.* [168]. Bayat *et al.* [169, 170] directly dispersed Fe_3O_4 nanoparticles in PAN solution and fabricated Fe_3O_4/carbon nanofiber. The maximum EMI SE of 67.9 dB was obtained for a 0.7-mm-thick carbon composite nanofiber mat containing 5 wt.% Fe_3O_4. The shielding was also found to be dominated by absorption of electromagnetic radiation.

With respect to screening against telltale infrared heat radiation (IR radiation), Al doped zirconia nanofibrous membranes were believed to be a promising infrared stealth material in the confrontation strategy field for defense, aircraft, satellites, etc. [113]. This material exhibited low infrared emissivity levels of 0.589 and 0.703 in the 3–5 and 8–14 μm wavebands, respectively.

Shielding efficiency of nanoparticles in air or liquid mainly relies on the geometry of the electrospun fiber mat. Fiber mat thickness was found the key parameter that strongly affects nanoparticle blocking/filtering results based on the studies on polyamide-66 nanofiber mat laminated/deposited on nonwoven substrates. In these studies, it was shown that the filtration efficiency could be increased to over 99% when the fiber mats were thick enough [171, 172]. By directly electrospinning PVDF/PTFE NP polymer solution into nanofiber, electret nanofiber membranes with numerous charges and desirable charge stability were successfully fabricated with PVDF as the matrix polymer and PTFE nanoparticles (PTFE NPs) as the charge enhancer. The resultant fibrous membrane exhibited a high filtration efficiency of 99.972% with a low pressure drop of 57 Pa, as well as a satisfactory quality factor of 0.14 Pa^{-1} and superior long-term service performance [173].

2.4.4 Wearables and Sensors

Wearables are smart electronic devices such as electronics and sensors that can be worn on the body. Wearables are usually associated with fabrics for the sake of lightweight, long-lasting, flexible, conformable, and wearable devices. Commercially available wearables made with flexible textiles are still rare. In research, reports on wearables made with nanofibers mainly fall under two categories: energy-harvesting and storage devices, and sensors.

A high-quality power source is essential for sustainable wearables. Besides use of batteries, a self-powered system that can harvest energy from multiple sources that are available in our personal and daily environments such as body movement, pressure, and friction is highly desirable. Various energy-harvesting systems have been successfully assembled. Piezoelectric materials are one of the most important multifunctional materials that have been extensively used for energy harvesting and sensor applications. Many piezoelectric materials, such as lead zirconate titanate (a piezoelectric ceramic, known as PZT) [174], PVDF [175], and P(VDF-TrFE) [46], have been electrospun into nanofibers for fabrication of nanogenerators. The type of piezoelectric material selected for a power-harvesting application can have a major influence on the harvester's functionality and performance [176]. PZT is widely used as a power-harvesting material with a high piezoelectric coefficient. However, this piezoceramic is extremely brittle and thus is difficult to manipulate during the fabrication process and is susceptible to fatigue crack growth when subjected to high-frequency cyclic loading. Wu *et al.* [174] successfully developed a suspending sintering technique for fabrication of flexible, dense, and tough PZT textile composed of aligned parallel nanowires from electrospun nanofibers. This PZT textile was then transferred onto a PDMS-coated PET film and made into a flexible and wearable nanogenerator. The maximum output voltage and current of the nanogenerator were reported to be 6 V and 45 nA, respectively. This nanogenerator was demonstrated to light a commercial LCD and power a ZnO nanowire UV sensor to detect UV light. Another common piezoelectric material is PVDF. Thank to the polymeric nature, PVDF exhibits considerable flexibility compared to PZT and other piezoelectric ceramics. PVDF nanogenerator can be easily fabricated with electrospun PVDF nanofibers. A single PVDF nanofiber has been deposited across a pair of electrodes using a near-field electrospinning process (direct writing) to harvest small mechanical vibrations [177]. The resulting nanogenerators exhibited a maximum voltage of 30 mV and current of 3 nA. The peak output voltage of randomly oriented PVDF nanofiber-based nanogenerator was reported to be 1 V [178]. By massively depositing aligned PVDF fibers, Fuh *et al.* [179] successfully increased the peak output voltage and current of PVDF nanofiber-based nanogenerator to 1.7 V and 300 nA, respectively. The mechanical-to-electrical energy conversion efficiency was shown further improvement through the increase in the β crystal phase of PVDF with high electric field involving needleless electrospinning. The voltage output of randomly oriented PADF nanofiber-based nanogenerator increased from just less than 1 V to about 2.6 V, as the current boosted from around 1.4 to 4.5 μA [180].

Another type of electrospun nanofiber-based power generator worth mentioning is triboelectric nanogenerator (TENG). Triboelectrification is one of the most common effects in our daily life, which is usually taken as a negative effect. Triboelectric nanogenerators are based on the conjunction of triboelectrification and electrostatic induction, and they can utilize the most common materials available in our daily life, such as PTFE, PDMS, PVC, etc. [181–183], Using PDMS-coated electrospun PVDF and PA6 coated PAN nanofiber membranes as the power-generating materials, Li *et al.* [184] fabricated a nanofibrous membrane constructed wearable triboelectric nanogenerator. With an effective device area of 16 cm^2 under gentle hand tapping, this nanogenerator demonstrated a current and voltage output of up to 110 μA and 540 V, respectively, and a capability of sustainably powering an electronic thermometer, an electronic watch, and 560 LEDs.

A considerable amount of work on electrospun nanofiber-based sensors, such as chemical sensors, optical sensors, and biosensors, has been reported [105, 108, 109]. The application of chemical sensors for wearables is rare when compared with mechanical sensors. Mechanical sensors are mainly based on the principle of change in the resistance/conductance of the incorporated conductive sensing component responding to external stimuli such as pressure and strength/strain [185]. Strain sensors are fabricated with conductive fibers [104, 186–191]. When the fiber or fiber network is stretched or elongated, the resistance/electric conductance will increase/decrease. The relationship between the elongation and the resistivity can be correlated; thus, the deformation of the strain sensor can be measured by measuring the resistivity of the sensor. Strain sensor, on the other hand, can be regarded/used as a pressure sensor as the sensor is elongated perpendicularly to the compressed direction. Pressure sensor can also be made with piezoelectric materials [185, 192]. When pressure is applied to the pressure sensor, an electric signal is measured to quantify the pressure. In most cases, pressure sensors are sensitive to mechanical deformation (bending). When a soft object is pressed by another soft object, the normal pressure cannot be measured independently from the mechanical stress [193]. To reduce the bending sensitivity, Lee *et al.* [193] used carbon nanotubes and graphene composite nanofibers to fabricate their pressure sensor. When compressed, these fibers change their relative alignment to accommodate bending deformation, thus reducing the strain in individual fibers. The sensor showed real-time (response time of ~20 ms), large-area, normal pressure monitoring under different, complex bending conditions.

In order to broaden utilization of displays in wearable electronics, transparent and flexible LEDs that can be integrated with fabrics suggest another field of wearables that requires development. Using keratin that is extracted from human hair, Park *et al.* [194] fabricated a transparent nanofiber layer for the cover of flexible consolidated polymer light-emitting diodes (PLEDs). The LED characteristic of this PLEDs was found maintained the device efficiencies similar to PLEDs without the cover layer. This research extends the application of electrospun nanofibers to transparent protection for wearable electronics.

2.4.5 Medical Care

The use of polymer nanofibers for biomedical and biotechnological applications has many intrinsic advantages [2]. Polymer nanofibers can provide a proper route to emulate or duplicate biosystems—a biomimetic approach. Cells live in a nano- or micro-featured environment and, therefore, they attach and organize well around fibers with diameters smaller than themselves. Many research studies [195–199] have shown evidences that apart from surface chemistry, the nanoscale surface features and topography have important effects on regulating cell behavior in terms of cell adhesion, activation, proliferation, alignment, and orientation [198]. Recently, electrospinning has been widely studied as an enabling technology to develop bioactive nanofibers for various biomedical applications including tissue engineering, drug delivery, and biosensors. From the point of view of medical textiles, wound dressing is highlighted as one of the most favorite applications of electrospun nanofibers.

Wound healing involves four processes, namely, homeostasis, inflammation, granulation tissue formation, and remodeling, which overlap in time [200–202]. Modern wound dressings are tailored to enable promoting recovery and preventing further harm from the wound more than covering it [203]. Dressings for wounds are designed to provide protection, removal of exudate, inhibition of exogenous microorganism invasion, accelerated healing, and improved appearance [204]. Electrospun nanofibrous membranes have higher air permeation [205], can prevent wound from infection [206] and dehydration [64, 204, 207], and promote wound healing [208–210] when combined with controlled drug release function. The air permeability and prevention of dehydration can be adjusted by design of fiber mat structure parameters such as the porosity and the thickness [3, 211, 212]. Protecting a wound from infection and inflammation is generally achieved by introducing antibacterial components such as Ag [64, 213–216], chitosan [207, 217–219], ZnO [220, 221], and antibiotic drugs [223–225] to nanofibers. For the need of accelerating wound healing, growth factors [209, 225, 226] are frequently incorporated. In the case of treatment of exudates, electrospun hydrogels and fibers with superabsorbents [227] have attracted more and more attention. Besides the capability of absorbing exudates, fibril hydrogels [217, 228–230] also possess advantage in preventing wound from dehydrating while keeping high breathability to wound. More detailed information can be found in references [231–234].

2.5 Challenges Facing Electrospinning

2.5.1 Enhancement of Mechanical Properties

Electrospinning is a technique that can produce continuous nanofilaments with diameters ranging from a few nanometers to micrometers by fast stretching polymer jet in high electric fields. Electrospun nanofibers have attracted wide attention for their numerous promising potential applications. However, industrialization of electrospun nanofibers has been forcibly restrained for their weaker mechanical properties.

The mechanical properties of electrospun polymeric nanofibers are strongly influenced by their molecular orientation and crystalline structures [235–236]. It has been reported that the degree of crystallinity and molecular orientation of electrospun nanofiber can be enhanced by reducing the diameter of the fibers [235, 237–240]. Therefore, the mechanical properties of electrospun nanofibers have been shown to improve through control of fiber diameter and improvement of fiber supermolecular structures during or after electrospinning [121, 241–243]. Papkov *et al.* [244] improved the elastic modulus of electrospun fibers from 0.36 to 48 GPa, and the tensile strength from 15 to 1750 MPa by reducing the fiber diameters from 2.8 μm to ~100 nm. Their data also showed that the strength and elastic modulus were further improved by annealing. The best recorded data far exceeded the properties of conventional PAN microfibers (250–400 MPa strength and 3–8 GPa modulus). Crosslinking [245] or welding [246] of nanofiber mats and reinforcing nanofibers with nanofillers [6] have also shown improvement effects. By using a highly viscous electrospinning dope, the weak points introduced during fiber solidification can be significantly eliminated as the evaporative solvent is significantly reduced. Research has proved that stronger fibers can be obtained by electrospinning of polymer solutions with higher viscosity [247]. However, as the viscosity increases, polymer molecular chains will be heavily entangled and can be hardly broken by electrostatic force; thus, the electrospinnability of such a dope is severely impaired [9, 248]. Aiming to enable the electrospinning of polymer dopes with high viscosity, sonication has been introduced to treat spinning solutions. Research studies have shown that application of sonication not only can improve the electrospinnability of high-viscosity solutions by reducing solution viscosity and surface tension, and increasing electric conductivity, but also can dramatically reduce fiber diameter, increase fiber density, and hence improve the mechanical properties [121, 249, 250]. However, excessive sonication can cause adverse effects. Systematic studies need to be conducted before sonication can be practically applied for fabrication of strong electrospun nanofibers.

2.5.2 Large-Scale Production

Another huge challenge that hinders the commercialization of electrospun nanofibers is the low productivity of electrospinning equipment. The output of conventional needle/nozzle-based electrospinning systems as described in Section 2.3 is typically 0.1–1.0 g h^{-1} by fiber weight or 1.0–5.0 mL

h^{-1} by flow rate, depending on the viscosity of the polymer solution [251, 252], which is too low to meet the strong market requirement.

Many efforts have been devoted to improve the output of electrospinning systems by multiplexing the electrospinning system either by stacking arrays of capillaries/syringes or operating a multiple-jet mode [253–258]. Significant improvements of output have been reported; however, these systems are still not robust enough for commercialization. One of the major issues is the interference among the electrically charged adjacent jets caused by electrostatic repulsion. Consequently, the inference distorts the paths of the jets resulting in non-uniform thickness of nanofiber mats. To solve the electrostatic repulsion problem, Kim *et al.* [259] applied a cylindrical electrode to a five-nozzle setup. This method is obviously only applicable to a nozzle array of limited amount. Moreover, fluid dripping and nozzle clogging are besetting multiple-jet electrospinning throughout the spinning process. Bottom-up multiple-jet system can avoid the solution dripping problem [260], but nozzle clogging caused by quick evaporation of solvent from the surface of a Taylor tone has not found a solution. Furthermore, cleaning hundreds of needles after each electrospinning process is also a tedious work that people have to face. The invention of rotary porous tube [261], rotary cone [262], and conical wire [263] has shown to be promising in terms of increasing productivity, but still has the unsolved dripping problem with an aggravated drawback of wide fiber diameter distribution.

To avoid the needle clogging and solution dripping problems, as well as the tedious needle cleaning, bottom-up needleless systems including multiple-spike [264], multiple-cleft [265], bubble [266], and rotary drum/wire (nanospider) [267] electrospinning systems have been proposed. In contrast to the needle-based multiple-jet electrospinning systems, there are no needles involved in these setups. The mechanism of these systems is based on electrohydrodynamics (EHD)-induced fluctuations on top of free surfaces of a thin layer of solution [265, 268]: above a certain critical value of the applied electric field intensity/field strength, the free surface of the thin solution layer starts to be self-organized in mesoscopic scale as particular waves with a characteristic wavelength. The amplitude of these waves quickly grows as the strength of the electric field increases until the electric force breaks the surface tension of the wave crests where nanofibers originate. In the multiple-spike multiple-cleft setup, a magnetic field and a metal spike or cleft complex are involved in assisting jet provoking. The rotary drum and bubble systems are, on the contrary, trying to initiate "wave jets" through the drum or by blowing bubbles. When compared with needle systems, the needleless electrospinning methods do not have the issues associated with needles. The problem facing needleless electrospinning is the large diameter variation induced by the free surface. Besides, the productivity of needleless electrospinning is still low due to the use of low viscosity/concentration polymer solutions where the solvent, taking a large portion of the electrospinning dope, evaporates into air during electrospinning with a very small portion of polymer left in the form of nanofiber.

2.5.3 Formation of Nanofiber-Based Yarn and Fabric

As electrospun nanofiber takes the form of a continuous linear shape, it is an instinct for people to try to transform the nanoscale material into macroscale structures by textile processes. Due to the low productivity of electrospinning systems, the work on fabrication of nanofiber yarns is mainly demonstrated in lab studies. Continuous nanofiber yarns have been fabricated by coating a core yarn/wire [269–272] with nanofibers or directly twisting nanofiber bundles [273–278]. Fennessey and Farris [273] twisted unidirectional aligned, structurally oriented PAN bundles into continuous yarns. The ultimate strength and modulus of the twisted yarns increase with increasing angle of twist to a maximum of 162±8.5 MPa and 5.9± 0.3 GPa, respectively, at an angle of 9.3°. Wang *et al.* [279] combined self-bundling electrospinning technique with post-treatments such as stretching and annealing under conditions similar to those used for conventional fibers to fabricate nanofiber yarns. The strength of the obtained nanofiber yarns approached to values equivalent to conventional fibers as the crystallinity and molecular orientation of PAN nanofibers were improved.

With regard to commercialization of electrospun nanofiber products, direct lamination of electrospun nanofiber mats with conventional fabrics is preferred as nanofiber mats possess the characteristics of high air permeability, programmable waterproofness, excellent sensibility, and high efficient filtration capability enabled protective performance. A few attempts for fabrication of nanofiber mat-laminated fabrics have been conducted. [280–282]. The durability of nanofiber mat laminates such as adhesion strength, abrasion resistance, tensile strength, and tearing strength was found to be the greatest challenge due to the weakness of electrospun nanofibers [283, 284]. Solvent-laminated samples were found to have lower adhesive force, which could only withstand 10 washes, while hot melt laminates have higher adhesive force, which decreases quickly after wash [285]. Rombaldoni *et al.* [286] explore the potential of low-temperature oxygen plasma treatment for enhancement of adhesion between polypropylene nonwoven and various nanofiber mats. The mean adhesion energy was found to have significantly increased from 0.58 and 0.39 J/m^2 to 4.80 and 0.89 J/m^2 for PEO and PA6 nanofibers, respectively. Chemical treatments on conventional fabrics for enhanced adhesion between the fabrics and PA6 nanofibers were also studied [287]. The treatments improved bonding of nanofibers toward fabrics, in particular, adhesion energy, for 60% and 51% on alkali-treated cotton fabrics and ethanol-treated nylon-66 fabrics, respectively. Nevertheless, for practical applications, adhesion should be at least one-order-of-magnitude greater than these reported values [287].

2.5.2 Other Issues

Due to the instability of electrospinning process, the uniformity of the diameter and the deposit spot of a single nanofiber are hardly controllable, and hence the performance of electrospun nanofiber products is still unreliable. Near-field electrospinning [177, 288–293] has been developed to deposit solid nanofibers in a direct, continuous, and controllable manner (see Figure 2.6). Near-field electrospinning uses microscale capillary and low electrostatic force for spinning and collects the fibers at the stable

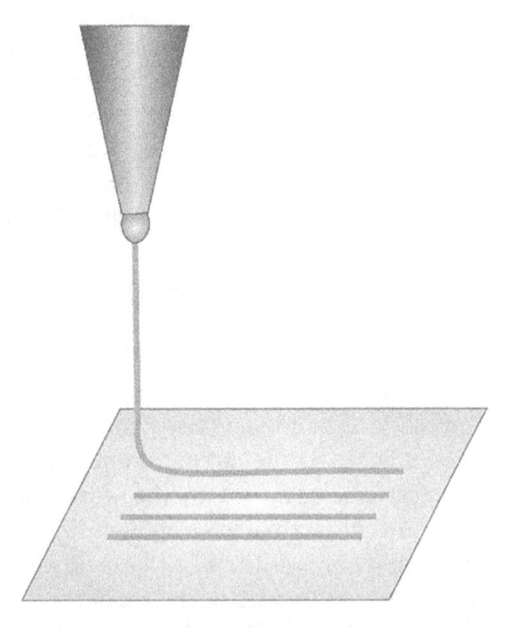

Figure 2.6 Schematic of near-field electrospinning. Note: The image has been prepared by the author.

liquid jet region by reducing the collection distance. However, the productivity of near-field spinning is much lower than large-scale electrospinning technology. Remarkable technological breakthrough needs to be made to take advantage of controllable near-field electrospinning technology and the extraordinary properties of electrospun nanofibers.

The potential applications of electrospun nanofibers have not yet been practically realized, though a vast variety of functional nanofibers have been created. For example, separation properties of electrospun nanofibers for different materials [294–299] have been extensively reported. However, systematic studies of the filtration properties such as pressure drop, durability, and washability of nanofiber-based filters are still vacant. Another instance are nanofiber-based sensors [43, 300–302]; the sensibility and the sensing capability have been intensively studied, yet commercially applicable sensors based on electrospun nanofibers have rarely been reported. Therefore, more practical and realistic studies should be conducted.

2.6 Future Outlook

As it assumes the universal form of the fundamental building blocks of almost every matter in this physical world, nanofiber has found huge potential in a wide range of applications. Research studies and publications on electrospun nanofibers are literally exploding. Nevertheless, the potential of nanofibers has not been realized and the practicability of nanofiber industry has not been accomplished yet. In this section, we will take a glance into the future of nanofibers from two aspects: the potential fabrication technologies of nanofibers, and the potential applications of nanofibers in the field of smart textiles.

2.6.1 Fabrication Technology

As we have mentioned, there are various nanofiber fabrication technologies that have been developed, among which only limited technologies can produce continuous nanofibers. Although electrospinning is regarded as one of the most promising techniques, there are some other technologies that are quite competitive. These technologies include but are not limited to composite spinning, melt blowing, CVD, and 3D printing.

Composite spinning has been applied more and more often in textile industry for mass production of microfibers. Though the fiber diameter is still beyond 1000 nm, composite spinning has demonstrated its potential for creating nanofibers. One of the challenges might be the limitation created by the harsh textile fabrication processes. As the filament becomes finer, its tensile strength gets weaker. If the filament is too fine, then it will be too weak to withstand texturizing, weaving, or knitting on conventional textile equipment. Eastman [303] has created an innovative way to generate finer fibers (Avra®) by protecting the ultrathin filament with a proprietary removable polymer: the filaments are extruded and held together by the proprietary removable polymer, enabling easy knitting or weaving, which is completely washed away in hot water after the fabric is made. This innovative work demonstrated that synthetic filament with a diameter below 1 μm might emerge to the world soon.

Melt blowing has proven to be capable of spinning nanofibers [92, 304]. Though the fibers are currently restricted to the nonwoven format, the productivity of melt blowing is much higher than electrospinning. Besides, there is still a possibility for proper yarn-collecting systems to be invented if needed.

CVD is one of the methods mainly involved for production of discrete inorganic nanomaterials such as nanoparticles, nanowires, nanotubes, and nanoplatelets among which CNT is a quite special product that has attracted many efforts in seeking of continuous yarn fabrication manners. Alan Windle's group showed the possibility of making continuous CNT yarns without an apparent limit to the length [305, 306] by mechanically drawing CNTs from the CVD synthesis zone of a furnace. The key requirements for continuous spinning are the rapid production of high-purity CNTs to form an aerogel in the furnace hot zone and the forcible removal of the product by continuous wind-up. Different levels of CNT orientation, fiber density, and mechanical properties of the yarn can be achieved through control of the process parameters including carrier gas [307], gas flow rates [307, 308], and winding rates [308, 309]. This direct spinning process allows one-step production of CNT yarns with potentially excellent properties [308]. It might not be applicable to organic materials but should still be regarded as a potential approach for continuous nanofiber yarns, especially when we are thinking of inorganic materials.

3D printing, also known as additive manufacturing (AM), has gained immense popularity as a fast prototyping technology. 3D printing technology is nearly completely distinct from traditional textile fabrication techniques [310]. The reported high-resolution 3D printing techniques are mainly based on two-photo polymerization (TPP) microfabrication (TPPMF) and soft nanolithography (SL). TPPMF has emerged as a promising 3D micro/nanoscale manufacturing tool for rapid and flexible fabrication of arbitrary and ultraprecise 3D structures with sub-100-nm resolution [311]. TPP process is initiated by three-order nonlinear absorption within the focal region. TPPMF is realized through continuous TPP by moving the focused beam in photoresist according to a computer-designed 3D route with the resolution beyond the optical diffraction limit [312–314]. Many typical and useful microstructures such as photonic crystals [315], mechanical devices [312, 316], and 3D hydrogels [314, 317] have been fabricated by TPPMF. SL provides low-cost and convenient access to micro- and nanoscale 3-D structures with well-defined and controllable surface structures [318, 319]. A typical SL can be divided into four major steps: 1) design of the pattern; 2) fabrication of the mask and the master; 3) fabrication of the soft stamp; and 4) fabrication of micro- and nanostructures with the stamp by printing, molding, embossing, decal transfer, phase-shifting edge lithography, and other procedures. SL has been widely applied for generation of 3D surfaces [320], microfluidic devices [321], protein and cell patterning [319, 322], micro-optics [323], microsensors [324], microactuators [325], etc. Reports on SL of nanofibers [326, 327] are rare, but there are already studies demonstrating the feasibility.

2.6.2 Applications Meet the Needs

Corresponding to Maslow's theory of hierarchy of needs [328], the functions of smart textiles can be divided into five levels: physiological satisfying, safety protecting, networking, entertaining and life facilitating, and performance enhancing (see Figure 2.7). Physiological satisfying satisfies all the

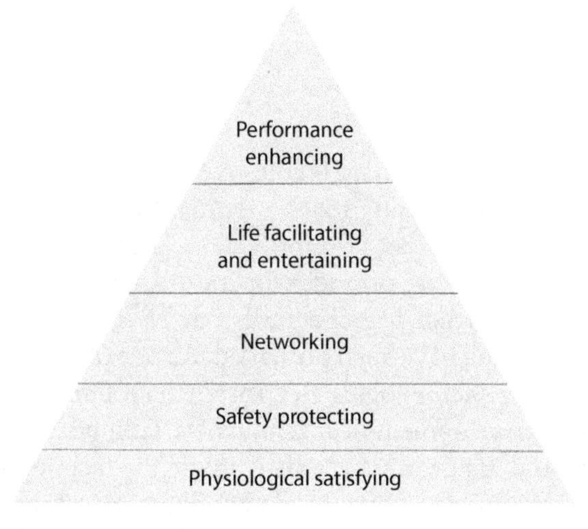

Figure 2.7 Five levels of applications that smart textiles serve for. Note: The image has been prepared by the author.

physiological comfort requirements that a human being requests from textiles such as warmth, cooling, and smoothness. Safety protecting protects wearers from environment harms such as rain, fire, bacteria, and radiation. Most of the designs of fundamental textiles and smart textiles are in accord with these two levels of the requirements. Once the physical needs are met, people feel eager to pursuit more on the spiritual level. So far, the remaining three levels of requirements are served by electronics and wearables. However, the desires for durable and flexible smart electronic components that can be integrated seamlessly with textiles are strong and overwhelming as people prefer something they can comfortably wear on without burden in comparison to something that they have to carry. Many wearables have been designed to meet these requirements. But, in principle, wearables currently available on the market are still "hardwares" [329] that are embedded in or simply boned to textiles. Disruptive wearables or smart textiles not only should fulfil their tasks as electronics but also should assume the flexible and lightweight nature of textiles. Therefore, development of flexible thread/wire-based electronic components such as sensors, batteries and supercapacitors, and displays is essential and critical. The emergence of nanofiber opens up great opportunities for quick realization of such an ambition. Performance enhancing, especially in the field of sportswear, can also be closely linked to actuators. By control of the external stimuli, actuators can simulate the actions of a muscle fiber by reversibly contracting, expanding, twisting, etc. Hence, an actuator is also called an artificial muscle [330]. Equipping with external skeleton or extra artificial muscles, human beings might be able to perform duties that are beyond a normal person's capability.

2.7 Conclusion

In conclusion, a variety of smart textiles have quickly emerged in textile industry boosted by the strong market demand for intelligent garments. Nanofiber-based textiles that can provide a pleasant microclimate by smartly regulating moisture or/and temperature, or protect wearers from ambient threats such as rain, fire, chemicals, bacteria, and radiations, or help with health monitoring have been presented. As one of the most promising nanofiber fabrication technologies, electrospun nanofibers have been intensively studied for development of smart textiles. Composite spinning, melt blowing, CVD, and 3D printing are also among the technologies that own huge potentials for mass production of nanofibers. However, there are still a few challenges confronting nanofiber technologies. The low productivity, high cost, and weak mechanical performance are severely hindering the commercialization of nanofibers and need to be solved before nanofiber-based textiles can be industrialized.

As the socioeconomic development is rapidly progressing, people's requirements for textiles are quickly progressing as well. They are more and more attracted to textiles that can help with networking, entertaining, and life-facilitating rather than those that can only provide physiological comfort. Therefore, development of flexible thread/fiber-based intelligent components such as sensors, batteries, supercapacitors, displays, and actuators has become critical for future smart textiles.

References

1. Rocco, M., William, R., and Alivisiatos, P., *Nanotechnology Research Directions: IWGN Workshop Report*, National Science and technology Council, 1999.
2. Ko, F.K. and Wan, Y., *Introduction to Nanofiber Materials*, Cambridge University Press, 2014.
3. Gibson, P., Schreuder-Gibson, H., and Rivin, D., Transport properties of porous membranes based on electrospun nanofibers. *Colloids Surf. Physicochem. Eng. Aspects*, 187–188, 469–481, 2001.
4. Nabet, B., *When Is Small Good? On Unusual Electronic Properties of Nanowires*, PA-19104, Editor, Philadelphia, 2002.
5. El-Aufy, A., Nabet, B., and Ko, F., Carbon nanotube reinforced (PEDT/PAN) nanocomposite for wearable electronics. *Polym. Prepr.*, 44, 134–135, 2003.
6. Ko, F., Gogotsi, Y., Ali, A., *et al.*, Electrospinning of continuous carbon nanotube-filled nanofiber yarns. *Adv. Mater.*, 15, 1161–1165, 2003.
7. Ko, F., Gandhi, M., and Karatzas, C., Carbon Nanotube Reinforced Spider Silk—A Model for the Next Generation of Super Strong and Tough Fibers. In *19th American Society for Composites Annual Technical Conference*, Atlanta, GA, 2004.
8. Mack, J., Viculis, L., Ali, A., *et al.*, Graphite Nanoplatelet Based Nanocomposites by the Electrospinning Process. In *17th Annual Conf. of the Am. Soc. for Composites*, Purdue University, 2003.
9. Wan, Y.Q., He, J.H., and Yu, J.Y., Carbon nanotube-reinforced polyacrylonitrile nanofibers by vibration-electrospinning. *Polym. Int.*, 56, 1367–1370, 2007.
10. Park, M., Jung, Y.-J., Kim, J., *et al.*, Synergistic effect of carbon nanofiber/nanotube composite catalyst on carbon felt electrode for high-performance all-vanadium redox flow battery. *Nano Lett.*, 13, 4833–4839, 2013.
11. Zhang, K., Zhang, L.L., Zhao, X., *et al.*, Graphene/polyaniline nanofiber composites as supercapacitor electrodes. *Chem. Mater.*, 22, 1392–1401, 2010.

12. Wu, Q., Xu, Y., Yao, Z., *et al.*, Supercapacitors based on flexible graphene/polyaniline nanofiber composite films. *ACS Nano*, 4, 1963–1970, 2010.
13. Ambrosi, A., Sasaki, T., and Pumera, M., Platelet graphite nanofibers for electrochemical sensing and biosensing: The influence of graphene sheet orientation. *Chem. Asian. J.*, 5, 266–271, 2010.
14. Huang, X., Hu, N., Gao, R., *et al.*, Reduced graphene oxide–polyaniline hybrid: Preparation, characterization and its applications for ammonia gas sensing. *J. Mater. Chem.*, 22, 22488–22495, 2012.
15. Yang, A., Tao, X., Wang, R., *et al.*, Room temperature gas sensing properties of SnO_2/multiwall-carbon-nanotube composite nanofibers. *Appl. Phys. Lett.*, 91, 133110–133113, 2007.
16. Liu, Y., Teng, H., Hou, H., *et al.*, Nonenzymatic glucose sensor based on renewable electrospun Ni nanoparticle-loaded carbon nanofiber paste electrode. *Biosens. Bioelectron.*, 24, 3329–3334, 2009.
17. Gibson, P. and Lee, C., *Application of nanofiber technology to nonwoven thermal insulation*, Army Soldier Systems Command Natick Ma, 2006.
18. Zhang, L., Zhong, Y., Cha, D., *et al.*, A self-cleaning underwater superoleophobic mesh for oil–water separation. *Sci. Rep.*, 3, 2326, 2013.
19. Im, J.S., Kim, J.G., Lee, S.H., *et al.*, Effective electromagnetic interference shielding by electrospun carbon fibers involving Fe_2O_3/$BaTiO_3$/MWCNT additives. *Mater. Chem. Phys.*, 124, 434–438, 2010.
20. Bayat, M., Yang, H., Ko, F., *et al.*, Electromagnetic interference shielding effectiveness of hybrid multifunctional Fe_3O_4/carbon nanofiber composite. *Polymer*, 55, 936–943, 2014.
21. Im, J.S., Kim, J.G., Lee, S.H., *et al.*, Enhanced adhesion and dispersion of carbon nanotube in PANI/PEO electrospun fibers for shielding effectiveness of electromagnetic interference. *Colloids Surf. Physicochem. Eng. Aspects*, 364, 151–157, 2010.
22. Fan, Z., Yan, J., Wei, T., *et al.*, Asymmetric supercapacitors based on graphene/MnO_2 and activated carbon nanofiber electrodes with high power and energy density. *Adv. Funct. Mater.*, 21, 2366–2375, 2011.
23. Chen, L.F., Huang, Z.H., Liang, H.W., *et al.*, Bacterial-cellulose-derived carbon nanofiber@ MnO_2 and nitrogen-doped carbon nanofiber electrode materials: An asymmetric supercapacitor with high energy and power density. *Adv. Mater.*, 25, 4746–4752, 2013.
24. Zhang, G., Zheng, J., Liang, R., *et al.*, α-MnO_2/carbon nanotube/carbon nanofiber composite catalytic air electrodes for rechargeable lithium-air batteries. *J. Electrochem. Soc.*, 158, A822–A827, 2011.
25. Ji, L. and Zhang, X., Fabrication of porous carbon/Si composite nanofibers as high-capacity battery electrodes. *Electrochem. Commun.*, 11, 1146–1149, 2009.
26. Fu, K., Yildiz, O., Bhanushali, H., *et al.*, Aligned carbon nanotube-silicon sheets: A novel nano-architecture for flexible lithium ion battery electrodes. *Adv. Mater.*, 25, 5109–5114, 2013.

27. Meng, C., Xiao, Y., Wang, P., *et al.*, Quantum-dot-doped polymer nanofibers for optical sensing. *Adv. Mater.*, 23, 3770–3774, 2011.
28. Li, L., Zhu, P., Peng, S., *et al.*, Controlled growth of CuS on electrospun carbon nanofibers as an efficient counter electrode for quantum dot-sensitized solar cells. *J. Phys. Chem. C*, 118, 16526–16535, 2014.
29. Sudhagar, P., Jung, J.H., Park, S., *et al.*, The performance of coupled (CdS: CdSe) quantum dot-sensitized TiO_2 nanofibrous solar cells. *Electrochem. Commun.*, 11, 2220–2224, 2009.
30. Fang, B., Kim, M., Fan, S.-Q., *et al.*, Facile synthesis of open mesoporous carbon nanofibers with tailored nanostructure as a highly efficient counter electrode in CdSe quantum-dot-sensitized solar cells. *J. Mater. Chem.*, 21, 8742–8748, 2011.
31. Li, M., Zhang, J., Zhang, H., *et al.*, Electrospinning: A facile method to disperse fluorescent quantum dots in nanofibers without Förster resonance energy transfer. *Adv. Funct. Mater.*, 17, 3650–3656, 2007.
32. Liu, Z., Sun, D.D., Guo, P., *et al.*, An efficient bicomponent TiO_2/SnO_2 nanofiber photocatalyst fabricated by electrospinning with a side-by-side dual spinneret method. *Nano Lett.*, 7, 1081–1085, 2007.
33. Yang, D., Liu, H., Zheng, Z., *et al.*, An efficient photocatalyst structure: TiO_2(B) nanofibers with a shell of anatase nanocrystals. *J. Am. Chem. Soc.*, 131, 17885–17893, 2009.
34. Wang, C., Shao, C., Liu, Y., *et al.*, Photocatalytic properties BiOCl and Bi_2O_3 nanofibers prepared by electrospinning. *Scripta Mater.*, 59, 332–335, 2008.
35. Zhang, Z., Shao, C., Li, X., *et al.*, Electrospun nanofibers of p-type NiO/n-type ZnO heterojunctions with enhanced photocatalytic activity. *ACS Appl. Mater. Interfaces*, 2, 2915–2923, 2010.
36. Lin, D., Wu, H., Zhang, R., *et al.*, Enhanced photocatalysis of electrospun Ag–ZnO heterostructured nanofibers. *Chem. Mater.*, 21, 3479–3484, 2009.
37. Chuangchote, S., Jitputti, J., Sagawa, T., *et al.*, Photocatalytic activity for hydrogen evolution of electrospun TiO_2 nanofibers. *ACS Appl. Mater. Interfaces*, 1, 1140–1143, 2009.
38. Song, M.Y., Kim, D.K., Ihn, K.J., *et al.*, Electrospun TiO_2 electrodes for dye-sensitized solar cells. *Nanotechnology*, 15, 1861, 2004.
39. Pham, Q.P., Sharma, U., and Mikos, A.G., Electrospinning of polymeric nanofibers for tissue engineering applications: A review. *Tissue Eng.*, 12, 1197–1211, 2006.
40. Cao, T., Li, Y., Wang, C., *et al.*, A facile *in situ* hydrothermal method to $SrTiO_3/TiO_2$ nanofiber heterostructures with high photocatalytic activity. *Langmuir*, 27, 2946–2952, 2011.
41. Olson, D.C., Piris, J., Collins, R.T., *et al.*, Hybrid photovoltaic devices of polymer and ZnO nanofiber composites. *Thin Solid Films*, 496, 26–29, 2006.
42. Kim, I.D., Hong, J.M., Lee, B.H., *et al.*, Dye-sensitized solar cells using network structure of electrospun ZnO nanofiber mats. *Appl. Phys. Lett.*, 91, 163109, 2007.

43. Li, Z., Zhang, H., Zheng, W., *et al.*, Highly sensitive and stable humidity nanosensors based on LiCl doped TiO_2 electrospun nanofibers. *J. Am. Chem. Soc.*, 130, 5036–5037, 2008.
44. Park, J.A., Moon, J.H., Lee, S.J., *et al.*, SnO_2–ZnO hybrid nanofibers-based highly sensitive nitrogen dioxides sensor. *Sensors Actuators B: Chem.*, 145, 592–595, 2010.
45. Persano, L., Dagdeviren, C., Su, Y., *et al.*, High performance piezoelectric devices based on aligned arrays of nanofibers of poly(vinylidenefluoride-co-trifluoroethylene). *Nat. Commun.*, 4, 1633, 2013.
46. Mandal, D., Yoon, S., and Kim, K.J., Origin of piezoelectricity in an electrospun poly(vinylidene fluoride-trifluoroethylene) nanofiber web-based nanogenerator and nano-pressure sensor. *Macromol. Rapid Commun.*, 32, 831–837, 2011.
47. Chen, X., Xu, S., Yao, N., *et al.*, 1.6 V nanogenerator for mechanical energy harvesting using PZT nanofibers. *Nano Lett.*, 10, 2133–2137, 2010.
48. Chang, J., Dommer, M., Chang, C., *et al.*, Piezoelectric nanofibers for energy scavenging applications. *Nano Energy*, 1, 356–371, 2012.
49. Bafqi, M.S.S., Bagherzadeh, R., and Latifi, M., Fabrication of composite PVDF-ZnO nanofiber mats by electrospinning for energy scavenging application with enhanced efficiency. *J. Polym. Res.*, 22, 130, 2015.
50. Lee, M., Chen, C.Y., Wang, S., *et al.*, A hybrid piezoelectric structure for wearable nanogenerators. *Adv. Mater.*, 24, 1759–1764, 2012.
51. Jin, G., Prabhakaran, M.P., Kai, D., *et al.*, Electrospun photosensitive nanofibers: Potential for photocurrent therapy in skin regeneration. *Photochem. Photobiol. Sci.*, 12, 124–134, 2013.
52. Jin, G., Li, J., and Li, K., Photosensitive semiconducting polymer-incorporated nanofibers for promoting the regeneration of skin wound. *Mater. Sci. Eng. C*, 70, 1176–1181, 2017.
53. Liu, S.P., Tan, L.J., Hu, W.L., *et al.*, Cellulose acetate nanofibers with photochromic property: Fabrication and characterization. *Mater. Lett.*, 64, 2427–2430, 2010.
54. Fu, G.-D., Xu, L.-Q., Yao, F., *et al.*, Smart nanofibers with a photoresponsive surface for controlled release. *ACS Appl. Mater. Interfaces*, 1, 2424–2427, 2009.
55. Fu, G., Xu, L., Yao, F., *et al.*, Smart nanofibers from combined living radical polymerization, "click chemistry", and electrospinning. *ACS Appl. Mater. Interfaces*, 1, 239–243, 2009.
56. Konosu, Y., Matsumoto, H., Tsuboi, K., *et al.*, Enhancing the effect of the nanofiber network structure on thermoresponsive wettability switching. *Langmuir*, 27, 14716–14720, 2011.
57. Ranganath, A.S., Ganesh, V.A., Sopiha, K., *et al.*, Investigation of wettability and moisture sorption property of electrospun poly (N-isopropylacrylamide) nanofibers. *MRS Advances*, 1, 1959–1964, 2016.

58. Chen, C., Wang, L., and Huang, Y., Morphology and thermal properties of electrospun fatty acids/polyethylene terephthalate composite fibers as novel form-stable phase change materials. *Sol. Energy Mater. Sol. Cells*, 92, 1382–1387, 2008.
59. McCann, J.T., Marquez, M., and Xia, Y., Melt coaxial electrospinning: A versatile method for the encapsulation of solid materials and fabrication of phase change nanofibers. *Nano Lett.*, 6, 2868–2872, 2006.
60. Chen, C., Wang, L., and Huang, Y., Electrospun phase change fibers based on polyethylene glycol/cellulose acetate blends. *Appl. Energy*, 88, 3133–3139, 2011.
61. Xing, Z.C., Chae, W.P., Baek, J.Y., *et al.*, *In vitro* assessment of antibacterial activity and cytocompatibility of silver-containing PHBV nanofibrous scaffolds for tissue engineering. *Biomacromolecules*, 11, 1248–1253, 2010.
62. Ma, Z., Kotaki, M., Inai, R., *et al.*, Potential of nanofiber matrix as tissue-engineering scaffolds. *Tissue Eng.*, 11, 101–109, 2005.
63. Unnithan, A.R., Barakat, N.A., Pichiah, P.T., *et al.*, Wound-dressing materials with antibacterial activity from electrospun polyurethane–dextran nanofiber mats containing ciprofloxacin HCl. *Carbohydr. Polym.*, 90, 1786–1793, 2012.
64. Rujitanaroj, P.-O., Pimpha, N., and Supaphol, P., Wound-dressing materials with antibacterial activity from electrospun gelatin fiber mats containing silver nanoparticles. *Polymer*, 49, 4723–4732, 2008.
65. Liu, L., Liu, Z., Bai, H., *et al.*, Concurrent filtration and solar photocatalytic disinfection/degradation using high-performance Ag/TiO_2 nanofiber membrane. *Water Res.*, 46, 1101–1112, 2012.
66. Maleki, M., Latifi, M., Amani-Tehran, M., *et al.*, Electrospun core–shell nanofibers for drug encapsulation and sustained release. *Polym. Eng. Sci.*, 53, 1770–1779, 2013.
67. Che, H., Huo, M., Peng, L., *et al.*, CO_2-responsive nanofibrous membranes with switchable oil/water wettability. *Angew. Chem. Int. Ed.*, 54, 8934–8938, 2015.
68. Ma, W., Samal, S.K., Liu, Z., *et al.*, Dual pH-and ammonia-vapor-responsive electrospun nanofibrous membranes for oil-water separations. *J. Membr. Sci.*, 537, 128–139, 2017.
69. Chen, M., Dong, M., Havelund, R., *et al.*, Thermo-responsive core–sheath electrospun nanofibers from poly(N-isopropylacrylamide)/polycaprolactone blends. *Chem. Mater.*, 22, 4214–4221, 2010.
70. Xu, X., Jiang, L., Zhou, Z., *et al.*, Preparation and properties of electrospun soy protein isolate/polyethylene oxide nanofiber membranes. *ACS Appl. Mater. Interfaces*, 4, 4331–4337, 2012.
71. Obaid, M., Barakat, N.A.M., Fadali, O.A., *et al.*, Stable and effective superhydrophilic polysulfone nanofiber mats for oil/water separation. *Polymer*, 72, 125–133, 2015.

72. Obaid, M., Fadali, O., Lim, B.-H., *et al.*, Super-hydrophilic and highly stable in oils polyamide–polysulfone composite membrane by electrospinning. *Mater. Lett.*, 138, 196–199, 2015.
73. Abdal-hay, A., Pant, H.R., and Lim, J.K., Super-hydrophilic electrospun nylon-6/hydroxyapatite membrane for bone tissue engineering. *Eur. Polym. J.*, 49, 1314–1321, 2013.
74. Wong, W.S., Nasiri, N., Rodriguez, A.L., *et al.*, Hierarchical amorphous nanofibers for transparent inherently super-hydrophilic coatings. *J. Mater. Chem. A*, 2, 15575–15581, 2014.
75. Chiou, N.-R., Lu, C., Guan, J., *et al.*, Growth and alignment of polyaniline nanofibres with superhydrophobic, superhydrophilic and other properties. *Nat. Nanotechnol.*, 2, 354, 2007.
76. Nuraje, N., Khan, W.S., Lei, Y., *et al.*, Superhydrophobic electrospun nanofibers. *J. Mater. Chem. A*, 1, 1929–1946, 2013.
77. Si, Y., Fu, Q., Wang, X., *et al.*, Superelastic and superhydrophobic nanofiber-assembled cellular aerogels for effective separation of oil/water emulsions. *ACS Nano*, 9, 3791–3799, 2015.
78. Dong, Y., Kong, J., Phua, S.L., *et al.*, Tailoring surface hydrophilicity of porous electrospun nanofibers to enhance capillary and push–pull effects for moisture wicking. *ACS Appl. Mater. Interfaces*, 6, 14087–14095, 2014.
79. Lim, H.S., Park, S.H., Koo, S.H., *et al.*, Superamphiphilic janus fabric. *Langmuir*, 26, 19159–19162, 2010.
80. Ahmed, F.E., Lalia, B.S., Hilal, N., *et al.*, Underwater superoleophobic cellulose/electrospun PVDF–HFP membranes for efficient oil/water separation. *Desalination*, 344, 48–54, 2014.
81. Ge, J., Zhang, J., Wang, F., *et al.*, Superhydrophilic and underwater superoleophobic nanofibrous membrane with hierarchical structured skin for effective oil-in-water emulsion separation. *J. Mater. Chem. A*, 5, 497–502, 2017.
82. Zhang, F., Zhang, W.B., Shi, Z., *et al.*, Nanowire-haired inorganic membranes with superhydrophilicity and underwater ultralow adhesive superoleophobicity for high-efficiency oil/water separation. *Adv. Mater.*, 25, 4192–4198, 2013.
83. Nishimoto, S. and Bhushan, B., Bioinspired self-cleaning surfaces with superhydrophobicity, superoleophobicity, and superhydrophilicity. *RSC Adv.*, 3, 671–690, 2013.
84. Wu, H., Krifa, M., and Koo, J.H., Flame retardant polyamide 6/nanoclay/intumescent nanocomposite fibers through electrospinning. *Text. Res. J.*, 84, 1106–1118, 2014.
85. Yin, X., Krifa, M., and Koo, J.H., Flame-retardant polyamide 6/carbon nanotube nanofibers: Processing and characterization. *J. Eng. Fabrics Fibers (JEFF)*, 10, 2015.
86. Doshi, J. and Reneker, D., Electrospinning process and applications of electrospun fibers. *J. Electrostatics*, 35, 151–160, 1995.

87. Reneker, D.H. and Chun, I., Nanometre diameter fibres of polymer, produced by electrospinning. *Nanotechnology*, 7, 216–223, 1996.
88. Zhou, Z., Lai, C., Zhang, L., *et al.*, Development of carbon nanofibers from aligned electrospun polyacrylonitrile nanofiber bundles and characterization of their microstructural, electrical, and mechanical properties. *Polymer*, 50, 2999–3006, 2009.
89. Yang, D., Lu, B., Zhao, Y., *et al.*, Fabrication of aligned fibrous arrays by magnetic electrospinning. *Adv. Mater.*, 19, 3702–3706, 2007.
90. Li, D., Wang, Y., and Xia, Y., Electrospinning of polymeric and ceramic nanofibers as uniaxially aligned arrays. *Nano Lett.*, 3, 1167–1171, 2003.
91. Desai, K., Kit, K., Li, J., *et al.*, Nanofibrous chitosan non-wovens for filtration applications. *Polymer*, 50, 3661–3669, 2009.
92. Barhate, R.S. and Ramakrishna, S., Nanofibrous filtering media: Filtration problems and solutions from tiny materials. *J. Membr. Sci.*, 296, 1–8, 2007.
93. Zhou, W., Bahi, A., Li, Y., *et al.*, Ultra-filtration membranes based on electrospun poly(vinylidene fluoride) (PVDF) fibrous composite membrane scaffolds. *RSC Adv.*, 3, 11614–11620, 2013.
94. Zeng, J., Xu, X., Chen, X., *et al.*, Biodegradable electrospun fibers for drug delivery. *J. Controlled Release*, 92, 227–231, 2003.
95. Yu, D.G., Zhu, L.M., White, K., *et al.*, Electrospun nanofiber-based drug delivery systems. *Health (N. Y.)*, 1, 67–75, 2009.
96. Leung, V., Hartwell, R., Faure, E., *et al.*, Electrospun nanofibrous scaffold for controll drug delivery. *June*, 13, 15, 2016.
97. Song, F., Wang, X.-L., and Wang, Y.-Z., Poly(N-isopropylacrylamide)/poly(ethylene oxide) blend nanofibrous scaffolds: Thermo-responsive carrier for controlled drug release. *Colloids Surf. B. Biointerfaces*, 88, 749–754, 2011.
98. Leung, V., Yang, H., Gupta, M., *et al.*, Alginate nanofibre based tissue engineering scaffolds by electrospinning. In *Society for the Advancement of Materials and Process Engineering 2010 Conference*, 2010.
99. Yoshimoto, H., Shin, Y.M., Terai, H., *et al.*, A biodegradable nanofiber scaffold by electrospinning and its potential for bone tissue engineering. *Biomaterials*, 24, 2077–2082, 2003.
100. Li, X., Xie, J., Yuan, X., *et al.*, Coating electrospun poly(ε-caprolactone) fibers with gelatin and calcium phosphate and their use as biomimetic scaffolds for bone tissue engineering. *Langmuir*, 24, 14145–14150, 2008.
101. Boland, E., Matthews, J., Pawlowski, K., *et al.*, Electrospinning collagen and elastin: Preliminary vascular tissue engineering. *Front. Biosci.*, 9, 1422–1432, 2004.
102. Li, W., Laurencin, C., Caterson, E., *et al.*, Electrospun nanofibrous structure: A novel scaffold for tissue engineering. *J. Biomed. Mater. Res.*, 60, 613–621, 2002.
103. Subramanian, A., Lin, H., Vu, D., *et al.*, Synthesis and evaluation of scaffolds prepared from chitosan fibers for potential use in cartilage tissue engineering. *Biomed. Sci. Instrum.*, 40, 117, 2004.

104. Sharma, T., Langevine, J., Naik, S., *et al.*, Aligned electrospun PVDF-TrFE nanofibers for flexible pressure sensors on catheter. In *Solid-State Sensors, Actuators and Microsystems, 2013 Transducers & Eurosensors XXVII: The 17th International Conference on*, IEEE, 2013.
105. Macagnano, A., Zampetti, E., and Kny, E., *Electrospinning for High Performance Sensors*, Springer, 2015.
106. Marx, S., Jose, M.V., Andersen, J.D., *et al.*, Electrospun gold nanofiber electrodes for biosensors. *Biosens. Bioelectron.*, 26, 2981–2986, 2011.
107. Servati, P., Soltanian, S., and Ko, F., Electrospun nanofiber based strain sensors for structural health monitoring. In *Structural Health Monitoring 2013: A Roadmap to Intelligent Structures: Proceedings of the Ninth International Workshop on Structural Health Monitoring, September 10–12, 2013*, DEStech Publications, Inc., 2013.
108. Ding, B., Wang, M., Wang, X., *et al.*, Electrospun nanomaterials for ultrasensitive sensors. *Mater. Today*, 13, 16–27, 2010.
109. Ding, B., Wang, M., Yu, J., *et al.*, Gas sensors based on electrospun nanofibers. *Sensors*, 9, 1609–1624, 2009.
110. Guo, F., Xu, X.X., Sun, Z.Z., *et al.*, A novel amperometric hydrogen peroxide biosensor based on electrospun Hb-collagen composite. *Colloids Surf. B. Biointerfaces*, 86, 140–145, 2011.
111. Shakir, I., Ali, Z., Bae, J., *et al.*, Conformal coating of ultrathin $Ni(OH)_2$ on ZnO nanowires grown on textile fiber for efficient flexible energy storage devices. *RSC Adv.*, 4, 6324–6329, 2014.
112. Ko, F.K. and Yang, H.J., Functional nanofibre: Enabling materials for the next generation SMART textiles. In *Textile Bioengineering and Informatics Symposium Proceedings, Vols 1 and 2*, pp. 1217–1228, 2008.
113. Mao, X., Bai, Y., Yu, J., *et al.*, Insights into the flexibility of ZrMxOy (M = Na, Mg, Al) nanofibrous membranes as promising infrared stealth materials. *Dalton Transactions*, 45, 6660–6666, 2016.
114. Test Method for Air Permeability of Textile Fabrics, ASTM D737-96, 1996.
115. Greiner, A. and Wendorff, J.H., Electrospinning: A fascinating method for the preparation of ultrathin fibers. *Angew. Chem. Int. Ed.*, 46, 5670–5703, 2007.
116. Kadam, V., Jadhav, A., Nayak, R., *et al.*, Air permeability and moisture management properties of electrospun nanofiber Membranes. *J. Fiber Bioeng. Inform.*, 9, 167–176, 2016.
117. Borhani, S., Etemad, S.G., and Ravandi, S.A.H., Dynamic heat and moisture transfer in bulky PAN nanofiber mats. *Heat Mass Transfer.*, 47, 807–811, 2011.
118. Sheikh, F.A., Zargar, M.A., Tamboli, A.H., *et al.*, A super hydrophilic modification of poly (vinylidene fluoride) (PVDF) nanofibers: By *in situ* hydrothermal approach. *Appl. Surf. Sci.*, 385, 417–425, 2016.
119. Kaufman, Y.J. and Fraser, R.S., The effect of smoke particles on clouds and climate forcing. *Science*, 277, 1636–1639, 1997.

120. Kathiravelu, G., Lucke, T., and Nichols, P., Rain drop measurement techniques: A review. *Water*, 8, 29, 2016.
121. Qiang, J., Wan, Y.Q., Yang, L.N., *et al.*, Effect of ultrasonic vibration on structure and performance of electrospun PAN fibrous membrane. *J. Nano Res.*, 23, 96–103, 2013.
122. Yoon, B. and Lee, S., Designing waterproof breathable materials based on electrospun nanofibers and assessing the performance characteristics. *Fiber. Polym.*, 12, 57–64, 2011.
123. Ah Hong, K., Sook Yoo, H., and Kim, E., Effect of waterborne polyurethane coating on the durability and breathable waterproofing of electrospun nanofiber web-laminated fabrics. *Text. Res. J.*, 85, 160–170, 2015.
124. Qu, M., Zhao, G., Cao, X., *et al.*, Biomimetic fabrication of lotus-leaf-like structured polyaniline film with stable superhydrophobic and conductive properties. *Langmuir*, 24, 4185–4189, 2008.
125. Cassie, A. and Baxter, S., Wettability of porous surfaces. *Trans. Faraday Soc.*, 40, 546–551, 1944.
126. Baker, E., Chemistry and morphology of plant epicuticular waxes. In *The Plant Cuticle*, C. DF, A. KL, *et al.* (Eds.), Academic Press, London, pp. 139–166, 1982.
127. Jeffree, C., The cuticle, epicuticular waxes and trichomes of plants, with reference to their structure, functions and evolution. In *Insects and the Plant Surface*, J. BE and S. SR (Eds.), Edward Arnold, London, pp. 23–64, 1986.
128. Barthlott, W., Scanning electron microscopy of the epidermal surface in plants. In *Scanning Electron Microscopy in Taxonomy and Functional Morphology. Systematics Association Special Volume*, pp. 69–94, 1990.
129. Gorji, M., Jeddi, A., and Gharehaghaji, A., Fabrication and characterization of polyurethane electrospun nanofiber membranes for protective clothing applications. *J. Appl. Polym. Sci.*, 125, 4135–4141, 2012.
130. Ge, J., Raza, A., Fen, F., *et al.*, Mechanically robust polyurethane microfibrous membranes exhibiting high air permeability. *J. Fiber Bioeng. Inform.*, 5, 411–421, 2012.
131. Han, H.R., Chung, S.E., and Park, C.H., Shape memory and breathable waterproof properties of polyurethane nanowebs. *Text. Res. J.*, 83, 76–82, 2013.
132. Gopinathan, J., Indumathi, B., Thomas, S., *et al.*, Morphology and hydroscopic properties of acrylic/thermoplastic polyurethane core–shell electrospun micro/nano fibrous mats with tunable porosity. *RSC Adv.*, 6, 54286–54292, 2016.
133. Jiang, L., Zhao, Y., and Zhai, J., A lotus-leaf-like superhydrophobic surface: A porous microsphere/nanofiber composite film prepared by electrohydrodynamics. *Angew. Chem.*, 116, 4438–4441, 2004.
134. Yasuhiro, M., *et al.*, Fabrication of a silver-ragwort-leaf-like superhydrophobic micro/nanoporous fibrous mat surface by electrospinning. *Nanotechnology*, 17, 5151, 2006.

135. Kang, M., Jung, R., Kim, H.-S., *et al.*, Preparation of superhydrophobic polystyrene membranes by electrospinning. *Colloids Surf. Physicochem. Eng. Aspects*, 313–314, 411–414, 2008.
136. Zhan, N., Li, Y., Zhang, C., *et al.*, A novel multinozzle electrospinning process for preparing superhydrophobic PS films with controllable bead-on-string/microfiber morphology. *J. Colloid Interface Sci.*, 345, 491–495, 2010.
137. Muthiah, P., Hsu, S.H., and Sigmund, W., Coaxially electrospun PVDF-Teflon AF and Teflon AF-PVDF core-sheath nanofiber mats with superhydrophobic properties. *Langmuir*, 26, 12483–12487, 2010.
138. Lee, M.S., Lee, T.S., and Park, W.H., Highly hydrophobic nanofibrous surfaces generated by poly(vinylidene fluoride). *Fiber. Polym.*, 14, 1271–1275, 2013.
139. Lim, J.-M., Yi, G.-R., Moon, J.H., *et al.*, Superhydrophobic films of electrospun fibers with multiple-scale surface morphology. *Langmuir*, 23, 7981–7989, 2007.
140. Yoon, Y.I., Moon, H.S., Lyoo, W.S., *et al.*, Superhydrophobicity of PHBV fibrous surface with bead-on-string structure. *J. Colloid Interface Sci.*, 320, 91–95, 2008.
141. Tang, H., Wang, H., and He, J., Superhydrophobic titania membranes of different adhesive forces fabricated by electrospinning. *J. Phys. Chem. C*, 113, 14220–14224, 2009.
142. Yoon, Y.I., Moon, H.S., Lyoo, W.S., *et al.*, Superhydrophobicity of cellulose triacetate fibrous mats produced by electrospinning and plasma treatment. *Carbohydr. Polym.*, 75, 246–250, 2009.
143. Ma, M., Gupta, M., Li, Z., *et al.*, Decorated electrospun fibers exhibiting superhydrophobicity. *Adv. Mater.*, 19, 255–259, 2007.
144. Birajdar, M.S. and Lee, J., Nanoscale bumps and dents on nanofibers enabling sonication-responsive wetting and improved moisture collection. *Macromol. Mater. Eng.*, 300, 1108–1115, 2015.
145. Hardman, S.J., Muhamad-Sarih, N., Riggs, H.J., *et al.*, Electrospinning superhydrophobic fibers using surface segregating end-functionalized polymer additives. *Macromolecules*, 44, 6461–6470, 2011.
146. Ahn, H.W., Park, C.H., and Chung, S.E., Waterproof and breathable properties of nanoweb applied clothing. *Text. Res. J.*, 81, 1438–1447, 2011.
147. Ge, J., Si, Y., Fu, F., *et al.*, Amphiphobic fluorinated polyurethane composite microfibrous membranes with robust waterproof and breathable performances. *RSC Adv.*, 3, 2248–2255, 2013.
148. Li, Y., Yang, F., Yu, J., *et al.*, Hydrophobic fibrous membranes with tunable porous structure for equilibrium of breathable and waterproof performance. *Adv. Mater. Interfaces*, 3, 2016.
149. Fang, J., Wang, H., Wang, X., *et al.*, Superhydrophobic nanofibre membranes: Effects of particulate coating on hydrophobicity and surface properties. *J. Text. I.*, 103, 937–944, 2012.

150. Jin, S., Park, Y., and Park, C.H., Preparation of breathable and superhydrophobic polyurethane electrospun webs with silica nanoparticles. *Text. Res. J.*, 86, 1816–1827, 2016.
151. Chen, M. and Besenbacher, F., Light-driven wettability changes on a photoresponsive electrospun mat. *ACS Nano*, 5, 1549–1555, 2011.
152. Sybilska, W. and Korycki, R., Analysis of thermal-insulating parameters in two-and three-layer textiles with semi-permeable membranes. *Fibers Text. East. Eur.*, 80–87, 2016.
153. Teo, W.E., *Thermal Insulation Properties of Electrospun Fibers.* http://electrospintech.com/thermalinsulation.html#.W7kYrXtKiUk, 2017.
154. Wang, B. and Wang, Y.D., Effect of fiber diameter on thermal conductivity of the electrospun carbon nanofiber mats. In *Advanced Materials Research*, Trans Tech Publ, 2011.
155. Sabetzadeh, N., Bahrambeygi, H., Rabbi, A., *et al.*, Thermal conductivity of polyacrylonitrile nanofibre web in various nanofibre diameters and surface densities. *Micro Nano Lett.*, 7, 662–666, 2012.
156. Kim, K.S. and Park, C.H., Thermal comfort and waterproof-breathable performance of aluminum-coated polyurethane nanowebs. *Text. Res. J.*, 83, 1808–1820, 2013.
157. Mondal, S., Phase change materials for smart textiles—An overview. *Appl. Therm. Eng.*, 28, 1536–1550, 2008.
158. Jegadheeswaran, S. and Pohekar, S.D., Performance enhancement in latent heat thermal storage system: A review. *Renew. Sust. Energ. Rev.*, 13, 2225–2244, 2009.
159. Sarier, N., Arat, R., Menceloglu, Y., *et al.*, Production of PEG grafted PAN copolymers and their electrospun nanowebs as novel thermal energy storage materials. *Thermochim. Acta*, 643, 83–93, 2016.
160. Lee, S., Developing UV-protective textiles based on electrospun zinc oxide nanocomposite fibers. *Fiber. Polym.*, 10, 295–301, 2009.
161. Vitchuli, N., Shi, Q., Nowak, J., *et al.*, Multifunctional ZnO/Nylon 6 nanofiber mats by an electrospinning–electrospraying hybrid process for use in protective applications. *Sci. Technol. Adv. Mater.*, 12, 055004, 2011.
162. Shi, Q., Vitchuli, N., Nowak, J., *et al.*, Multifunctional and durable nanofiber-fabric-layered composite for protective application. *J. Appl. Polym. Sci.*, 128, 1219–1226, 2013.
163. Sundarrajan, S., Venkatesan, A., and Ramakrishna, S., Fabrication of nanostructured self-detoxifying nanofiber membranes that contain active polymeric functional groups. *Macromol. Rapid Commun.*, 30, 1769–1774, 2009.
164. Lu, A.X., McEntee, M., Browe, M.A., *et al.*, MOFabric: Electrospun nanofiber mats from PVDF/UiO-66-NH2 for chemical protection and decontamination. *ACS Appl. Mater. Interfaces*, 9, 13632–13636, 2017.
165. Jiang, S., Ma, B.C., Reinholz, J., *et al.*, Efficient nanofibrous membranes for antibacterial wound dressing and UV protection. *ACS Appl. Mater. Interfaces*, 8, 29915–29922, 2016.

166. Kim, H.R., Fujimori, K., Kim, B.S., *et al.*, Lightweight nanofibrous EMI shielding nanowebs prepared by electrospinning and metallization. *Composites Sci. Technol.*, 72, 1233–1239, 2012.
167. Kim, H.R., Kim, B.S., and Kim, I.S., Fabrication and EMI shielding effectiveness of Ag-decorated highly porous poly (vinyl alcohol)/Fe_2O_3 nanofibrous composites. *Mater. Chem. Phys.*, 135, 1024–1029, 2012.
168. Zhu, Y., Zhang, J.C., Zhai, J., *et al.*, Multifunctional carbon nanofibers with conductive, magnetic and superhydrophobic properties. *Chemphyschem*, 7, 336–341, 2006.
169. Bayat, M., Yang, H., and Ko, F., Electrical and magnetic properties of Fe_3O_4/ carbon composite nanofibres. In *SAMPE*, Seattle, 2010.
170. Bayat, M., Yang, H., and Ko, F., Electromagnetic properties of electrospun Fe_3O_4/carbon composite nanofibers. *Polymer*, 52, 1645–1653, 2011.
171. Faccini, M., Vaquero, C., and Amantia, D., Development of protective clothing against nanoparticle based on electrospun nanofibers. *J. Nanomater.*, 2012, 18, 2012.
172. Shahrabi, S., Gharehaghaji, A.A., and Latifi, M., Fabrication of electrospun polyamide-66 nanofiber layer for high-performance nanofiltration in clean room applications. *J. Ind. Text.*, 45, 1100–1114, 2016.
173. Wang, S., Zhao, X., Yin, X., *et al.*, Electret polyvinylidene fluoride nanofibers hybridized by polytetrafluoroethylene nanoparticles for high-efficiency air filtration. *ACS Appl. Mater. Interfaces*, 8, 23985–23994, 2016.
174. Wu, W., Bai, S., Yuan, M., *et al.*, Lead zirconate titanate nanowire textile nanogenerator for wearable energy-harvesting and self-powered devices. *ACS Nano*, 6, 6231–6235, 2012.
175. Fang, J., Wang, X., and Lin, T., Electrical power generator from randomly oriented electrospun poly(vinylidene fluoride) nanofibre membranes. *J. Mater. Chem.*, 21, 11088–11091, 2011.
176. Anton, S.R. and Sodano, H.A., A review of power harvesting using piezoelectric materials (2003–2006). *Smart Mater. Struct.*, 16, R1, 2007.
177. Chang, C., Tran, V.H., Wang, J., *et al.*, Direct-write piezoelectric polymeric nanogenerator with high energy conversion efficiency. *Nano Lett.*, 10, 726–731, 2010.
178. Gheibi, A., Latifi, M., Merati, A.A., *et al.*, Piezoelectric electrospun nanofibrous materials for self-powering wearable electronic textiles applications. *J. Polym. Res.*, 21, 469, 2014.
179. Fuh, Y.K., Chen, S.Y., and Ye, J.C., Massively parallel aligned microfibers-based harvester deposited via *in situ*, oriented poled near-field electrospinning. *Appl. Phys. Lett.*, 103, 033114, 2013.
180. Fang, J., Niu, H., Wang, H., *et al.*, Enhanced mechanical energy harvesting using needleless electrospun poly(vinylidene fluoride) nanofibre webs. *Energy Environ. Sci.*, 6, 2196–2202, 2013.
181. Wang, Z.L., Triboelectric nanogenerators as new energy technology for self-powered systems and as active mechanical and chemical sensors. *ACS Nano*, 7, 9533–9557, 2013.

182. Wang, Z.L., Triboelectric nanogenerators as new energy technology and self-powered sensors—Principles, problems and perspectives. *Faraday Discuss.*, 176, 447–458, 2015.
183. Li, T., Xu, Y., Willander, M., *et al.*, Lightweight triboelectric nanogenerator for energy harvesting and sensing tiny mechanical motion. *Adv. Funct. Mater.*, 26, 4370–4376, 2016.
184. Li, Z., Shen, J., Abdalla, I., *et al.*, Nanofibrous membrane constructed wearable triboelectric nanogenerator for high performance biomechanical energy harvesting. *Nano Energy*, 36, 341–348, 2017.
185. Wang, Y., Zheng, J., Ren, G., *et al.*, A flexible piezoelectric force sensor based on PVDF fabrics. *Smart Mater. Struct.*, 20, 045009, 2011.
186. Tiwari, M.K., Yarin, A.L., and Megaridis, C.M., Electrospun fibrous nanocomposites as permeable, flexible strain sensors. *J. Appl. Phys.*, 103, 044305, 2008.
187. Liu, N., Fang, G., Wan, J., *et al.*, Electrospun PEDOT: PSS–PVA nanofiber based ultrahigh-strain sensors with controllable electrical conductivity. *J. Mater. Chem.*, 21, 18962–18966, 2011.
188. Soltanian, S., Rahmanian, R., Gholamkhass, B., *et al.*, Highly stretchable, sparse, metallized nanofiber webs as thin, transferrable transparent conductors. *Adv. Energy Mater.*, 3, 1332–1337, 2013.
189. Soltanian, S., Servati, A., Rahmanian, R., *et al.*, Highly piezoresistive compliant nanofibrous sensors for tactile and epidermal electronic applications. *J. Mater. Res.*, 30, 121, 2015.
190. Sun, B., Long, Y.-Z., Liu, S.-L., *et al.*, Fabrication of curled conducting polymer microfibrous arrays via a novel electrospinning method for stretchable strain sensors. *Nanoscale*, 5, 7041–7045, 2013.
191. Kim, I., Lee, E.G., Jang, E., *et al.*, Characteristics of polyurethane nanowebs treated with silver nanowire solutions as strain sensors. *Text. Res. J.*, 0040517517697647, 2017.
192. Ren, G., Cai, F., Li, B., *et al.*, Flexible pressure sensor based on a poly (VDF-TrFE) nanofiber web. *Macromol. Mater. Eng.*, 298, 541–546, 2013.
193. Lee, S., Reuveny, A., Reeder, J., *et al.*, A transparent bending-insensitive pressure sensor, 11, 472, 2016.
194. Park, M., Lee, K.S., Shim, J., *et al.*, Environment friendly, transparent nanofiber textiles consolidated with high efficiency PLEDs for wearable electronics. *Org. Electron.*, 36, 89–96, 2016.
195. Flemming, R., Murphy, C., Abrams, G., *et al.*, Effects of synthetic micro-and nano-structured surfaces on cell behavior. *Biomaterials*, 20, 573–588, 1999.
196. Desai, T., Micro- and nanoscale structures for tissue engineering constructs. *Med. Eng. Phys.*, 22, 595–606, 2000.
197. Curtis, A. and Wilkinson, C., Nanotechniques and approaches in biotechnology. *Mater. Today*, 4, 22–28, 2001.
198. Zhang, Y., Lim, C., Ramakrishna, S., *et al.*, Recent development of polymer nanofibers for biomedical and biotechnological applications. *J. Mater. Sci. Mater. Med.*, 16, 933–946, 2005.

199. Gerardo-Nava, J., Führmann, T., Klinkhammer, K., *et al.*, Human neural cell interactions with orientated electrospun nanofibers *in vitro. Nanomedicine*, 4, 11–30, 2009.
200. Martin, P., Wound healing—Aiming for perfect skin regeneration. *Science*, 276, 75–81, 1997.
201. Singer, A.J. and Clark, R.A., Cutaneous wound healing. *New Engl. J. Med.*, 341, 738–746, 1999.
202. Broughton, G., 2nd, Janis, J.E., and Attinger, C.E., The basic science of wound healing. *Plast. Reconstr. Surg.*, 117, 12S–34S, 2006.
203. Gokarneshan, N., Rachel, D.A., Rajendran, V., *et al.*, *Emerging Research Trends in Medical Textiles*, Springer, 2015.
204. Khil, M.S., Cha, D.I., Kim, H.Y., *et al.*, Electrospun nanofibrous polyurethane membrane as wound dressing. *J. Biomed. Mater. Res., Part B*, 67, 675–679, 2003.
205. Liu, X., Lin, T., Gao, Y., *et al.*, Antimicrobial electrospun nanofibers of cellulose acetate and polyester urethane composite for wound dressing. *J. Biomed. Mater. Res., Part B*, 100, 1556–1565, 2012.
206. Duan, Y.Y., Jia, J., Wang, S.H., *et al.*, Preparation of antimicrobial poly (ϵ-caprolactone) electrospun nanofibers containing silver-loaded zirconium phosphate nanoparticles. *J. Appl. Polym. Sci.*, 106, 1208–1214, 2007.
207. Jayakumar, R., Prabaharan, M., Kumar, P.S., *et al.*, Biomaterials based on chitin and chitosan in wound dressing applications. *Biotechnol. Adv.*, 29, 322–337, 2011.
208. Barrientos, S., Stojadinovic, O., Golinko, M.S., *et al.*, Growth factors and cytokines in wound healing. *Wound Repair Regen.*, 16, 585–601, 2008.
209. Schneider, A., Wang, X., Kaplan, D., *et al.*, Biofunctionalized electrospun silk mats as a topical bioactive dressing for accelerated wound healing. *Acta Biomater.*, 5, 2570–2578, 2009.
210. Dai, X.-Y., Nie, W., Wang, Y.-C., *et al.*, Electrospun emodin polyvinylpyrrolidone blended nanofibrous membrane: A novel medicated biomaterial for drug delivery and accelerated wound healing. *J. Mater. Sci. Mater. Med.*, 23, 2709–2716, 2012.
211. Lee, S. and Obendorf, S.K., Use of electrospun nanofiber web for protective textile materials as barriers to liquid penetration. *Text. Res. J.*, 77, 696–702, 2007.
212. Gibson, P., Schreuder-Gibson, H., and Pentheny, C., Electrospinning technology: Direct application of tailorable ultrathin membranes. *J. Coat. Fabr.*, 28, 63–72, 1998.
213. Hong, K.H., Preparation and properties of electrospun poly (vinyl alcohol)/silver fiber web as wound dressings. *Polym. Eng. Sci.*, 47, 43–49, 2007.
214. Melaiye, A., Sun, Z., Hindi, K., *et al.*, Silver (I)– imidazole cyclophane gem-diol complexes encapsulated by electrospun tecophilic nanofibers: Formation of nanosilver particles and antimicrobial activity. *J. Am. Chem. Soc.*, 127, 2285–2291, 2005.

215. Abdelgawad, A.M., Hudson, S.M., and Rojas, O.J., Antimicrobial wound dressing nanofiber mats from multicomponent (chitosan/silver-NPs/polyvinyl alcohol) systems. *Carbohydr. Polym.*, 100, 166–178, 2014.
216. Hong, K.H., Park, J.L., Sul, I.H., *et al.*, Preparation of antimicrobial poly (vinyl alcohol) nanofibers containing silver nanoparticles. *J. Polym. Sci., Part B: Polym. Phys.*, 44, 2468–2474, 2006.
217. Zhou, Y., Yang, D., Chen, X., *et al.*, Electrospun water-soluble carboxyethyl chitosan/poly (vinyl alcohol) nanofibrous membrane as potential wound dressing for skin regeneration. *Biomacromolecules*, 9, 349–354, 2007.
218. Ignatova, M., Starbova, K., Markova, N., *et al.*, Electrospun nano-fibre mats with antibacterial properties from quaternised chitosan and poly (vinyl alcohol). *Carbohydr. Res.*, 341, 2098–2107, 2006.
219. Cai, Z., Mo, X.M., Zhang, K.H., *et al.*, Fabrication of chitosan/silk fibroin composite nanofibers for wound-dressing applications. *Int. J. Mol. Sci.*, 11, 3529–3539, 2010.
220. Shalumon, K., Anulekha, K., Nair, S.V., *et al.*, Sodium alginate/poly (vinyl alcohol)/nano ZnO composite nanofibers for antibacterial wound dressings. *Int. J. Biol. Macromol.*, 49, 247–254, 2011.
221. Sudheesh Kumar, P., Lakshmanan, V.-K., Anilkumar, T., *et al.*, Flexible and microporous chitosan hydrogel/nano ZnO composite bandages for wound dressing: *In vitro* and *in vivo* evaluation. *ACS Appl. Mater. Interfaces*, 4, 2618–2629, 2012.
222. Kataria, K., Gupta, A., Rath, G., *et al.*, *In vivo* wound healing performance of drug loaded electrospun composite nanofibers transdermal patch. *Int. J. Pharm.*, 469, 102–110, 2014.
223. Thakur, R.A., Florek, C.A., Kohn, J., *et al.*, Electrospun nanofibrous polymeric scaffold with targeted drug release profiles for potential application as wound dressing. *Int. J. Pharm.*, 364, 87–93, 2008.
224. Merrell, J.G., McLaughlin, S.W., Tie, L., *et al.*, Curcumin-loaded poly(ϵ-caprolactone) nanofibres: Diabetic wound dressing with anti-oxidant and anti-inflammatory properties. *Clin. Exp. Pharmacol. Physiol.*, 36, 1149–1156, 2009.
225. Choi, J.S., Leong, K.W., and Yoo, H.S., *In vivo* wound healing of diabetic ulcers using electrospun nanofibers immobilized with human epidermal growth factor (EGF). *Biomaterials*, 29, 587–596, 2008.
226. Xie, Z., Paras, C.B., Weng, H., *et al.*, Dual growth factor releasing multifunctional nanofibers for wound healing. *Acta Biomater.*, 9, 9351–9359, 2013.
227. Lalani, R. and Liu, L., Electrospun zwitterionic poly(sulfobetaine methacrylate) for nonadherent, superabsorbent, and antimicrobial wound dressing applications. *Biomacromolecules*, 13, 1853–1863, 2012.
228. Uppal, R., Ramaswamy, G.N., Arnold, C., *et al.*, Hyaluronic acid nanofiber wound dressing—Production, characterization, and *in vivo* behavior. *J. Biomed. Mater. Res., Part B*, 97B, 20–29, 2011.

229. Kamoun, E.A., Chen, X., Eldin, M.S.M., *et al.*, Crosslinked poly (vinyl alcohol) hydrogels for wound dressing applications: A review of remarkably blended polymers. *Arabian J. Chem.*, 8, 1–14, 2015.
230. Růžičková, J., Velebný, V., Novák, J., *et al.*, Hyaluronic acid based nanofibers for wound dressing and drug delivery carriers. In *Intracellular Delivery II: Fundamentals and Applications*, Prokop, A., Iwasaki, Y., Harada, A. (Eds.), Springer Netherlands, Dordrecht, pp. 417–433, 2014.
231. Zahedi, P., Rezaeian, I., Ranaei-Siadat, S.O., *et al.*, A review on wound dressings with an emphasis on electrospun nanofibrous polymeric bandages. *Polym. Adv. Technol.*, 21, 77–95, 2010.
232. Abrigo, M., McArthur, S.L., and Kingshott, P., Electrospun nanofibers as dressings for chronic wound care: Advances, challenges, and future prospects. *Macromol. Biosci.*, 14, 772–792, 2014.
233. Rieger, K.A., Birch, N.P., and Schiffman, J.D., Designing electrospun nanofiber mats to promote wound healing—A review. *J. Mater. Chem. B*, 1, 4531–4541, 2013.
234. Jayakumar, R. and Nair, S., *Biomedical Applications of Polymeric Nanofibers*, Springer Science & Business Media, 2012.
235. Lim, C.T., Tan, E.P.S., and Ng, S.Y., Effects of crystalline morphology on the tensile properties of electrospun polymer nanofibers. *Appl. Phys. Lett.*, 92, 141908, 2008.
236. Arinstein, A., Burman, M., Gendelman, O., *et al.*, Effect of supramolecular structure on polymer nanofibre elasticity. *Nat Nano*, 2, 59–62, 2007.
237. Wong, S.C., Baji, A., and Leng, S.W., Effect of fiber diameter on tensile properties of electrospun poly(ε-caprolactone). *Polymer*, 49, 4713–4722, 2008.
238. Pai, C.-L., Boyce, M.C., and Rutledge, G.C., Mechanical properties of individual electrospun PA 6(3)T fibers and their variation with fiber diameter. *Polymer*, 52, 2295–2301, 2011.
239. Tan, E.P.S. and Lim, C.T., Physical properties of a single polymeric nanofiber. *Appl. Phys. Lett.*, 84, 1603–1605, 2004.
240. Naraghi, M., Arshad, S.N., and Chasiotis, I., Molecular orientation and mechanical property size effects in electrospun polyacrylonitrile nanofibers. *Polymer*, 52, 1612–1618, 2011.
241. Gandhi, M., Yang, H., Shor, L., *et al.*, Post-spinning modification of electrospun nanofiber nanocomposite from *Bombyx mori* silk and carbon nanotubes. *Polymer*, 50, 1918–1924, 2009.
242. Cao, Q., Wan, Y., Qiang, J., *et al.*, Effect of sonication treatment on electrospinnability of high-viscosity PAN solution and mechanical performance of microfiber mat. *Iranian Polym. J.*, 1–7, 2014.
243. Lai, C., Zhong, G., Yue, Z., *et al.*, Investigation of post-spinning stretching process on morphological, structural, and mechanical properties of electrospun polyacrylonitrile copolymer nanofibers. *Polymer*, 52, 519–528, 2011.

244. Papkov, D., Zou, Y., Andalib, M.N., *et al.*, Simultaneously strong and tough ultrafine continuous nanofibers. *ACS Nano*, 7, 3324–3331, 2013.
245. Leung, V., Hartwell, R., Elizei, S.S., *et al.*, Postelectrospinning modifications for alginate nanofiber-based wound dressings. *J. Biomed. Mater. Res., Part B*, 102, 508–515, 2014.
246. Li, H., Zhu, C., Xue, J., *et al.*, Enhancing the mechanical properties of electrospun nanofiber mats through controllable welding at the cross points. *Macromol. Rapid Commun.*, 38, 2017.
247. He, J.H., Wan, Y.Q., and Xu, L., Nano-effects, quantum-like properties in electrospun nanofibers. *Chaos, Solitons Fractals*, 33, 26–37, 2007.
248. Thompson, C.J., Chase, G.G., Yarin, A.L., *et al.*, Effects of parameters on nanofiber diameter determined from electrospinning model. *Polymer*, 48, 6913–6922, 2007.
249. Wan, Y.Q., He, J.H., Wu, Y., *et al.*, Vibrorheological effect on electrospun polyacrylonitrile (PAN) nanofibers. *Mater. Lett.*, 60, 3296–3300, 2006.
250. Cao, Q., Wan, Y., Qiang, J., *et al.*, Effect of sonication treatment on electrospinnability of high-viscosity PAN solution and mechanical performance of microfiber mat. *Iran. Polym. J.*, 23, 947–953, 2014.
251. Deitzel, J., Kleinmeyer, J., BeckTan, N., *et al.*, *Generation of polymer nanofibers through electrospinning*, Army Research Lab Aberdeen Proving Ground Md, 1999.
252. Salem, D.R., Electrospinning of nanofibers and the charge injection method. In *Nanofibers and Nanotechnology in Textiles*, P.J. Brown and K. Stevens (Eds.), Woodhead Publishing Limited, Cambridge, England, 2007.
253. Yamashita, Y., Ko, F., Miyake, H., *et al.*, Establishment of nanofiber preparation technique by electrospinning. *Sen'i Gakkaishi*, 64, 24–28, 2008.
254. Varabhas, J.S., Chase, G.G., and Reneker, D.H., Electrospun nanofibers from a porous hollow tube. *Polymer*, 49, 4226–4229, 2008.
255. Varesano, A., Carletto, R.A., and Mazzuchetti, G., Experimental investigations on the multi-jet electrospinning process. *J. Mater. Process. Technol.*, 209, 5178–5185, 2009.
256. Theron, S.A., Yarin, A.L., Zussman, E., *et al.*, Multiple jets in electrospinning: Experiment and modeling. *Polymer*, 46, 2889–2899, 2005.
257. Yang, Y., Jia, Z., Li, Q., *et al.*, *Electrospun Uniform Fibres with a Special Regular Hexagon Distributed Multi-Needles System*, IOP Publishing, 2008.
258. Tomaszewski, W. and Szadkowski, M., Investigation of electrospinning with the use of a multi-jet electrospinning head. *Fibers Text. East. Eur.*, 13, 22–26, 2005.
259. Kim, G.H., Cho, Y.S., and Kim, W.D., Stability analysis for multi-jets electrospinning process modified with a cylindrical electrode. *Eur. Polym. J.*, 42, 2031–2038, 2006.
260. Varesano, A., Rombaldoni, F., Mazzuchetti, G., *et al.*, Multi-jet nozzle electrospinning on textile substrates: Observations on process and nanofibre mat deposition. *Polym. Int.*, 2010.

261. Dosunmu, O.O., Chase, G.G., Kataphinan, W., *et al.*, Electrospinning of polymer nanofibers from multiple jets on a porous tubular surface. *Nanotechnology*, 17, 1123, 2006.
262. Lu, B., Wang, Y., Liu, Y., *et al.*, Superhigh-throughput needleless electrospinning using a rotary cone as spinneret. *Small*, 6, 1612–1616, 2010.
263. Wang, X., Niu, H., Lin, T., *et al.*, Needleless electrospinning of nanofibers with a conical wire coil. *Polym. Eng. Sci.*, 49, 1582–1586, 2009.
264. Yarin, A.L. and Zussman, E., Upward needleless electrospinning of multiple nanofibers. *Polymer*, 45, 2977, 2004.
265. Lukas, D., Sarkar, A., and Pokorny, P., Self-organization of jets in electrospinning from free liquid surface: A generalized approach. *J. Appl. Phys.*, 103, 084309–084307, 2008.
266. Liu, Y. and He, J., Bubble electrospinning for mass production of nanofibers. *Int. J. Nonlinear Sci. Numer. Simul.*, 8, 393, 2007.
267. Petrik, S. and Maly, M. *Production* nozzle-less electrospinning nanofiber technology. In *2009 Fall MRS Symposium*, Boston, MA, 2009.
268. Wu, D., Huang, X., Lai, X., *et al.*, High throughput tip-less electrospinning via a circular cylindrical electrode. *J. Nanosci. Nanotechnol.*, 10, 4221–4226, 2010.
269. Dabirian, F., Ravandi, S.A.H., Hinestroza, J.P., *et al.*, Conformal coating of yarns and wires with electrospun nanofibers. *Polym. Eng. Sci.*, 52, 1724–1732, 2012.
270. He, J.X., Zhou, Y.M., Wu, Y.C., *et al.*, Nanofiber coated hybrid yarn fabricated by novel electrospinning-airflow twisting method. *Surf. Coat. Technol.*, 258, 398–404, 2014.
271. Zhou, Y., He, J., Wang, H., *et al.*, Continuous nanofiber coated hybrid yarn produced by multi-nozzle air jet electrospinning. *J. Text. I.*, 108, 783–787, 2017.
272. Liu, C.K., He, H.J., Sun, R.J., *et al.*, Preparation of continuous nanofiber core-spun yarn by a novel covering method. *Mater. Des.*, 112, 456–461, 2016.
273. Fennessey, S.F. and Farris, R.J., Fabrication of aligned and molecularly oriented electrospun polyacrylonitrile nanofibers and the mechanical behavior of their twisted yarns. *Polymer*, 45, 4217–4225, 2004.
274. Teo, W. and Ramakrishna, S., Electrospun fibre bundle made of aligned nanofibres over two fixed points. *Nanotechnology*, 16, 1878, 2005.
275. Ali, U., Zhou, Y., Wang, X., *et al.*, Direct electrospinning of highly twisted, continuous nanofiber yarns. *J. Text. I.*, 103, 80–88, 2012.
276. Ali, U., Zhou, Y., Wang, X., *et al.*, Electrospinning of continuous nanofiber bundles and twisted nanofiber yarns. In *Nanofibers—Production, Properties and Functional Applications*, Lin, T. (Ed.), InTech, 2011.
277. He, J.X., Qi, K., Zhou, Y.M., *et al.*, Fabrication of continuous nanofiber yarn using novel multi-nozzle bubble electrospinning. *Polym. Int.*, 63, 1288–1294, 2014.

278. Mokhtari, F., Salehi, M., Zamani, F., *et al.*, Advances in electrospinning: The production and application of nanofibres and nanofibrous structures. *Text. Prog.*, 48, 119–219, 2016.
279. Wang, X., Zhang, K., Zhu, M., *et al.*, Enhanced mechanical performance of self-bundled electrospun fiber yarns via post-treatments. *Macromol. Rapid Commun.*, 29, 826–831, 2008.
280. Lee, S. and Kay Obendorf, S., Developing protective textile materials as barriers to liquid penetration using melt-electrospinning. *J. Appl. Polym. Sci.*, 102, 3430–3437, 2006.
281. Bagherzadeh, R., Latifi, M., Najar, S.S., *et al.*, Transport properties of multi-layer fabric based on electrospun nanofiber mats as a breathable barrier textile material. *Text. Res. J.*, 82, 70–76, 2012.
282. Lee, S. and Obendorf, S.K., Transport properties of layered fabric systems based on electrospun nanofibers. *Fiber. Polym.*, 8, 501–506, 2007.
283. Knizek, R., Karhankova, D., and Fridrichova, L., Two and three layer lamination of nanofiber. *World Acad. Sci. Eng. Technol. Int. J. Chem. Molec. Nucl. Mater. Metall. Eng.*, 9, 1266–1269, 2017.
284. Heinisch, T., Bajzík, V., Knížek, R., *et al.*, Effect of the process of lamination microporous nanofiber membrane on the evaporative resistance of the two-layer laminate. In *Advanced Materials Research*, Trans Tech Publ, 2013.
285. Sumin, L., Kimura, D., Yokoyama, A., *et al.*, The effects of laundering on the mechanical properties of mass-produced nanofiber web for use in wear. *Text. Res. J.*, 79, 1085–1090, 2009.
286. Rombaldoni, F., Mahmood, K., Varesano, A., *et al.*, Adhesion enhancement of electrospun nanofiber mats to polypropylene nonwoven fabric by low-temperature oxygen plasma treatment. *Surf. Coat. Technol.*, 216, 178–184, 2013.
287. Varesano, A., Rombaldoni, F., Tonetti, C., *et al.*, Chemical treatments for improving adhesion between electrospun nanofibers and fabrics. *J. Appl. Polym. Sci.*, 131, 2014.
288. Maynor, B.W., Filocamo, S.F., Grinstaff, M.W., *et al.*, Direct-writing of polymer nanostructures: Poly (thiophene) nanowires on semiconducting and insulating surfaces. *J. Am. Chem. Soc.*, 124, 522–523, 2002.
289. Sun, D., Chang, C., Li, S., *et al.*, Near-field electrospinning. *Nano Lett.*, 6, 839–842, 2006.
290. Pu, J., Yan, X., Jiang, Y., *et al.*, Piezoelectric actuation of direct-write electrospun fibers. *Sens. Actuators, A*, 164, 131–136, 2010.
291. Lee, J., Lee, S.Y., Jang, J., *et al.*, Fabrication of patterned nanofibrous mats using direct-write electrospinning. *Langmuir*, 28, 7267–7275, 2012.
292. Theron, A., Zussman, E., and Yarin, A., Electrostatic field-assisted alignment of electrospun nanofibres. *Nanotechnology*, 12, 384, 2001.
293. Dalton, P.D., Joergensen, N.T., Groll, J., *et al.*, Patterned melt electrospun substrates for tissue engineering. *Biomed. Mater.*, 3, 034109, 2008.

294. Gopal, R., Kaur, S., Feng, C.Y., *et al.*, Electrospun nanofibrous polysulfone membranes as pre-filters: Particulate removal. *J. Membr. Sci.*, 289, 210–219, 2007.
295. Aussawasathien, D., Teerawattananon, C., and Vongachariya, A., Separation of micron to sub-micron particles from water: Electrospun nylon-6 nanofibrous membranes as pre-filters. *J. Membr. Sci.*, 315, 11–19, 2008.
296. Lee, M.W., An, S., Latthe, S.S., *et al.*, Electrospun polystyrene nanofiber membrane with superhydrophobicity and superoleophilicity for selective separation of water and low viscous oil. *ACS Appl. Mater. Interfaces*, 5, 10597–10604, 2013.
297. Gopal, R., Kaur, S., Ma, Z., *et al.*, Electrospun nanofibrous filtration membrane. *J. Membr. Sci.*, 281, 581–586, 2006.
298. Haider, S. and Park, S.Y., Preparation of the electrospun chitosan nanofibers and their applications to the adsorption of Cu (II) and Pb (II) ions from an aqueous solution. *J. Membr. Sci.*, 328, 90–96, 2009.
299. Kirichenko, V., Filatov, Y., and Budyka, A., Electrospinning of micro- and nanofibers: Fundamentals in separation and filtration processes. *Int. J. Multiscale Comput. Eng.*, 8, 2010.
300. Kim, I.-D., Rothschild, A., Lee, B.H., *et al.*, Ultrasensitive chemiresistors based on electrospun TiO_2 nanofibers. *Nano Lett.*, 6, 2009–2013, 2006.
301. Zhang, Y., He, X., Li, J., *et al.*, Fabrication and ethanol-sensing properties of micro gas sensor based on electrospun SnO_2 nanofibers. *Sensors Actuators B: Chem.*, 132, 67–73, 2008.
302. Aussawasathien, D., Dong, J.-H., and Dai, L., Electrospun polymer nanofiber sensors. *Synth. Met.*, 154, 37–40, 2005.
303. Eastman launches Avra™ Performance Fibers, introducing a polyester fiber with a unique combination of size, shape, and performance to the active apparel market. http://www.eastman.com/Company/News_Center/2016/Pages/Eastman-launches-Avra-Performance-Fibers.aspx, 2017.
304. Ellison, C.J., Phatak, A., Giles, D.W., *et al.*, Melt blown nanofibers: Fiber diameter distributions and onset of fiber breakup. *Polymer*, 48, 3306–3316, 2007.
305. Singh, C., Shaffer, M.S.P., and Windle, A.H., Production of controlled architectures of aligned carbon nanotubes by an injection chemical vapour deposition method. *Carbon*, 41, 359–368, 2003.
306. Li, Y.L., Kinloch, I.A., and Windle, A.H., *Direct Spinning of Carbon Nanotube Fibers from Chemical Vapor Deposition Synthesis*, American Association for the Advancement of Science, pp. 276–278, 2004.
307. Motta, M., Moisala, A., Kinloch, I.A., and Windle, A.H., High performance fibres from 'dog bone' carbon nanotubes. *Adv. Mater.*, 19, 3721–3726, 2007.
308. Li, Y.L., Kinloch, I.A., and Windle, A.H., Direct spinning of carbon nanotube fibers from chemical vapor deposition synthesis. *Science*, 304, 276–278, 2004.
309. Koziol, K., Vilatela, J., Moisala, A., *et al.*, High-performance carbon nanotube fiber. *Science*, 318, 1892, 2007.

310. Wang, B.Z. and Chen, Y., The Effect of 3D Printing Technology on the Future Fashion Design and Manufacturing. In *Applied Mechanics and Materials*, Trans Tech Publ, 2014.
311. Mao, M., He, J., Li, X., *et al.*, The emerging frontiers and applications of high-resolution 3D printing. *Micromachines*, 8, 113, 2017.
312. Kawata, S., Sun, H.-B., Tanaka, T., *et al.*, Finer features for functional microdevices. *Nature*, 412, 697, 2001.
313. Farsari, M. and Chichkov, B.N., Materials processing: Two-photon fabrication. *Nat. Photonics*, 3, 450, 2009.
314. Xing, J.F., Zheng, M.L., and Duan, X.M., Two-photon polymerization microfabrication of hydrogels: An advanced 3D printing technology for tissue engineering and drug delivery. *Chem. Soc. Rev.*, 44, 5031–5039, 2015.
315. Sakellari, I., Kabouraki, E., Gray, D., *et al.*, Quantum dot based 3D photonic devices. In *Advanced Fabrication Technologies for Micro/Nano Optics and Photonics X*, International Society for Optics and Photonics, 2017.
316. Wang, W.K., Sun, Z.B., Zheng, M.L., *et al.*, Magnetic nickel–phosphorus/polymer composite and remotely driven three-dimensional micromachine fabricated by nanoplating and two-photon polymerization. *J. Phys. Chem. C*, 115, 11275–11281, 2011.
317. Torgersen, J., Qin, X.H., Li, Z., *et al.*, Hydrogels for two-photon polymerization: A toolbox for mimicking the extracellular matrix. *Adv. Funct. Mater.*, 23, 4542–4554, 2013.
318. Xia, Y. and Whitesides, G.M., Soft lithography. *Annu. Rev. Mater. Sci.*, 28, 153, 1998.
319. Qin, D., Xia, Y., and Whitesides, G.M., Soft lithography for micro-and nanoscale patterning. *Nat. Protoc.*, 5, 491, 2010.
320. Wang, F., Li, S., and Wang, L., Fabrication of artificial super-hydrophobic lotus-leaf-like bamboo surfaces through soft lithography. *Colloids Surf. Physicochem. Eng. Aspects*, 513, 389–395, 2017.
321. Kamei, K.-I., Mashimo, Y., Koyama, Y., *et al.*, 3D printing of soft lithography mold for rapid production of polydimethylsiloxane-based microfluidic devices for cell stimulation with concentration gradients. *Biomed. Microdevices*, 17, 36, 2015.
322. Whitesides, G.M., Ostuni, E., Takayama, S., *et al.*, Soft lithography in biology and biochemistry. *Annu. Rev. Biomed. Eng.*, 3, 335–373, 2001.
323. Huang, Y., Paloczi, G.T., Yariv, A., *et al.*, Fabrication and replication of polymer integrated optical devices using electron-beam lithography and soft lithography. *J. Phy. Chem. B*, 108, 8606–8613, 2004.
324. Tokonami, S., Shiigi, H., and Nagaoka, T., Micro-and nanosized molecularly imprinted polymers for high-throughput analytical applications. *Anal. Chim. Acta*, 641, 7–13, 2009.
325. Buguin, A., Li, M.-H., Silberzan, P., *et al.*, Micro-actuators: When artificial muscles made of nematic liquid crystal elastomers meet soft lithography. *J. Am. Chem. Soc.*, 128, 1088–1089, 2006.

326. Hung, A.M. and Stupp, S.I., Simultaneous self-assembly, orientation, and patterning of peptide–amphiphile nanofibers by soft lithography. *Nano Lett.*, 7, 1165–1171, 2007.
327. Mele, E., Lezzi, F., Polini, A., *et al.*, Enhanced charge-carrier mobility in polymer nanofibers realized by solvent-resistant soft nanolithography. *J. Mater. Chem.*, 22, 18051–18056, 2012.
328. Maslow, A.H., A theory of human motivation. *Psychol. Rev.*, 50, 370, 1943.
329. Pantelopoulos, A. and Bourbakis, N.G., A survey on wearable sensor-based systems for health monitoring and prognosis. *IEEE Trans. Syst. Man. Cybern. C Appl. Rev.*, 40, 1–12, 2010.
330. Madden, J.D., Vandesteeg, N.A., Anquetil, P.A., *et al.*, Artificial muscle technology: Physical principles and naval prospects. *IEEE J. Oceanic Eng.*, 29, 706–728, 2004.

3

Nanosols for Smart Textiles

Boris Mahltig

Hochschule Niederrhein, Faculty of Textile and Clothing Technology, Mönchengladbach, Germany

Abstract

Nanosols are innovative coating agents useable for the functionalization and finishing of textile materials. Nanosols are based on inorganic particles in nanometer scale produced by sol–gel technology, which are stabilized in solvents – mainly ethanol, iso-propanol or water. One main advantage of this type of coating agent is that the agents can be modified with simple methods in a broad range, and by this, a variety of new and smart functions can be realized on textile surfaces. However, for application on textiles, there is also a main challenge; this is the realization of sol–gel systems using water as the main solvent. Due to stability reasons, nanosols are often prepared with organic solvents, which are from the point of flammability, safety reasons and cost often not wished.

This chapter will introduce the preparation and modification of nanosols especially dedicated to the application on textiles to realize functional and smart materials. Examples of the realization of water-based sol–gel coating agents are given. Special focus is set on photocatalytic/light responsive materials, antimicrobial systems and controlled release systems.

Keywords: Sol–gel technology, inorganic nanoparticle coating, photocatalysis, antimicrobial, controlled release

3.1 Introduction

The sol–gel technology is probably one of the most successful and promising methods to produce new, advantageous and functional materials in all areas of life, science and industry. The number of new materials realized

Email: mahltig@gmx.de

Nazire D. Yilmaz (ed.) Smart Textiles, (91–110) © 2019 Scrivener Publishing LLC

and developed by sol–gel method is nearly uncountable. As well, the fields of applications and realized functions are manifold [1–5].

To give here an impression on the broad field of sol–gel technology, just take a look on the recent sol–gel conference held last September 2017 in Liège (Belgium). This conference takes place every second year, and in 2017, more than 300 contributions from different fields of sol–gel technology were presented by scientists from all over the world. The idea of functional sol–gel coatings is mainly to create an inorganic coating with a certain functional property and to apply it onto a certain substrate (e.g. metal, glass, textile, wood, leather, etc.) [6–11].

By this, the functional property is transferred to the coated substrate. This created inorganic sol–gel coating can be self-functional from its own nature such as the metal oxide coating, e.g. to improve the abrasion resistance of glass fiber fabrics [3]. However, in most cases, the function is introduced by special additives carrying the functional properties. A simple example can be given with colored sol–gel coatings. Mixed metal oxides of spinel type can be realized by sol–gel method and contain a self-inherent colorant [12]. In comparison, an uncolored silica coating realized by sol–gel method can be modified by incorporation of a dye stuff that carries the color as a functional property [13–21].

Because of the many different functions, which can be applied by sol–gel coatings, it is suitable to place the functions in several specific main groups, as shown in Figure 3.1.

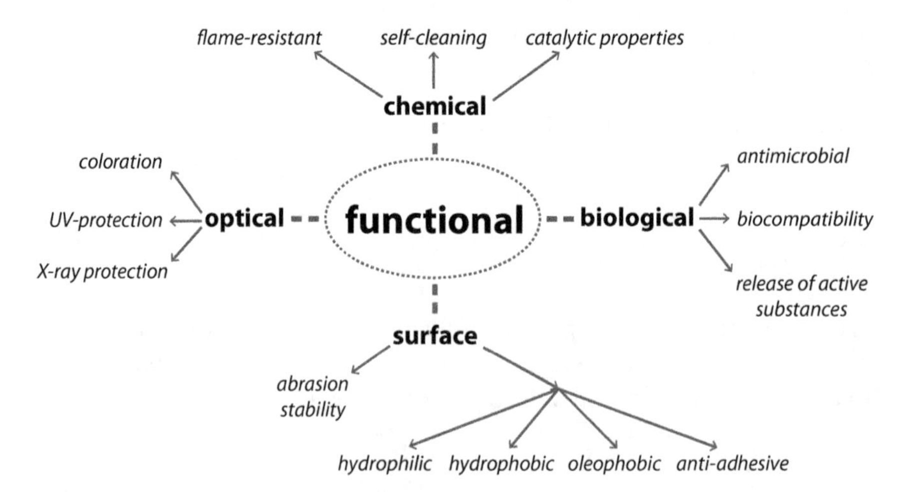

Figure 3.1 Schematic overview of possible functions that can be achieved by nanosols on textiles. The functions are divided into four main groups – optical-, chemical-, biological- and surface-functional. Note: This schematic drawing has been created by the author.

Optical functions summarize all properties modifying the interaction with light or in a broader view with any type of electromagnetic radiation. To this field count the ordinary coloration but also functions as UV-protection, IR-absorption or X-ray protection properties [15–17, 22–24]. The chemical functionalization is related to all types of coatings modifying the chemical properties of the coated substrates. This can be a coating introducing flame-retardant properties [3, 25–30]. Also, the application of catalytic properties is possible [30, 31]. In this field, many developments have been achieved with photocatalytic coatings often based on titania or zinc oxide materials [33–38].

The biological functionalization is related to any property interacting with living organisms. Main field in this area is the antimicrobial function acting against microorganisms [39–44]. Sol–gel coatings can be also modified to realize coatings with biocompatibility [3, 45–48].

The term "surface functional" summarizes all coatings modifying the interaction of the fiber surface to other media. This can be an enhanced abrasion stability, stabilizing the textile surface against mechanical influences in contact with other materials [3].

Besides, repellent coatings are surface functional. Water-repellent hydrophobic coatings can be realized on textiles by sol–gel method as well as oleophobic coatings [46, 49–51]. A repellent modification can be also described as antiadhesive. Such antiadhesive coatings can be used to realize antiadhesive wound bandages, which are less sticky to the healing wound [8, 52].

A hydrophilic coating increases the ability of the coated textile substrate for water uptake. The content of water on a textile substrate is directly related to its antistatic properties, because a static charging is conducted away by the uptaken water [53–55]. This approach can be broadened in a double functional sol–gel coating combining the water-repellent and antistatic properties. This combination is especially challenging, because two contradictive properties (hydrophobic and hydrophilic) are combined and realized in one single application [56, 57].

3.2 Preparation of Nanosols as Coating Agents

Nanosols are solutions containing particles with small diameters in the range of a few nanometers up to 80 nm. The nanosol particles are often built up by semimetal oxide or metal oxide compounds. These solutions are meta-stable, meaning that the particles would agglomerate and precipitate or gelate because of their small size [45]. The driving force of agglomeration is the large surface area of the small particles compared to their small particle volume. In case of nanosols, particle agglomeration is prevented by different influences, e.g. surface charges or sterical reasons [58]. Another and more traditional definition of nanosols could be the description as colloidal solution.

Nanosols can be used and applied as liquid coating agents on textile substrates by different methods such as padding, dipping or spraying. After application on the substrate, the solvent evaporates and the remaining nanoparticles aggregate forming a three-dimensional network. This network forms the final sol–gel coating. A further thermal treatment leads to condensation of the coating and finally to the formation of a ceramic coating (Figure 3.2) [3, 59].

The preparation and application of a silica-based SiO_2-nanosol is schematically depicted in Figure 3.2. This silica sol is shown as an example, because most reported nanosols are related to silicon-based systems. Other prominent examples for materials building up nanosols are titania (TiO_2), zirconia (ZrO_2), zinc oxide (ZnO) or barium titanate ($BaTiO_3$) [3, 23].

In case of SiO_2 sols, the preparation starts by hydrolysis of metal alkoxide compounds. In Figure 3.2, the example tetraethoxysilane (TEOS) is given. TEOS has the advantage of medium reactivity, compared to tetramethoxysilane (TMOS) with higher and tetrapropoxysilane (TPOS) with lower reactivity [58]. The hydrolysis can be performed under acidic or alkaline conditions and leads to the formation of silanol groups Si–OH in the first step. A following condensation of silanol groups leads to the formation of nanosized SiO_2 particles and the nanosol is formed.

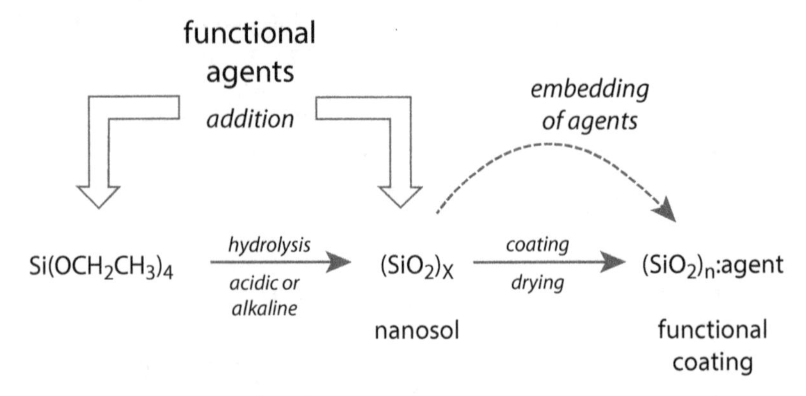

Figure 3.2 Schematic drawing of the production route of functional nanosols and the following realization of functional coatings. The shown picture is related to a nanosol based on semimetal oxide – here silicon dioxide (silica; SiO_2). Note: This schematic drawing has been created by the author.

To functionalize nanosols, agents containing the desired functions are added to the nanosol or to the precursors before hydrolysis is started (Figure 3.2). These agents are embedded into the sol–gel coating during the coating and drying processes. This embedding is also named as physical modification of the sol–gel coating. A chemical modification is performed, if a covalent bond between the SiO_2 matrix of the coating and the functional agent is also formed, besides the simple embedding [3, 45].

3.3 Application on Textiles

In a first and easy view, a nanosol agent can be applied on any kind of substrate to gain the wished function. However, each kind of substrate has its own demands to realize an optimal sol–gel coating.

Textile substrates contain higher flexibility compared to other substrates, e.g. wood, stone or glass. This flexibility is a typical property essential for many textile applications, so the flexibility should not be negatively influenced by the functional coating. Other properties such as air permeability also have to be kept even after the coating application. To achieve flexible coatings, nanosols should be applied in low concentrations [3, 43, 60].

Another important point is the stability of the coatings in case of usage, which is related to abrasion stability and washing stability. For this purpose, anchoring of the coating to the textile surface is required. For industrial applications, nanosols, which are free from flammable or toxic organic solvents, are often required.

Table 3.1 summarizes some basic nanosols and their preparation parameters. These sols can be used after further modification for textile treatments. All of these sols contain water as the solvent. The used silane precursors contain hydrophobic properties and are, for this reason, not mixable with water, so a strong stirring has to be performed during preparation to force the components together to an emulsion. After the hydrolysis of the silane precursor is finished and the SiO_2 particles are formed, a clear, transparent and homogeneous solution is obtained. The reaction time can be as long as two days. After the reaction, it has to be confirmed that no precipitation occurs in the sol. Such a precipitation is a hint for an insufficient sol formation and an unstable recipe. The stability of these nanosols varies in a broad range. The mentioned silica sol is stable at room temperature for several weeks. The duration until the gelation process is increased, if the compound 3-glycidyloxypropyltriethoxysilane (GLYEO) is added as the second precursor [42]. In comparison, the mentioned alkaline silica sol is stable against gelation for a period of one year or even longer (Table 3.1).

Table 3.1 Overview of recipes for some aqueous silica sols. The utilized silane precursors are tetraethoxysilane (TEOS) and 3-glycidyloxypropyltriethoxysilane (GLYEO). Note: The table has been arranged by the author.

Sols	Components	Comments	References
Acidic SiO_2 sol – type 1	TEOS as silane precursor, water as solvent and HCl or HNO_3 as acid to promote the hydrolysis	Acidic SiO_2 sol	[42]
Acidic SiO_2 sol – type 2	TEOS and GLYEO as silane precursors, water as solvent and HCl or HNO_3 as acid to promote the hydrolysis	Acidic SiO_2 sol with epoxy component	[42]
Alkaline SiO_2 sol	TEOS as silane precursor, water as solvent and triethanolamine as base to promote the hydrolysis	Alkaline SiO_2 sol	[40]

3.4 Nanosols and Smart Textiles

Nanosols that are used to realize functional materials are described in the following subsections. The application of these sols onto textile substrates is in their first steps or even when the sol formation has not started yet. However, the perspective functional properties are highly attractive, worth to mention here and to highlight the chances for realization of functional and advantageous textiles in the future.

3.4.1 Photocatalytic and Light Responsive Materials

Light responsive materials are systems exhibiting any kind of action in case of exhibition to light. In case of textile materials, the light responsive action that is discussed mostly in literature is the photocatalytic activity. In this field of photocatalytic materials, the photooxidation promoted by catalytic TiO_2 species is the most prominent one.

For application onto textiles, titania sols in aqueous solutions are often wanted as the finishing agents for textiles. However, the reactivity of titanium precursors with water is higher compared to silane precursors, so it is quite challenging to prepare water-based TiO_2 sol coating agents.

One change to prepare photoactive TiO_2 particles in aqueous solutions is to use a special pH-regime and to stabilize titanium alkoxides in the presence of amino compounds. For this purpose, compounds like triethanolamine and polymers containing amino groups can be used [33]. A prominent example here is polyvinylamine, which is also presented in Figure 3.3. Polyvinylamine is also well known under the trade name Lupamin (BASF product).

After preparation of the water-based titania sols, a thermal treatment is performed on the liquid recipes to transfer the previously formed amorphous TiO_2 into the photoactive anatase modification. This thermal treatment can be done under ambient pressure in reflux or under solvothermal conditions with high pressure in an autoclave device. The presence of crystalline anatase is determined by X-ray diffraction (XRD) measurements. By this analytical method, the formation of brookite as second crystalline modification of TiO_2 is also determined. Obviously the brookite is formed as well in small amounts during the mentioned preparation technique. The prepared TiO_2 sols can be used as coating agents for textile treatments. The photoactivity of the coated textiles is measured by color changes of dye solutions under illumination with UV A light. The related measurement procedure is described in the literature [31, 32]. The photoactivity A [%] is given in relation to the color change during a reference measurement with an analogous but uncoated textile. The measurement results are given in Table 3.2.

The method used to determine the photoactivity is simple and fast. However, it has to be kept in mind that the determined photoactivity is related to the used dye stuff [32]. Also the type and number of used UV light sources and their distance have significant influence on the measurement result. For this, a comparison of percentage values of photoactivity is only

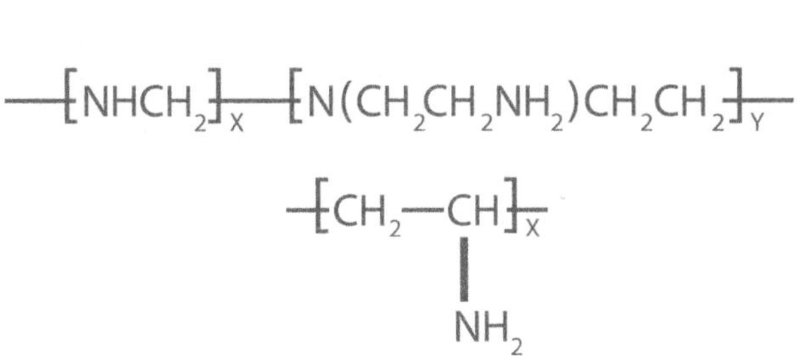

$$-[NHCH_2]_X-[N(CH_2CH_2NH_2)CH_2CH_2]_Y-$$

$$-[CH_2-CH(NH_2)]_X-$$

Figure 3.3 Chemical structures of two polymeric amino compounds, useful as additives for preparation of photocatalytic active and water-based titania sols. On top: polyethylenimine (PEI); below: polyvinylamine. Note: The schematic drawings have been created by the author.

Table 3.2 Selection of several crystalline titania sols prepared under the presence of different amino compounds. After coating onto polyester fabrics, the photoactivity A [%] is determined by the decomposition of the dyes Rhodamine B and Methylenblue in an aqueous solution under illumination with UV A light. Note: The table has been arranged by the author.

				Photoactivity A [%]	
TiO_2-Sol No.	**Amino compound**	**Concentration of amino compound [mol/l]**	**pH**	**With Rhodamin B**	**With Methylenblue**
1	Triethanolamine	0.2	3.46	76	24
2	Polymer – PEI	9.4	1.81	73	35
3	Polymer – PEI	4.7	1.30	89	46
4	Polymer – PEI	2.4	1.26	89	59
5	Polyvinylamine	18.9	1.11	96	46
6	Polyvinylamine	9.4	1.12	92	55
7	Polyvinylamine	4.7	1.08	96	66
8	Polyvinylamine	2.4	1.11	96	66
Reference pure polyester fabric	Titania sol prepared analogously but without addition of any amino compound			49	<10

possible if the measurements are performed with the same dye stuff, same dye concentration and similar measurement arrangement. Nevertheless, measurement results for the TiO_2 coatings onto polyester fabrics are given in Table 3.2; some main results can be summarized. The addition of the amino compounds increases the photoactivity. The addition of polymers containing amino groups especially leads to effective results.

The preparation of TiO_2 sols can be modified by addition of water-soluble silver salts, e.g. silver nitrate, silver acetate or silver lactate. By this, crystalline elementary silver particles are also formed similar to the formation of crystalline anatase (Figure 3.4). Formation of the metallic silver during the thermal treatment is probably the result of reduction of the silver ions by the added amino compounds [40]. The silver-modified TiO_2 sol was investigated using transmission electron microscopy (TEM) (Figure 3.4). The formed crystalline areas of anatase exhibited diameters around 7 nm. In comparison, the formed silver particles were significantly larger possessing diameters around 30 nm.

Besides pure TiO_2 coatings, combinations of TiO_2 and SiO_2 are also used in sols to prepare photocatalytic coatings on textile substrates. Gregori *et al.* [61] report on suitable aqueous and alcohol-based suspensions and solutions. The same group of researchers recently presented an interesting modification of TiO_2/SiO_2 systems by using gold nanoparticles and gold nanospheres [62, 63]. Even by addition of only 1 wt% gold particles, the photocatalytic degradation rate is doubled in the case of illumination with UV light. The often aimed activity with visible light is not introduced by the addition of gold particles [62, 63].

In case of light responsive materials, coatings leading to coloration by interference effects are also interesting and fascinating. Such interference effects can be achieved by application of hybrid organic/inorganic materials [64]. Interference colored materials like opals can be realized by coatings of highly ordered SiO_2 spheres with diameters of several hundred nanometers [65]. This approach is quite close to sol–gel techniques. A recently developed approach in this field contains structurally colored multilayer films made of a silk-based material. This approach is inspired from nature, where specific species of beetles receive their color via interference effects [66].

The last example related to light responsive materials given here are carbon particles used for fluorescence effects. Fluorescence effects on textiles are demanded for working clothes or clothes with high visibility. Usually fluorescence textiles are prepared by application of organic fluorescence dyes. Manifold types of fluorescence dyes are developed, and the application on several types of textiles such as polyester, polyamide or cotton is

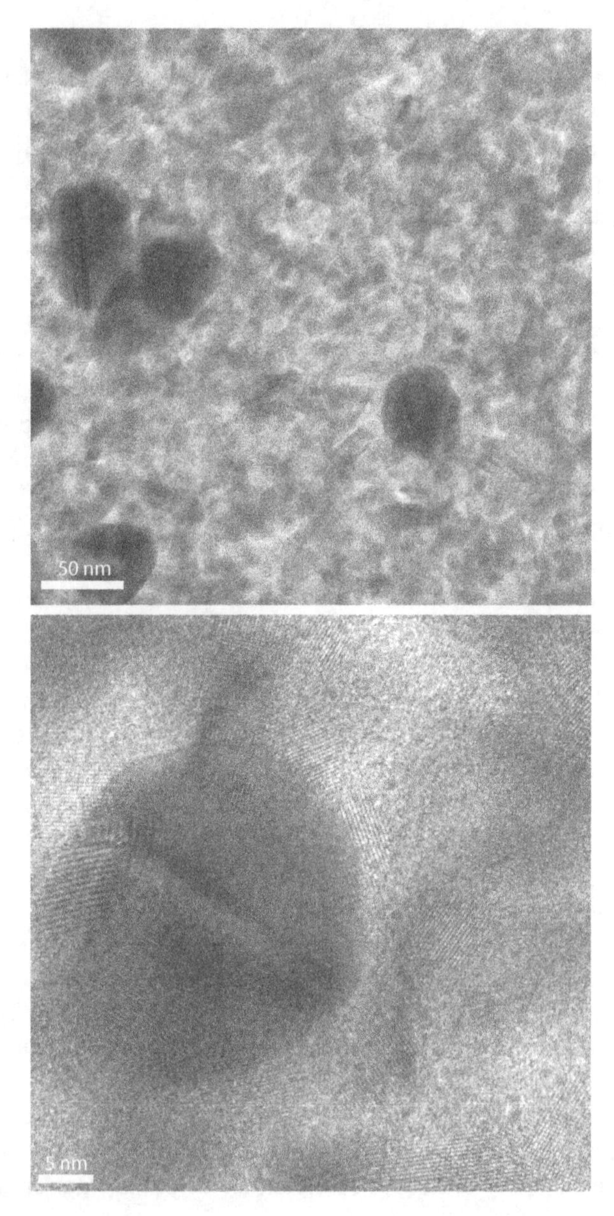

Figure 3.4 Electron microscopic images of crystalline TiO_2 sol coatings with crystalline silver particles (dark circles around 30 nm diameter) – images taken by transmission electron microscopy (TEM). Note: The image has been created by the author.

easily performed. However, one main disadvantage of organic dye stuffs could be their limited stability under light exposure. The exposure to light can also cause photochemical reactions leading to the decomposition of dyes [3, 37].

In contrast, carbon nanoparticles (as well named as carbon nanodots) exhibit high light stability. An excellent review on carbon nanodots embedded in mesoporous materials is given by Innocenzi *et al.* [67]. Such mesoporous materials can be realized by the sol–gel approach. The shown preparation of bulk materials can be transferred, in a next step, to the development of sol–gel-based coating agents used for the application of thin inorganic mesoporous coatings with embedded carbon nanodots for textile treatments.

Carbon nanodots can be functionalized on their surface by introduction of amino groups. Epoxy modified silane compounds like GLYEO can be covalently bonded to these amino groups. By this, the chemical modification of silica sol coatings with carbon nanodots is possible, and photoluminescence coatings for textile finishing can be realized easily. One advantage for the development of these coating recipes is the good water solubility of the modified carbon nanodots [68].

Alternative to SiO_2 coatings, carbon nanodots can also be embedded in ZnO coatings [69]. In contrast to SiO_2, ZnO is a semi-conductive material and can be activated by UV light. The electron band structure of zinc oxide can be modified via embedding of carbon nanodots leading to narrowing of the ZnO band gap. By this, an activation with visible light can be introduced [69].

3.4.2 Antimicrobial and Bioactive Systems

Most applications on antimicrobial sol–gel coatings are probably related to recipes using the embedding of silver ions, silver compounds or metallic silver particles [39–41, 70–72].

An application developed especially for the antimicrobial treatment of cotton is reported by Xing *et al.* [73]. This group uses sol–gel coating agents prepared from silica water glass modified with silver nitrate. An advantage of this recipe is its higher water content compared to earlier reported recipes containing significant amounts of organic solvents. Besides silver-containing systems, copper-containing coatings are also investigated for antimicrobial sol–gel coatings as an alternative [42, 74]. For further information on the broad field of antimicrobial sol–gel coatings, several excellent review papers can be referred to [8, 75, 76].

One interesting approach is sol–gel coatings releasing antibacterial acting gases [3]. These antibacterial sol–gel coatings act against bacteria, even if there is no contact to the coated surface or no intermediate liquid medium. A possible process in this field is the use of nitric oxide (NO) releasing sol–gel systems [77, 78]. However, two points have to be kept in mind. First, the released gases can also influence the human health. Second, the release into the gas phase can exhaust the antimicrobial depot after short duration. By view on both limitations, it is clear that these releasing systems would find main applications as textile packaging materials.

Just have a view on other more innovative and forecasting bioactive sol–gel materials [79]. Sol–gel materials can be set to the interface of inorganic chemistry and biology by bioencapsulation of biomolecules, e.g. proteins, enzymes, polysaccharides, lipids and nucleic acids. The resulting materials are often also named as biocers (from bioceramics). Potential applications for biocers are the development of new biosensors, bioreactors, filter systems for waste water cleaning or drug release systems [79–82].

Especially attractive is the embedding of enzymes in sol–gel coatings. Enzymes are highly active biomolecules working as biocatalysts. By fixation of these enzymes on textile surfaces, their specific catalytic properties are transferred to the coated textile substrates. A simple example is the embedding of the enzyme lysozyme in sol–gel coatings applied onto polyester foils [83]. Lysozyme presents antimicrobial activity, and by its introduction into the coating recipe, the antimicrobial properties are transferred to the coated polyester substrates. The advantageous part here is the realization of a biobased antibacterial coating, which could be a big issue also for marketing reasons.

The realization of biocatalytic coatings prepared by sol–gel encapsulation of enzymes is described in references [84, 85]. For this, silica sol coatings are mainly used. The use of water-based recipes without organic solvents or solvents at lowest concentration is absolutely necessary to avoid the denaturation of the embedded enzymes. The thermal fixation performed after the coating process is also limited to moderate temperatures to avoid any denaturation. If these demands are fulfilled, the enzyme containing coating agents can be used to functionalize textiles with biocatalytic functions. Applications can be found as additives in cleaning agents or in bioreactors.

An interesting approach is to stabilize the embedded biomolecules against thermal denaturation [86]. Some proteins have the property to take up and bond bivalent metal ions as Ca^{2+}. In case of bonding to Ca^{2+}, these proteins can receive an increased stability against higher temperatures but also under exposition to the organic solvent ethanol [86]. This is of course a very attractive approach for preparation of biocer coatings with sol–gel technology.

It is a further step to immobilize and embed complete cells of microorganisms in sol–gel coatings; this is done for algae, bacteria and cyanobacteria

[3, 87]. The final idea of this approach is to realize coatings containing living bioorganisms working as a kind of bioreactors. By embedding cyanobacteria, the development of novel photobioreactors is aimed [87].

It is of course a smart idea to apply these coatings onto textile substrates and to realize these textile-based bioreactors, which can be modified in various constructions and geometry by textile technology processes. However, even if this is a very attractive idea, the main difference in embedding a biomolecule like protein compared to a living cell should be kept in mind. In contrast to proteins, living cells need ambient temperature conditions, humidity and nutrients to survive. For this, some efforts have to be made to keep the functional properties of coatings containing living cells even after the coating process is finished.

3.4.3 Controlled Release Systems

Active agents, perfumes, fragrances or even medical substances can be impregnated onto textile substrates. In contact with water or other solvents, these substances are released into the solvent. In this situation, the rate of release depends on the solubility of the substances in the present solvent and on the adhesive properties of the textile surface. Usually the release is faster than wished for the aimed application, meaning that the concentration in the solvent decreases too fast and the depot function of the textile diminishes soon. In many cases, it is useful to decelerate the rate of releasing by a coating. Sol–gel coatings can be used also for embedding of active substances or drugs, and the velocity of release is no longer determined by the solubility of the drug [88–90].

Now the release is determined or, in other words, controlled by the type of coating used for the embedding. The addition of penetration agents such as sorbitol can be used to introduce to the silica coatings capillary channels that enable the embedded drug molecules to escape from the coating. By adjusting the concentration of sorbitol added to the sol–gel recipe, the rate of release can be controlled [88–90]. Such mesoporous silica materials can be used for release of DNA fragments as well, e.g. for the modification of living cells [91, 92]. The controlled release of antibiotics is useful for the preparation of antimicrobial coatings [93].

3.5 Summary

Nanosols are versatile and multifunctional tools to functionalize textile substrates. The realized new functions can lead to completely new materials and applications, which is never expected for textiles in the traditional world. These new functional textiles can be best described by the term "smart textiles."

The functional properties that can be realized on textiles with nanosol technology could be best described in a schematic overview distinguishing between optical, chemical, biological or surface functionalities. Many of the actually mentioned and summarized properties could be also understood as effects gained with traditional textile finishing, e.g. coloration, dyeing or water-repellent effect. All these functions can be realized as well by nanosol treatment of the textile. However, one aim of this chapter is to emphasize possible and extraordinary functional properties, which can be especially realized by nanosol application. In the section of optical functionalization, nanosols can be used to realize photocatalytic and light responsive textile materials. Of high potential are especially nanosol recipes containing carbon quantum dots (C-dots) with fluorescence properties. These C-dots are supposed to be innovative fluorescence materials with high lightfastness and low toxicity. Also innovative biofunctional textiles can be realized. These biofunctions are especially related to the embedding of biopolymers, enzymes and complete cells onto textile surface by using nanosol coatings. Prospective applications are found for textiles as carrier of enzymatic properties, e.g. as part of bioreactors or for waste water cleaning processes. Other potential applications could be developed in the fields of biosensors, bioreactors or drug release systems.

Acknowledgements

The author would like to thank Prof. Dr. T. Textor (Fachhochschule Reutlingen, Germany) for many helpful discussions and long-term cooperation. For help with electron microscopic measurements, many thanks are owed to M. Reibold (TU Dresden, Germany). All product and company names mentioned in this chapter may be trademarks of their respective owners, also without labeling.

References

1. Hench, L.L. and West, J.K., The sol–gel process. *Chem. Rev.*, 90, 33, 1990.
2. Brinker, C.J. and Scherer, G.W., *Sol–gel Science: The Physics and Chemistry of Sol–gel Processing*, Academic Press, Boston, 1993.

3. Mahltig, B. and Textor, T., *Nanosols & Textiles*, World Scientific, Singapore, 2008.
4. Dimitriev, Y., Ivanova, Y. and Iordanova, R., History of sol–gel science and technology. *J. Univ. Chem. Technol. Metall.*, 43, 181, 2008.
5. Ismail, W.N.W., Sol–gel technology for innovative fabric finishing—A review. *J. Sol–gel Sci. Technol.*, 78, 698, 2016.
6. Jentzsch, A., Böttcher, H., Haufe, H., Mahltig, B. and Richter H., Oberfläche für Maschinenteile einer Druckmaschine, German Patent DE10209296A1, 2002.
7. Jentzsch, A., Böttcher, H., Haufe, H., Mahltig, B. and Richter H., Feuchtwalze für Feuchtwerke in Druckmaschinen, German Patent DE10209297A1, 2002.
8. Mahltig, B., Haufe, H. and Böttcher, H., Functionalization of textiles by inorganic sol–gel coatings. *J. Mater. Chem.*, 15, 4385, 2005.
9. Mahltig, B., Swaboda, C., Roessler, A. and Böttcher, H., Functionalising wood by nanosol application. *J. Mater. Chem.*, 18, 3180, 2008.
10. Mahltig, B., Vakuumimprägnierung poröser Metallwerkstücke mit anorganischen Nanosolen - eine Methode zur Metallversiegelung. *Vakuum in Forschung und Praxis*, 22, 2, 25, 2010.
11. Mahltig, B., Vossebein, L., Ehrmann, A., Cheval, N. and Fahmi, A., Modified silica sol coatings for surface enhancement of leather. *Acta Chim. Slov.*, 59, 331, 2012.
12. Stangar, U.L., Orel, B. and Krajnc, M., Preparation and spectroscopic characterization of blue $CoAl_2O_4$ coatings. *J. Sol–gel Sci. Technol.*, 26, 771, 2003.
13. Kobayashi, Y., Imai, Y. and Kurokawa, Y., Preparation of a transparent alumina film doped with organic dye by sol–gel process. *J. Mater. Sci. Lett.*, 7, 1148, 1988.
14. Seckin, T., Gültek, A. and Kartaca, S., The grafting of Rhodamine B onto sol–gel derived mesoporous silicas. *Dyes Pigments*, 56, 51, 2003.
15. Mahltig, B., Knittel, D., Schollmeyer, E. and Böttcher, H., Incorporation of triarylmethane dyes into sol–gel matrices deposited on textiles. *J. Sol–gel Sci. Technol.*, 31, 293, 2004.
16. Mahltig, B., Böttcher, H., Knittel, D. and Schollmeyer, E., Light fading and wash fastness of dyed nanosol-coated textiles. *Textile Res. J.*, 74, 521, 2004.
17. Mahltig, B. and Textor, T., Combination of silica sol and dyes on textiles. *J. Sol–gel Sci. Technol.*, 39, 111, 2006.
18. Yin, Y., Wang, C. and Wang, C., An evaluation of the dyeing behavior of sol–gel silica doped with direct dyes. *J. Sol–gel Sci. Technol.*, 48, 308, 2008.
19. Onar, N., Leaching and fastness behavior of cotton fabrics dyed with different type of dyes using sol–gel process. *J. Appl. Polym. Sci.*, 109, 97, 2008.
20. Ribeiro, L.S., Pinto, T., Monteiro, A., Soares, O.S.G.P., Pereira, C., Freire, C. and Pereira, M.F.R., Silica nanoparticles functionalized with a thermochromic dye for textile applications. *J. Mater. Sci.*, 48, 5085, 2013.
21. Kartini, I., Ilmi, I. and Kunarti, E.S., Wash fastness improvement of malachite green-dyed cotton fabrics coated with nanosol composites of silica-titania. *Bull. Mater. Sci.*, 37, 1419, 2014.

22. Rauch, K., Dieckmann, U., Böttcher, H. and Mahltig B., UV-schützende transparente Beschichtungen für technische Anwendungen, German Patent DE102004027075A1, 2004.
23. Mahltig, B. and Haufe H., Fasermaterial zur Verminderung von Röntgenexposition, German Patent, DE102010056132A1, 2010.
24. Mahltig, B., Textor, T. and Akcakoca Kumbasar, P., Photobactericidal and photochromic textile materials realized by embedding of advantageous dye using sol–gel technology. *Celal Bayar Univ. J. Sci.*, 11, 306, 2015.
25. Alongi, J., Ciobanu, M. and Malucelli, G., Sol–gel treatments for enhancing flame retardancy and thermal stability of cotton fabrics optimization of the process and evaluation of the durability. *Cellulose*, 18, 167, 2011.
26. Chapple, S.A. and Ferg, E., The influence of precursor ratios on the properties of cotton coated with a sol–gel flame retardant. *AATCC Rev.*, 6, 36, 2006.
27. Vasiljevic, J., Tomsic, B., Jerman, I., Orel, B., Jaksa, G. and Simoncic, B., Novel multifunctional water- and oil-repellent, antibacterial, and flame-retardant cellulose fibres created by sol–gel process. *Cellulose*, 21, 2611, 2014.
28. Alongi, J., Colleoni, C., Rosace, G. and Malucelli, G., Sol–gel derived architectures for enhancing cotton flame retardancy: Effect of pure and phosphorus-doped silica phases. *Polym. Degrad. Stab.*, 99, 92, 2014.
29. Yaman, N., Preparation and flammability properties of hybrid materials containing phosphorous compounds via sol–gel process. *Fibers Polym.*, 10, 413, 2009.
30. Alongi, J. and Malucelli, G., Thermal stability, flame retardancy and abrasion resistance of cotton and cotton–linen blends treated by sol–gel silica coatings containing alumina micro- or nano-particles. *Polym. Degrad. Stab.*, 98, 1428, 2013.
31. Mahltig, B., Gutmann, E., Meyer, D.C., Reibold, M., Dresler, B., Günther, K., Faßler, D. and Böttcher, H., Solvothermal preparation of metalized titania sols for photocatalytic and antimicrobial coatings. *J. Mater. Chem.*, 17, 2367, 2007.
32. Böttcher, H., Mahltig, B., Sarsour, J. and Stegmaier, T., Qualitative investigations of the photocatalytic dye destruction by TiO_2-coated polyester fabrics. *J. Sol–gel Sci. Technol.*, 55, 177, 2010.
33. Mahltig, B. and Kim Y.H., Anatas-haltiges wasserbasiertes Beschichtungsmittel und dessen Anwendung zur Herstellung von photoaktiven Textilien, German Patent, DE102010009002.6, 2010.
34. Mahltig, B., Gutmann, E. and Meyer, D.C., Solvothermal preparation of nanocrystalline anatase containing TiO_2 and TiO_2/SiO_2 coating agents for application of photocatalytic treatments. *Mater. Chem. Phys.*, 127, 285, 2011.
35. Mahltig, B., Haufe, H., Kim, C.W., Kang, Y.S., Gutmann, E., Leisegang, T. and Meyer, D.C., Manganese/TiO_2 composites prepared and used for photocatalytic active textiles. *Croat. Chem. Acta*, 86, 143, 2013.
36. Mahltig, B. and Miao, H., Microwave assisted preparation of photoactive TiO_2 on textile substrates. *J. Coat. Technol. Res.*, 14, 721, 2017.

37. Mahltig, B., *Textiles: Advances in Research and Applications*, Nova Science Publishers Inc., New York, 2018.
38. Ibanescu Busila, M., Musat, V., Textor, T., Badilita, V. and Mahltig, B., Photocatalytic and antimicrobial Ag/ZnO nanocomposites for functionalization of textile fabrics. *J. Alloys Compd.*, 610, 244, 2014.
39. Haufe, H., Thron, A., Fiedler, D., Mahltig, B. and Böttcher, H., Biocidal nanosol coatings. *Surf. Coat. Int. B: Coat. Trans.*, 88, 55, 2005.
40. Mahltig, B., Gutmann, E., Meyer, D.C., Reibold, M., Bund, A. and Böttcher, H., Thermal preparation and stabilization of crystalline silver particles in SiO_2-based coating solutions. *J. Sol–gel Sci. Technol.*, 49, 202, 2009.
41. Mahltig, B., Gutmann, E., Reibold, M., Meyer, D.C. and Böttcher, H., Synthesis of Ag and Ag/SiO_2 sols by solvothermal method and their bactericidal activity. *J. Sol–gel Sci. Technol.*, 51, 204, 2009.
42. Mahltig, B., Fiedler, D., Fischer, A. and Simon, P., Antimicrobial coatings on textiles – modification of sol–gel layers with organic and inorganic biocides. *J. Sol–gel Sci. Technol.*, 55, 269, 2010.
43. Mahltig, B. and Fischer, A., Inorganic/organic polymer coatings for textiles to realize water repellent and antimicrobial properties – a study under respect to textile comfort. *J. Polym. Sci. B: Polymer Physics*, 48, 1562, 2010.
44. Grethe, T., Haase, H., Natarajan, H.S., Limandoko, N. and Mahltig, B., Coating process for antimicrobial textile surfaces derived from a polyester dyeing process. *J. Coat. Technol. Res.*, 12, 1133, 2015.
45. Mahltig, B. and Böttcher, H., Refining of textiles by nanosol coating. *Melliand Textilber. (English edition)*, 83, E50, 2002.
46. Tomsic, B., Simoncic, B., Orel, B., Cerne, L., Tavcer, P.F., Zorko, M., Jerman, I., Vilcnik, A. and Kovac, J., Sol–gel coating of cellulose fibres with antimicrobial and repellent properties. *J. Sol–gel Sci. Technol.*, 47, 44, 2008.
47. Vilcnik, A., Jerman, I., Vuk, A.S., Kozelj, M., Orel, B., Tomsic, B., Simoncic, B. and Kovac, J., Structural properties and antibacterial effects of hydrophobic and oleophobic sol–gel coatings for cotton fabrics. *Langmuir*, 25, 5869, 2009.
48. Simoncic, B., Tomsic, B., Cerne, L., Orel, B., Jerman, I., Kovac, J., Zerjav, M. and Simoncic, A., Multifunctional water and oil repellent and antimicrobial properties of finished cotton: Influence of sol–gel finishing procedure. *J. Sol–gel Sci. Technol.*, 61, 340, 2012.
49. Mahltig, B. and Böttcher, H., Modified silica sol coatings for water-repellent textiles. *J. Sol–gel Sci. Technol.*, 27, 43, 2003.
50. Yu, M., Gu, G., Meng, W.D. and Qing, F.L., Superhydrophobic cotton fabric coating based on a complex layer of silica nanoparticles and perfluorooctylated quaternary ammonium silane coupling agent. *Appl. Surf. Sci.*, 253, 3669, 2007.
51. Boukhriss, A., Boyer, D., Hannache, H., Roblin, J.-P., Mahiou, R., Cherkaoui, O., Therias, S. and Gmouth, S., Sol–gel based water repellent coatings for textiles. *Cellulose*, 22, 1415, 2015.

52. Mahltig, B., Böttcher, H., Langen, G. and Meister M., Antiadhäsive Beschichtung zur Ausrüstung von Wundverbänden, German Patent, DE1024987A1, 2002.
53. Behr, D., Was versteht man unter elektrischer Aufladung. *Wirkerei und Strickereitechnik*, 41, 7, 1991.
54. Mahltig, B., Günther, K., Askani, A., Bohnert, F., Brinkert, N., Kyosev, Y., Weide, T., Krieg, M. and Leisegang, T., X-ray protective organic/inorganic fiber – along the textile chain from fiber production to clothing application. *J. Text. Inst.*, 108, 1975, 2017.
55. Erdumlu, N. and Ozipek, B., Investigation of regenerated bamboo fibre and yarn characteristics. *Fibres Text. East. Europe*, 16, 43, 2008.
56. Textor, T. and Mahltig, B., Nanosols for preparation of antistatic coatings simultaneously yielding water and oil repellent properties for textile treatment. *Mater. Technol.*, 25, 74, 2010.
57. Textor, T. and Mahltig, B., A sol–gel-based surface treatment for preparation of water repellent antistatic textiles. *Appl. Surf. Sci.*, 256, 1668, 2010.
58. Wright, J.D. and Sommerdijk, N.A.J.M., *Sol–gel Materials Chemistry and Applications*, CRC Press, Boca Raton, 2000.
59. Ciriminna, R., Fidalgo, A., Pandarus, V., Beland, F., Ilharco, L.M. and Pagliaro, M., The sol–gel route to advanced silica-based materials and recent applications. *Chem. Rev.*, 113, 6592, 2013.
60. Li, Y. and Cai, Z., Effect of acid-catalyzed sol–gel silica coating on the properties of cotton fabric. *J. Text. Inst.*, 103, 1099, 2012.
61. Gregori, D., Guillard, C., Chaput, F. and Parola S., Method for depositing a photocatalytic coating and related coatings, textile materials and use in photocatalysis, US Patent, US20160040353A1, 2013.
62. Levchuk, I., Sillanpää, M., Guillard, C., Gregori, D., Chateau, D., Chaput, F., Lerouge, F. and Parola, S., Enhanced photocatalytic activity through insertion of plasmonic nanostructures into porous TiO_2/SiO_2 hybrid composite films. *J. Catalysis*, 342, 117, 2016.
63. Levchuk, I., Sillanpää, M., Guillard, C., Gregori, D., Chateau, D. and Parola, S., TiO_2/SiO_2 porous composite thin films: Role of TiO_2 areal loading and modification with gold nanospheres on the photocatalytic activity. *Appl. Surf. Sci.*, 383, 367, 2016.
64. Parola, S., Julian-Lopez, B., Carlos, L.D. and Sanchez, C., Optical properties of hybrid organic-inorganic materials and their applications. *Adv. Funct. Mater.*, 26, 6506, 2016.
65. Marlow, F., Parvin Sharifi, M., Brinkmann, R. and Mendive, D., Opale: Status und Perspektiven. *Angewandte Chemie*, 121, 6328, 2009.
66. Colusso, E., Perotto, G., Wang, Y., Sturaro, M., Omenetto, F. and Martucci, A., Bioinspired stimuli-responsive multilayer film made of silk-titanate nanocomposites. *J. Mater. Chem. C*, 16, 3924, 2017.
67. Innocenzi, P., Malfatti, L. and Carboni, D., Graphene and carbon nanodots in mesoporous materials: An interactive platform for functional applications. *Nanoscale*, 7, 12759, 2015.

68. Suzuki, K., Malfatti, L., Takahashi, M., Carboni, D., Messina, F., Tokudome, Y., Takemoto, M. and Innocenzi, P., Design of carbon dots photoluminescence through organo-functional silane grafting for solid-state emitting devices. *Sci. Rep.*, 5469, 1, 2017.
69. Suzuki, K., Malfatti, L., Carboni, D., Loche, D., Casula, M., Moretto, A., Maggini, M., Takahashi, M. and Innocenzi, P., Energy transfer induced by carbon quantum dots in porous zinc oxide nanocomposite films. *J. Phys. Chem. C*, 119, 2837, 2015.
70. Kawashita, M., Tsuneyama, S., Miyaji, F., Kokubo, T., Kozuka, H. and Yamamoto, K., Antibacterial silver-containing silica glass prepared by sol–gel method. *Biomaterials*, 21, 393, 2000.
71. Kokkoris, M., Trapalis, C.C., Kossionides, S., Vlastou, R., Nsouli, B., Grötzschel, R., Spartalis, S., Kordas, G. and Paradellis, T., RBS and HIRBS studies of nanostructured $AgSiO_2$ sol–gel thin coatings. *Nucl. Instr. Methods Phys. Res. B*, 188, 67, 2002.
72. Stobie, N., Duffy, B., McCormack, D.E., Colreavy, J., Hidalgo, M., McHale, P. and Hinder, S.J., Prevention of *Staphylococcus epidermidis* biofilm formation using a low-temperature silver-doped phenyltriethoxysilane sol–gel coating. *Biomaterials*, 29, 963, 2008.
73. Xing, Y., Yang, X. and Dai, J., Antimicrobial finishing of cotton textile based on water glass by sol–gel method. *J. Sol–gel Sci. Technol.*, 43, 187, 2007.
74. Trapalis, C.C., Kokkoris, M., Perdikakis, G. and Kordas, G., Study of antibacterial composite Cu/SiO_2 thin coatings. *J. Sol–gel Sci. Technol.*, 26, 1213, 2003.
75. Gao, Y. and Cranston, R., Recent advances in antimicrobial treatments of textiles. *Text. Res. J.*, 78, 60, 2008.
76. Simoncic, B. and Tomsic, B., Structures of novel antimicrobial agents for textiles - a review. *Text. Res. J.*, 80, 1721, 2010.
77. Dobmeier, K.P. and Schoenfisch, M.H., Antibacterial properties of nitric oxide-releasing sol–gel microarrays. *Biomacromolecules*, 5, 2493, 2004.
78. Nablo, B.J., Rothrock, A.R. and Schoenfisch, M.H., Nitric oxide-releasing sol–gels as antibacterial coatings for orthopedic implants. *Biomaterials*, 26, 917, 2005.
79. Avnir, D., Coradin, T., Lev, O. and Livage, J., Recent bio-applications of sol–gel materials. *J. Mater. Chem.*, 16, 1013, 2006.
80. Fiedler, D., Hager, U., Franke, H., Soltmann, U. and Böttcher, H., Algae biocers: Astaxanthin formation in sol–gel immobilised living microalgae. *J. Mater. Chem.*, 17, 261, 2007.
81. Raff, J., Soltmann, U., Matys, S., Selenska-Pobell, S., Böttcher, H. and Pompe, W., Biosorption of uranium and copper by biocers. *Chem. Mater.*, 15, 240, 2003.
82. Soltmann, U., Raff, J., Selenska-Pobell, S., Matys, S., Pompe, W. and Böttcher, H., Biosorption of heavy metals by sol–gel immobilized *Bacillus sphaericus* cells, spores and S-layers. *J. Sol–gel Sci. Technol.*, 26, 1209, 2003.

83. Corradini, C., Alfieri, I., Cavazza, A., Lantano, C., Lorenzi, A., Zucchetto, N. and Montenero, A., Antimicrobial films containing lysozyme for active packaging obtained by sol–gel technique. *J. Food Eng.*, 119, 580, 2013.
84. Brennan, J.D., Biofriendly sol–gel processing for the entrapment of soluble and membrane-bound proteins: Toward novel solid-phase assays for high-throughput screening. *Acc. Chem. Res.*, 40, 827, 2007.
85. Pierre, A.C., The sol–gel encapsulation of enzymes. *Biocatal. Biotransformation*, 22, 145, 2004.
86. Zhen, L., Flora, K. and Brennan, J.D., Improving the performance of a sol–gel entrapped metal-binding protein by maximizing protein thermal stability before entrapment. *Chem. Mater.*, 10, 3974, 1998.
87. Rooke, J.C., Leonard, A. and Su, B.-L., Targeting photobioreactors: Immobilisation of cyanobacteria within porous silica gel using biocompatible methods. *J. Mater. Chem.*, 18, 1333, 2008.
88. Barbe, C., Bartlett, J., Kong, L., Finnie, K., Lin, H.Q., Larkin, M., Calleja, S., Bush, A. and Calleja, G., Silica particles: A novel drug-delivery system. *Adv. Mater.*, 16, 1, 2004.
89. Böttcher, H., Slowik, P. and Süß, W., Sol–gel carrier systems for controlled drug delivery. *J. Sol-gel Sci. Technol.*, 13, 277, 1998.
90. Radin, S. and Ducheyne, P., Controlled release of vancomycin from thin sol–gel films on titanium alloy fracture plate material. *Biomaterials*, 28, 1721, 2007.
91. Slowing, I.I., Vivero-Escoto, J.L., Wu, C.-W. and Lin, V.S.-Y., Mesoporous silica nanoparticles as controlled release drug delivery and gene transfection carriers. *Adv. Drug Deliv. Rev.*, 60, 1278, 2008.
92. Trewyn, B.G., Slowing, I.I., Giri, S., Chen, H.-T. and Lin, V.S.Y., Synthesis and functionalization of a mesoporous silica nanoparticle based on the sol–gel process and applications in controlled release. *Acc. Chem. Res.*, 40, 846, 2007.
93. Gaetano, F.D., Ambrosio, L., Raucci, M.G., Marotta, A. and Catauro, M., Sol–gel processing of drug delivery materials and release kinetics. *J. Mater. Sci. Mater. Med.*, 16, 261, 2005.

4

Responsive Polymers for Smart Textiles

Eri Niiyama[1,2], **Ailifeire Fulati**[1,2] **and Mitsuhiro Ebara**[1,2,3]*

[1]*International Center for Materials Nanoarchitectonics (WPI-MANA), National Institute for Materials Science (NIMS), Tsukuba, Ibaraki, Japan*
[2]*Graduate School of Pure and Applied Sciences, University of Tsukuba, Tsukuba, Ibaraki, Japan*
[3]*Graduate School of Tokyo University of Science, Katsushika-ku, Tokyo, Japan*

Abstract

Fibrous materials are presenting increasing significance nowadays in a variety of applications such as textiles, nonwovens, composite materials, etc. Especially, nanofibers have recently gained much prominence in biological and medical applications because of the rather large surface area and high porosity provided by their nanoscale features. Furthermore, nanofibers with "smart" or "stimuli-responsive" surfaces are of great interest for such applications as "on–off" switchable control of permeability, wettability, and/or swelling/deswelling behavior, by applying external stimuli. Because the nanofibers have an extremely larger external surface area, the meshes or mats electrospun from smart polymers display much quicker response times than the corresponding bulk materials such as hydrogels. This chapter provides a literature overview of fibrous biomaterials, particularly stimuli-responsive nanofibers, which are composed of polymers responsive to stimuli, including temperature, light, pH, or electric/magnetic field.

Keywords: Nanofibers, electrospinning, nonwoven, temperature-responsive polymers, hyperthermia, cancer therapy

4.1 Classification of Stimuli-Responsive Polymers

This chapter focuses on the fabrication, characterization, and applications of stimuli-responsive fiber-based materials, which are composed of polymers responsive either solely or multiply to temperature, light, or electric/

**Corresponding author*: EBARA.Mitsuhiro@nims.go.jp

Nazire D. Yilmaz (ed.) Smart Textiles, (111–126) © 2019 Scrivener Publishing LLC

magnetic field. Stimuli-responsive polymers could be classified as either physical or chemical stimuli-responsive ones (Figure 4.1). The physical stimuli such as temperature, electric, or magnetic fields will affect the level of various energy sources and alter molecular interactions at critical onset points. Physical stimuli are sometimes favorable because they allow local and remote control. On the other hand, the appearance of numerous bioactive molecules is tightly controlled to maintain a normal metabolic balance via the feedback system called homeostasis in human body. Therefore, chemical or biochemical stimulus such as pH, ionic factors, or biomolecules has been considered as another important stimulus. Some systems have been developed to combine two or more stimuli-responsive mechanisms into one polymer system, the so-called dual- or multi-responsive polymer systems. One of the most widely used stimuli-responsive polymers is poly(*N*-isopropylacrylamide) (PNIPAAm) and its derivatives. A PNIPAAm solution, which undergoes a sharp yet reversible phase transition from monophasic below a specific temperature to biphasic above it, generally exhibits the so-called lower critical solution temperature (LCST) behavior. PNIPAAm has an LCST of 32°C in aqueous solution. There have been several studies on the electrospun PNIPAAm nanofibers [1–3]. These temperature-responsive nanofibers show much quicker response times than the corresponding bulk materials because the nanofibers have an extremely larger external surface area (Figure 4.2). However, one of the major challenges in the development of PNIPAAm nanofibers is that they are not stable in water and disperse easily; therefore, an incorporation of non-soluble components or cross-linkable moieties into PNIPAAm nanofibers is required. Therefore, careful choice of polymers and rational design of nanofiber are crucial to fabricate dynamic and responsive fibers. Also, use of different precursor materials makes it possible to fabricate fibers,

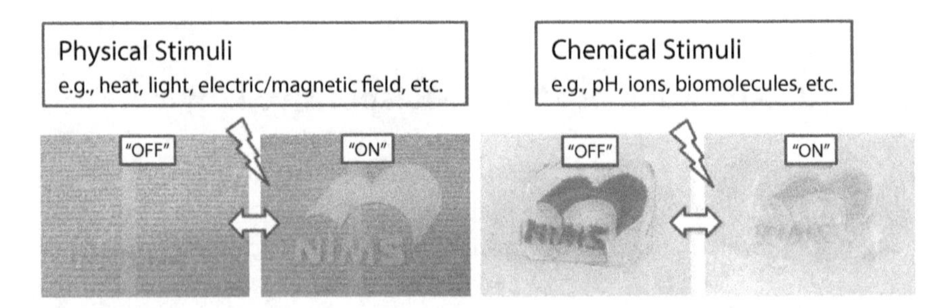

Figure 4.1 Classification of stimuli-responsive polymers on the basis of stimuli (physical stimuli: heat, light, and electric/magnetic field; chemical stimuli: pH, ions, and biomolecules). Note: The figure has been prepared by the author.

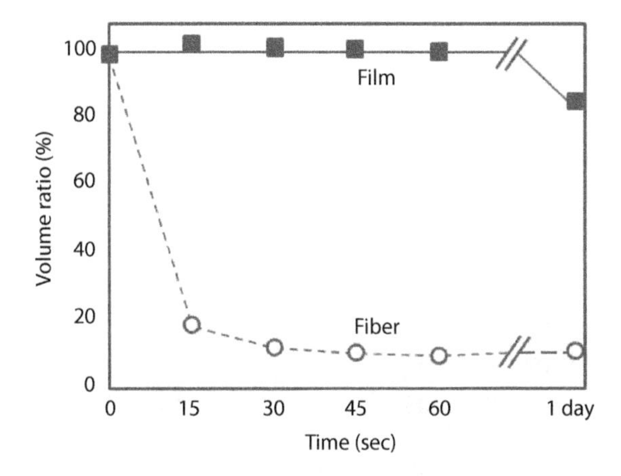

Figure 4.2 Shrinking behaviors for PNIPAAm film and fiber in response to temperature change from 20 to 40°C. The fiber shows much quicker response times than the corresponding film due to an extremely larger external surface area. Note: The figure has been arranged by the author.

which respond to other stimuli such as pH, light, temperature, or magnetic, that could better meet the demands of the desired applications. The following sections will review some of the varieties of stimuli-responsive nanofibers and their key fabrication methods as well as some of their applications in biomedical fields.

4.2 Fiber Fabrication

Polymeric nanofibers can be processed by diverse techniques such as phase separation, self-assembly, electrospinning, drawing, and microfluidic devices (Figure 4.3).

Phase separation is a method that has long been used to fabricate porous polymer fibrous membranes or sponges by inducing the separation of a polymer solution into two different phases, namely, the polymer-poor phase (low polymer concentration) and the polymer-rich phase (high polymer concentration). This enables the preparation of a three-dimensional nanofibrous structure with interconnected pores.

Self-assembly by a bottom-up method for the preparation of nanofibers from polymers, peptides, and macromolecules is a versatile and powerful technique to construct well-defined nanostructures. This is accomplished by spontaneous and automatic organization of molecules into desired structures through various types of intermolecular interactions [4]. Lipid

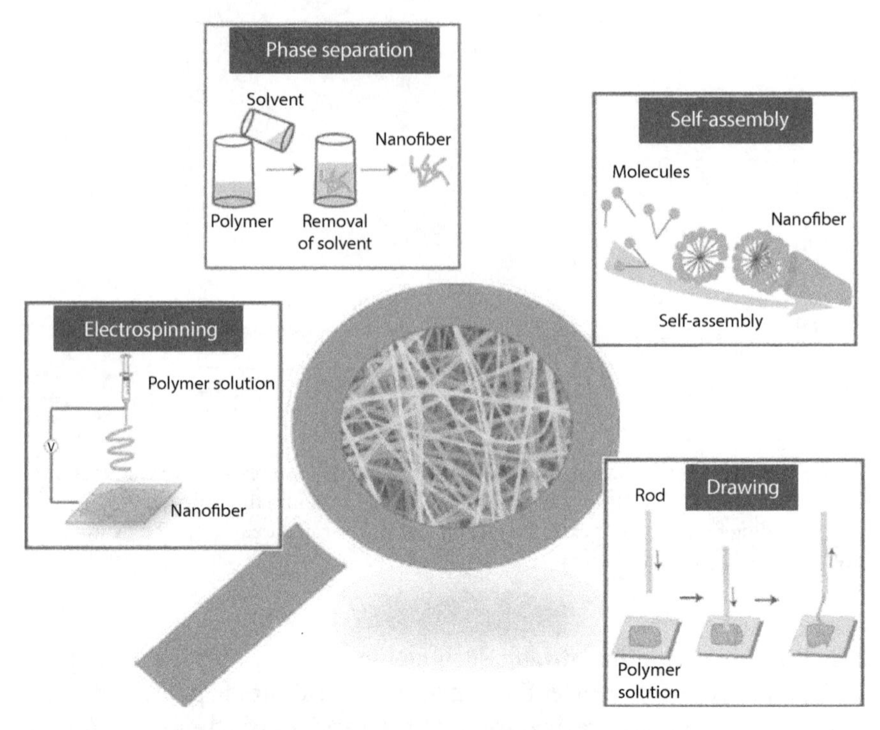

Figure 4.3 Fabrication methods of nanofibers such as phase separation, self-assembly, electrospinning, and drawing. Note: The figure has been arranged by the author.

membranes, which control cellular processes, assemble from a hydrophobic tail group and a hydrophilic head group [5]. This natural organization of life is driven by noncovalent forces. It is known that polymer fibers and liquid crystals (LCs) all self-assemble in solutions based on the same principles that drive natural molecular assembly [4]. Recently, Saito *et al.* [6] have developed a stimuli-responsive self-assembled system from lyotropic LCs composed of cyclic ethynylhelicene oligomers. Through careful control of temperature, the LCs were able to be dynamically changed from anisotropically aligned fibers to turbid gels. It is thought that these systems have potential biological applications in mimicking actin. Bitton *et al.* [7] have used peptide amphiphiles in combination with hyaluronic acid in dynamic self-assembly of hierarchical nanofibers under electrostatic control. Heparin was found to drive the self-assembly process by the formation of a dense diffusion barrier.

Electrospinning is considered to be a simple technique to produce micro-sized or nano-sized fibers. Nevertheless, controlling fiber alignment by electrospinning has not been a simple task [8]. In electrospinning, the

"Taylor cone" of a polymer solution droplet forms at the end of a capillary tip when electrical forces are applied [9]. When the electric field reaches a critical level at which the repulsive electric force overcomes the surface tension force, a charged jet of the solution is ejected from the tip of the "Taylor cone." As the jet diameter decreases when the jet flies to the collector, the radial forces from charged ions exceed the cohesive forces of the jet solution, causing it to split into many fibers. Furthermore, these divided fibers repel each other, leading to chaotic trajectories and bending instability. At the same time, the solvent is evaporated and the polymer solidifies on the collector. Thus, continuous fibers are laid to form a nonwoven sheet. With the increased control over spatial alignment and fiber diameters, electrospinning will play a key role in the development of future smart fibers. A recent example where control over the fiber diameter by tuning the electrospinning conditions had an influence on the fiber properties was shown using PNIPAAm/polystyrene [10]. Here electrospun fiber mats were fabricated with superhydrophilic and superhydrophobic properties. The large surface area of the fibers and the temperature-responsive PNIPAAm contributed to the fast wettability switching of these electrospun fiber mats.

Drawing is an optimized method for the fabrication of single fibers using a viscous polymer solution with volatile organic solvents. A continuous long linear fiber can be obtained by the drawing method, and the fiber diameter relies on the size of the needle (micropipette), polymer solution flow rate, and temperature, which affect the viscosity of the polymer and the evaporation rate of the solvent. The fabrication of fibers by the drawing process has been understood since Carothers *et al.* [11] reported it in the 1930s. Yuan *et al.* [12] have recently shown that 3D structures can be directly written using a programmable micro-milling machine. These precisely fabricated biodegradable arrays can be used for endothelial cell growth and are an exciting precursor to a functional microvascular network. Polymer fibers for microfluidic devices are also not a new concept, but recently, Yildirim *et al.* [13] have shown that microfluidic devices can be assembled from drawn polymer fibers, with the microfluidic channels incorporated into the individual fiber. Microfluidic devices fabricated by this method have the potential to be applied in a high-throughput, low-cost, point-of-care analysis, which can be used for early-stage detection of diseases in the field.

The fabrication of fibers using microfluidic devices is also being explored. By controlling the internal morphology of nanofibers using expansion flow in a coaxial microfluidic channel, a broad range of fibers with chemical and mechanical properties have been realized [14]. The control over fiber morphology has implications for tissue engineering, as cell proliferation is known to be dependent on the scaffold morphology. Often one of the limitations of other techniques used to fabricate fibers is the choice of materials. Yu *et al.* [15] have recently combined droplet microfluidics with wet spinning to fabricate hybrid bamboo biomimetic fibers, which have multiple functionalities. Calcium alginate was used to bind polymer microspheres composed of poly(lactic-*co*-glycolic acid) (PLGA) into a bamboo-like architecture. The authors also demonstrated the flexibility of their method by incorporating hydrophobic droplets and cell microspheroids into the calcium alginate fiber. The advantage of this method is the flexibility and the choice of materials that can be used in the fabrication of new fibers.

4.3 Biomedical Application

4.3.1 Sensors

In this section, applications of stimuli-responsive fiber-based biomaterials will be discussed with some examples. The first example is the application for sensors. The role of sensors is to transform physical or chemical responses into signals on the basis of the targeted application. In particular, polymeric electrospun nanofibers have been investigated as sensors of gases, chemicals, optical materials, and biomaterials. It is considered that highly sensitive sensors can be assembled using nanofibers that possess high surface area and porosity. Here, polymeric fibers for fabricating sensors will be discussed in detail. Light, electrochemical signal, mass, magnetism, and thermal effects can all be measured by sensors. The number and variety of sensors existing today are astounding [16, 17]. Stimuli-responsive polymers and their combination with existing technologies can lead to new sensors with improved sensitivity.

Recently, Wang *et al.* [18, 19] have reported the fabrication of spider-web-like nanonets as a humidity sensor. These spider-web-like nanonets were fabricated by electrospinning onto a quartz crystal microbalance (QCM). The QCM-based fibrous sensors presented the most obvious decrease in frequency, shift in the relative humidity (RH) range of 2–95%, and the response time of the sensors gradually decreased with increasing RH in the chamber. For a flat film sensor, the response decreased with increasing RH ranging from 2% to 80%; however, it decreased to an immeasurable value once the RH exceeded 80%, which could be attributed to the mass loading of moisture. For fiber-coated sensors, however, the response showed an initial linear decrease in frequency corresponding to an increase in RH up to 50%, and shifted to the opposite direction at a higher RH (>50%), indicating that the moisture starts to be released from fibers at a higher RH. Among biosensors, glucose sensors have been extensively studied using diverse materials. In particular, Heo *et al.* [20, 21, 22] developed a hydrogel fibrous glucose sensor that has the following significant advantages: 1. The fibers can remain at the implantation site for an extended period, whereas microbeads disperse from the implantation site. 2. The fibers can be implanted at a readily controllable fluorescence intensity by cutting them into a specified length, thereby enabling stable and repeatable sensing. 3. The fibers can be easily and non-surgically removed from the body. An *in vivo* glucose fiber sensor implanted under the ear skin of mice remained at the implantation site for an extended period because the increased contact area with the subcutaneous tissue decreased the mobility of the subcutaneous implants. Immediately after the implantation, the fiber sensor was visible through the ear skin of over 100 μm thickness. For glucose sensing, Heo and coworkers injected glucose to temporarily increase the glucose concentration to 300 mg dL^{-1}, and insulin to decrease the glucose concentration to below 140 mg dL^{-1}. As a result, the fluorescence intensity of hydrogel fibers constantly tracked the fluctuations in blood glucose concentration for two up-and-down cycles. For long-term monitoring, after about 4 months from implantation, the fluorescence intensity of hydrogel fibers responded to blood glucose concentration fluctuations in one up-and-down cycle. These results indicate that the hydrogel fibers maintained their sensing functionality *in vivo* for a long period without inducing inflammatory reactions.

4.3.2 Drug Delivery Systems (DDSs)

DDSs are quite interesting applications that have evolved over the years, long before sustained DDSs were developed. Recently, many researchers have focused on controlled DDSs because these have become possible with the use of smart materials with remotely controllable properties. Therefore, drugs can be released to targeted regions on demand.

In this regard, temperature-responsive polymer-grafted chitosan (CTS) nanofibers were reported [23]. First, CTS-graft-PNIPAAm copolymer was also synthesized by using 1-ethyl-3-(3-dimethylaminopropyl)-carbodiimide and N-hydroxysuccinimide (NHS) as grafting agents to graft carboxyl-terminated PNIPAAm chains onto the CTS biomacromolecules. And then, CTS-g-PNIPAAm with or without bovine serum albumin (BSA) was fabricated into nanofibers through electrospinning using poly(ethylene oxide) (PEO) as a fiber-forming facilitating additive. The CTS-g-PNIPAAm/PEO nanofibers showed a pH- and temperature-dependent swelling/deswelling behavior. The drug release study showed that the nanofibers provided controlled release of the entrapped protein.

Zhang and Yarin [24] introduced two types of smart nanofibers for the study of the release of a dye rhodamine 610 chloride (rhodamine B) as a drug model. The electrospun nanofibers were fabricated from poly (NIPAAm-*co*-methylmethacrylate (MMA)) and poly(NIPAAm-*co*-MMA-*co*-acrylic acid (AAc)) copolymers, which are temperature-responsive and pH/temperature dual-responsive copolymers, respectively. In the case of a dye release study using poly(NIPAAm-*co*-MMA) nanofibers, a relatively low cumulative release rate (on the order of 1%) was observed below LCST, which resembled that of pure PMMA nanofibers. By the time the temperature crossed the LCST, however, the corresponding cumulative release rate had rapidly reached about 10% and saturated at about 12%. Moreover, the release rate demonstrated the largest thermal response. It is emphasized that poly(NIPAAm-*co*-MMA) nanofibers exhibited a positive thermoresponsive release profile, that is, a higher release rate when the nanofibers shrink above LCST than when the nanofibers swell below LCST.

Cui *et al.* [25] fabricated pH-responsive nanofibers for a local drug delivery system by introducing acid-labile acetal groups into a biodegradable backbone. The drug (paracetamol)-incorporated nanofibers were prepared by electrospinning [25]. The profile of drug release from the electrospun nanofibers prepared from an acid-labile polymer was evaluated in buffer solutions of different pHs (4.0, 5.5, and 7.0). In the absence of an acid group in the polymer, there were no significant differences in the profile among the buffer solutions of different pHs. Following the introduction of an acid group, significantly different drug release profiles were observed. The total amounts of the drug released were about 67% and 78% after incubation in pH 5.5 and pH 4.0 buffer solutions, respectively, and only 26% after incubation in pH 7.4. Moreover, the amount of the drug burst-released depended on the contents of acid-labile segments and the polymer nanofibers incubated in pH 4.0 medium. Additionally, when the pHs of buffer solutions were 5.5 and 4.0, the amount of drug released from pH-sensitive nanofibers increased owing to the pH-induced structural changes of the polymeric nanofibers and the degradation of the matrix polymer.

Kim *et al.* [26] fabricated novel temperature-responsive nanofibers for “on–off” drug release systems (Figure 4.4). They synthesized the nanofibers using the temperature-responsive polymer, poly(NIPAAm-*co*-*N*-hydroxylmethylacrylamide(HMAAm)), and fluorescein isothiocyanate (FITC)-dextran was directly embedded into these nanofibers as the drug model.

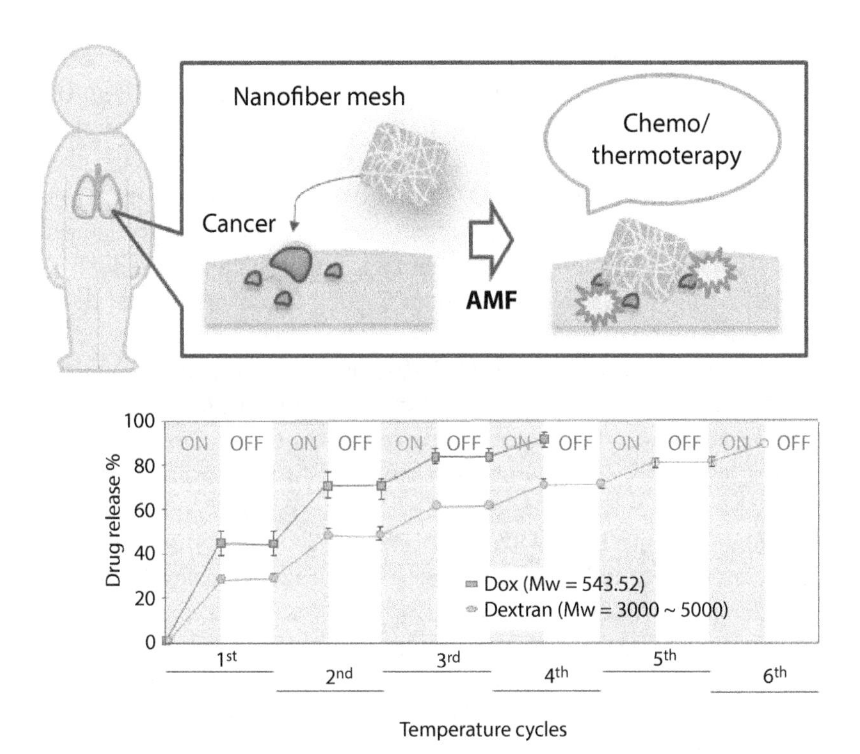

Figure 4.4 On–off drug release strategy using temperature-responsive nanofiber mesh. By incorporating magnetic nanoparticles, the nanofibers shrink in response to the alternating magnetic field (AMF) application. This nanofiber demonstrates the synergic effect of hyperthermia and chemotherapy for cancer cell therapy. Note: The figure has been arranged by the author.

The prepared nanofibers showed "on–off" switchable swelling–shrinking behavior in response to temperature alternation cycles upon crossing the LCST; correspondingly, the dextran release profile showed the "on–off" switchable behavior. After the first heating, approximately 30% of the loaded dextran was released from the nanofibers. The release stopped after cooling below LCST, but the release restarted upon the second heating. In this system, the dextran is released by it being squeezed out of the collapsed interconnected cross-linked polymer network. The release of dextran stops upon cooling because of the suppressed diffusion of the dextran molecules, which have high molecular weight. Almost all of the dextran was released from the nanofibers after six temperature cycles. This kind of "on–off" drug release system can release a certain amount of a drug within a short time after an off period that can be programmed according to the circadian rhythm of the disease being treated. They also showed temperature- and magnetic-field-responsive electrospun nanofibers containing an anticancer drug (doxorubicin, DOX) and magnetic nanoparticles (MNPs). Upon alternating magnetic field (AMF) application, the temperature-responsive nanofibers shrank in response to the increased temperature triggered by MNPs [27]. The incorporated DOX was released from nanofibers owing to the hyperthermic effect. These nanofibers demonstrated the synergic effect of hyperthermia and chemotherapy for cancer cell therapy with the hyperthermic treatment time being less than 5 min, which can reduce the side effects on normal tissues or cells.

On the other hand, Fu *et al.* [28] studied the development of a novel photoresponsive "on–off" release system for a prodrug, based on host–guest interaction on the photoresponsive and cross-linked nanofiber surface. The nanofibers with stimuli-responsive surfaces were electrospun from the block copolymer via controlled radical polymerization, and then the surfaces were modified with photosensitive 4-propargyloxyazobenzene by "click chemistry." Followed by UV irradiation at a wavelength of 365 nm, the prodrug was released quickly from nanofiber surfaces, as the photosensitive group transformed from the trans to the cis form configuration. In the dark, however, there was almost no release of the prodrug from the nanofiber surfaces, or between the host and the guest, that is, between the prodrug and the surfaces. Furthermore, this system showed a quick response and controllable release of the prodrug, as revealed by the multistep "on–off" release profile under UV irradiation. The concentration of the prodrug in solution increases gradually in the next 10 min upon UV exposure. The concentration of the prodrug ceases to increase upon removal of the UV irradiation after 20 min. The concentration of the prodrug remains almost constant in the next 20 min, in the absence of UV irradiation. When UV irradiation is resumed after that, the concentration of the prodrug in solution increases again.

4.3.3 Cell Application

The stimuli-responsive nanofiber meshes have been also explored as cell manipulation/storage systems. Maeda *et al.* [1] envision the temperature-responsive nanofiber meshes in a cell storage application. The cryopreservation of mammalian cells was demonstrated without loss of viability during freezing process by using a PNIPAAm mesh because dehydrated PNIPAAm chains suppressed the formation of large extracellular ice crystals during the freeze/thaw process (Figure 4.5). Sur *et al.* [29], on the other hand, prepared peptide amphiphile nanofiber matrices by incorporation of a photolabile artificial amino acid to control bioactivity. The peptide

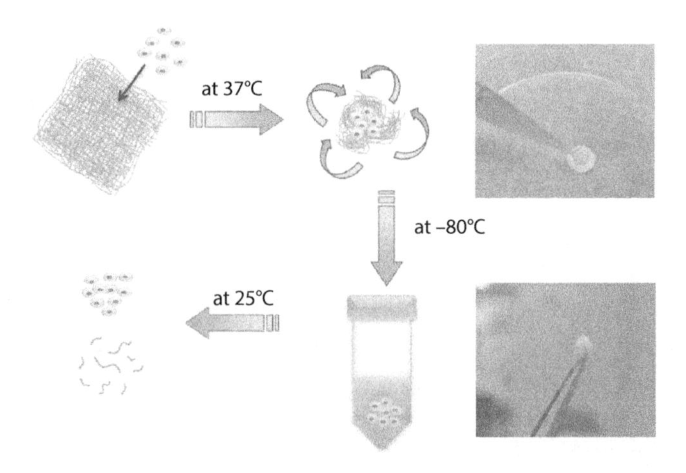

Figure 4.5 Temperature-modulated manipulation of cells using PNIPAAm nanofiber mesh for cryopreservation application. Note: The figure has been arranged by the author.

amphiphiles self-assembled into cylindrical nanofibers. Cell adhesion was dynamically controlled by rapid photolytic removal of the RGDS peptide from the nanofiber. This dynamic temporal control of cell–material interactions can become an important component in the design of new artificial matrices based on nanostructures for regenerative medicine research studies.

Fukunaga *et al.* [30] and Sawada *et al.* [31] developed calcium ion (Ca^{2+})-responsive hydrogels composed of designed β-sheet peptides. As the novel designed peptide, E1Y9, has a glutamic acid residue to interact with Ca^{2+}, the peptide in the sol-state self-assembled into hydrogels in the presence of Ca^{2+}. The hydrogels showed a high cell-adhesive ability that was similar in magnitude to fibronectin. Thus, the novel peptide-based nanofiber hydrogels can facilitate development studies for 3D cell cultures for tissue engineering.

On the other hand, the authors have innovated a nanofiber mesh for the removal of toxins from the blood, which they are hopeful to incorporate into wearable blood purification systems for kidney failure patients [32]. We made our nanofiber mesh using two components: a blood-compatible primary matrix polymer made from polyethylene-co-vinyl alcohol, or EVOH, and several different forms of zeolites. Zeolites have microporous structures capable of adsorbing toxins such as creatinine from the blood. Different zeolites have different pore sizes, meaning they can be used to selectively adsorb specific solutes. Our result demonstrated that a 16-g mesh is enough to remove all the creatinine produced in one day by the human body. Although the new design is still in its early stages and not yet ready for production, nanofiber-based biomaterials will soon be a feasible, compact, and cheap alternative to dialysis for kidney failure patients across the world.

Kim *et al.* [33] prepared a temperature-responsive fibrous hydrogel that was used as a cell capture and release membrane. The fibrous hydrogel captured and released cells by self-wrapping, encapsulation, and shrinking in response to temperature changes. In the cell culture medium at 37°C, a droplet of cells was dropped on a fiber web, and the web immediately started to fold up and wrap around the droplet. The folding of the fiber web was very fast and was completed within 30 s. After 10 min at 37°C, the fiber web was transferred to a refrigerator at 4°C and allowed to swell for another 10 min. The fiber web became transparent with a hydrogel-like morphology, and the fibrous structure was maintained. When the web was heated again to 37°C, the hydrogel-like fibrous web shrank and became opaque. After three cycles of temperature alternations, almost all of the cells (>95%) seeded on the web were released, whereas only a few cells were released upon swelling during the cooling from 37°C to 4°C. Live/dead assay of the cells released from the web showed that almost all of the cells were alive, and the condition and proliferation of the cells were determined. The most interesting application of magnetic-field-responsive fibers is in the hyperthermic treatment of cancer cells. Huang *et al.* [34] reported MNP-incorporated polystyrene (PS-MNP) nanofibers that showed the hyperthermic effect on cultured SKOV-3 cells. When the PS-MNP nanofibers were exposed to the AMF for 10 min or longer, all of the cancer cells were destroyed, and the association of the cancer cells with the nanofibers was also demonstrated. Here, improved hyperthermic smart fibers are introduced.

4.4 Filters

Owing to the high specific surface area and high porosity of nanofibers, they have been developed as filter media, which are very useful for the separation or purification of not only waste water but also biomolecules. Furthermore, filtration has been improved and new types of nanofibers have been developed, such as hollow fibers. pH-responsive hollow fibers for protein antifouling were developed by Zou *et al.* [35]. The hollow fibers were prepared by a dry-wet spinning technique based on a liquid–liquid phase separation technique. Poly(MMA-*co*-AAc-*co*-vinylpyrrolidone (VP)) terpolymer was synthesized to modify polyestersulfone hollow fiber membranes. When the pH changed from 2.0 to 11.0 in pH-responsive filtration, the fluxes increased under acid conditions owing to the increased hydrophilicity and showed pH dependence, that is, fluxes decreased with increasing pH. These results indicated that the flux variation increased with the increase in the amount of the terpolymer, and the pH sensitivity was caused by the dissociation of AAc in the terpolymers. In a further study by the same group, they used another type of terpolymer (poly (St-*co*-AAc-*co*-VP)), and two mechanisms underlying the functions of the pH-sensitive flat-sheet membranes (FSMs) and hollow fiber membranes were clarified, namely, the pore size change theory and electroviscous effect [36]. Firstly, regarding the pore size change theory, the swelling–shrinking effect of the ionized–deionized AAc chain was deemed to be the main reason for the water flux change and solute rejection of the pore-filled pH-sensitive FSM. Secondly, regarding the electroviscous effect, the carboxylic acids of AAc could dissociate to carboxylate ions at pH 12.0 to provide a high charge density in the copolymer, resulting in the swelling of the terpolymer. Furthermore, the solution flowing through the pores in the membranes was affected by the electroviscous effect during the filtration. The electroviscous effect is a physical phenomenon that occurs when an electrolyte solution passes through a narrow capillary or pore with charged surfaces.

4.5 Conclusion

Fibers produced from various types of responsive polymers at various production stages have already been explored for a wide range of applications in diverse fields, including nanotechnology, textiles, industry, fuel cells, tissue engineering, regenerative medicine, and biomaterials. These fibers, particularly well-defined nanofibers, have astounding features compared with other types of materials, for instance, high specific surface area, high porosity, and biomimetic properties, providing promising potential applications. Therefore, fibers have been explored as cell culture scaffolds, DDSs, and other biomedical applications. Among these types of fibers, those that combined well with stimuli-responsive media are called smart fibers. They can be turned on and off remotely ("on–off" controllable) by applying external stimuli while the fiber characteristics remained. The applications of smart fibers are not limited *in vitro* or *in vivo*. Thus, the smart fibers have gained much attention from researchers because their remotely controllable characteristics can be used to match the circadian rhythm of a disease, which could be a good candidate for the treatment of cancer, damaged tissues, or chronic diseases. Unfortunately, the most serious limitation of fibers is their lack of direct injectability into the body.

Hence, they are normally used in the form of mats, sheets, or webs of bulk size. Nevertheless, mat- or sheet-type smart fibers offer an opportunity to develop their unique applications, for example, dressings or anti-adhesion membranes for wounds. Further functionalization of smart fibers could also be used for the separation, purification, and preservation of biomolecules or cells. Taking together all these merits of smart fibers, they can be utilized as key tools in a wide range of applications with switchable properties in response to external remote control.

References

1. Maeda, T., Kim, Y.J., Aoyagi, T. and Ebara, M., The design of temperature-responsive nanofiber meshes for cell storage applications. *Fibers*, 5, 13, 2017.
2. Okuzaki, H., Kobayashi, K. and Yan, H., Non-woven fabric of poly(N-isopropylacrylamide) nanofibers fabricated by electrospinning. *Synthetic Metals*, 159, 2273, 2009.
3. Bhardwaj, N. and Kundu, S.C., Electrospinning: A fascinating fiber fabrication technique. *Biotechnol. Adv.*, 28, 325, 2010.
4. Kato, T., Self-assembly of phase segregated liquid crystal structures. *Science*, 295, 2414, 2002.
5. Ringsdorf, H., Schlarb, B. and Venzmer, J., Molecular architecture and function of polymeric oriented systems: Models for the study of organization, surface recognition, and dynamics of biomembranes. *Angew. Chem. Int. Ed. Engl.*, 27, 113, 1988.
6. Saito, N., Kanie, K., Matsubara, M., Muramatsu, A. and Yamaguchi, M., Dynamic and reversible polymorphism of self-assembled lyotropic liquid crystalline systems derived from cyclic bis(ethynylhelicene) oligomers. *J. Am. Chem. Soc.*, 137, 6594, 2015.
7. Bitton, R., Chow, L.W., Zha, R.H., Velichko, Y.S., Pashuck, E.T. and Stupp, S.I., Electrostatic control of structure in self-assembled membranes. *Small*, 10, 500, 2014.
8. Matabola, K.P. and Moutloali, R.M., The influence of electrospinning parameters on the morphology and diameter of poly(vinyledene fluoride) nanofibers- effect of sodium chloride. *J. Mater. Sci.*, 48, 5475, 2013.
9. Dahlin, R.L., Kasper, F.K. and Mikos, A.G., Polymeric nanofibers in tissue engineering. *Tissue Eng. Part B Rev.*, 17, 349, 2011.
10. Muthiah, P., Boyle, T.J. and Sigmund, W., Thermally induced, rapid wettability switching of electrospun blended polystyrene/poly(N-isopropylacrylamide) nanofiber mats. *Macromol. Mater. Eng.*, 298, 1251, 2013.
11. Carothers, W.H. and Hill, J.W., Studies of polymerization and ring formation. XV. Artificial fibers from synthetic linear condensation. *J. Am. Chem. Soc.*, 54, 1579, 1932.

12. Yuan, H., Cambron, S.D. and Keynton, R.S., Prescribed 3-D direct writing of suspended micron/sub-micron scale fiber structures via a robotic dispensing system. *J. Vis. Exp.*, e52834, 2015.
13. Yildirim, A., Yunusa, M., Ozturk, F.E., Kanik, M. and Bayindir, M., Surface textured polymer fibers for microfluidics. *Adv. Funct. Mater.*, 24, 4569, 2014.
14. Daniele, M.A., Boyd, D.A., Adams, A.A. and Ligler, F.S., Microfluidics microfluidic strategies for design and assembly of microfibers and nanofibers with tissue engineering and regenerative medicine applications. *Adv. Healthc. Mater.*, 4, 11, 2014.
15. Yu, Y., Wen, H., Ma, J., Lykkemark, S., Xu, H. and Qin, J., Flexible fabrication of biomimetic bamboo-like hybrid microfibers. *Adv. Mater.*, 26, 2494, 2014.
16. Singh, S. and Gupta, B.D., Fabrication and characterization of a surface plasmon resonance based fiber optic sensor using gel entrapment technique for the detection of low glucose concentration. *Sens. Actuators B*, 177, 589, 2013.
17. Wang, S., Chao, D., Berda, E.B., Jia, X., Yang, R., Wang, X., Jiang, T. and Wang, C., Fabrication of electroactive oligoaniline functionalized poly(amic acid) nanofibers for application as an ammonia sensor. *RSC Adv.*, 3, 4059, 2013.
18. Wang, X., Ding, B., Yu, J., Wang, M. and Pan, F., A highly sensitive humidity sensor based on a nanofibrous membrane coated quartz crystal microbalance. *Nanotechnology*, 21, 055502, 2010.
19. Wang, X., Ding, B., Yu, J. and Wang, M., Highly sensitive humidity sensors based on electro-spinning/netting a polyamide 6 nano-fiber/net modified by polyethyleneimine. *J. Mater. Chem.*, 21, 16231, 2011.
20. Heo, Y.J., Shibata, H., Okitsu, T., Kawanishi, T. and Takeuchi, S., Long-term *in vivo* glucose monitoring using fluorescent hydrogel fibers. *PNAS*, 108, 13399, 2011.
21. Shibata, H., Heo, Y.J., Okitsu, T., Matsunaga, Y., Kawanishi, T. and Takeuchi, S., Injectable hydrogel microbeads for fluorescence-based *in vivo* continuous glucose monitoring. *PNAS*, 107, 17894, 2010.
22. Heo, Y.J. and Takeuchi, S., Towards smart tattoos: Implantable biosensors for continuous glucose monitoring. *Adv. Healthc. Mater.*, 2, 43, 2013.
23. Yuan, H.H., Li, B.Y., Liang, K., Lou, X.X. and Zhang, Y.Z., Regulating drug release from pH- and temperature-responsive electrospun CTS-g-PNIPAAm/poly(ethylene oxide) hydrogel nanofibers. *Biomed. Mater.*, 9, 055001, 2014.
24. Zhang, Y. and Yarin, A.L., Stimuli-responsive copolymers of n-isopropyl acrylamide with enhanced longevity in water for micro- and nanofluidics, drug delivery and non-woven applications. *J. Mater. Chem.*, 19, 4732, 2009.
25. Cui, W., Qi, M., Li, X., Huang, S., Zhou, S. and Weng, J., Electrospun fibers of acid-labile biodegradable polymers with acetal groups as potential drug carriers. *Int. J. Pharm.*, 361, 47, 2008.
26. Kim, Y.J., Ebara, M. and Aoyagi, T., Temperature-responsive electrospun nanofibers for 'on-off' switchable release of dextran. *Sci. Technol. Adv. Mater.*, 13, 064203, 2012.

27. Kim, Y.J., Ebara, M. and Aoyagi, T., A smart hyperthermia nanofiber with switchable drug release for inducing cancer apoptosis. *Adv. Funct. Mater.*, 23, 5753, 2013.
28. Fu, G.D., Xu, L.Q., Yao, F., Li, G.L. and Kang, E.T., Smart nanofibers with a photoresponsive surface for controlled release. *ACS Appl. Mater. Interfaces*, 1, 2424, 2009.
29. Sur, S., Matson, J.B., Webber, M.J., Newcomb, C.J. and Stupp, S.I., Photodynamic control of bioactivity in a nanofiber matrix. *ACS Nano*, 6, 10776, 2012.
30. Fukunaga, K., Tsutsumi, H. and Mihara, H., Self-assembling peptide nanofibers promoting cell adhesion and differentiation. *Biopolymers*, 100, 731, 2013.
31. Sawada, T., Tsuchiya, M., Takahashi, T., Tsutsumi, H. and Mihara, H., Cell-adhesive hydrogels composed of peptide nanofibers responsive to biological ions. *Polym. J.*, 44, 651, 2012.
32. Namekawa, K., Schereiber, M.T., Aoyagi, T. and Ebara, M., Fabrication of zeolite–polymer composite nanofibers for removal of uremic toxins from kidney failure patients. *Biomater. Sci.*, 2, 674, 2014.
33. Kim, Y.J., Ebara, M. and Aoyagi, T., A smart nanofiber web that captures and releases cells. *Angew. Chem. Int. Ed. Engl.*, 51, 10537, 2012.
34. Huang, C., Soenen, S.J., Rejman, J., Trekker, J., Chengxun, L., Lagae, L., Ceelen, W., Wilhelm, C., Demeester, J., De, S. and Stefaan, C., Magnetic electrospun fibers for cancer therapy. *Adv. Funct. Mater.*, 22, 2479, 2012.
35. Zou, W., Huang, Y., Luo, J., Liu, J. and Zhao, C., Poly (methyl methacrylate–acrylic acid–vinyl pyrrolidone) terpolymer modified polyethersulfone hollow fiber membrane with pH sensitivity and protein antifouling property. *J. Membrane Sci.*, 358, 76, 2010.
36. Cheng, C., Ma, L., Wu, D., Ren, J., Zhao, W., Xue, J., Sun, S. and Zhao, C., Remarkable pH-sensitivity and anti-fouling property of terpolymer blended polyethersulfone hollow fiber membranes. *J. Membrane Sci.*, 378, 369, 2011.

5

Nanowires for Smart Textiles

Jizhong Song

Institute of Optoelectronics & Nanomaterials, MIIT Key Laboratory of Advanced Display Materials and Devices, College of Materials Science and Engineering, Nanjing University of Science and Technology, Nanjing, China

Abstract
Textile-based devices have gained significant attention because of their excellent flexibility, high sustainability, light weight, comfort, and wide-ranging applications in wearable electronic devices, robotic sensory skins, and biomedical devices. Nanowires (NWs) with high electrical performance and high-aspect ratio can be easily assembled into high-quality network structures, which expands their applications in highly flexible textile-based devices. The combination of different NWs and textiles also renders NWs highly applicable in integrated smart devices. In this chapter, recent progress in high-performance smart textiles based on NW materials is comprehensively described, with emphasis on typically explored NWs including metals, metal oxides, conducting polymers, sulfide, and other semiconductors. Meanwhile, perspectives on future research of NW-based textiles are also discussed.

Keywords: Metal nanowires, metal oxide nanowires, sulfide nanowires, energy storage, optoelectronic devices, solar cells, nanogenerators, photodetectors, lithium batteries

5.1 Introduction

A textile [1] is "a thin, flexible sheet of material with sufficient strength and tear resistance for use in clothing, interior applications, and other protective functions." These requirements, as well as the ability of a fabric to recover its original shape after bending or crumpling, markedly distinguish fabric from paper and plastic films. Staple or filament fibers can be twisted into yarns, after which the yarns can be woven, knitted, or felted

Email: songjizhong@njust.edu.cn

Nazire D. Yilmaz (ed.) Smart Textiles, (127–176) © 2019 Scrivener Publishing LLC

into textiles as shown in Figure 5.1a. Textile development reflects human civilization to a certain extent. Various leaves were connected by our ancestors to form the early models of textiles. Natural materials, such as silk and cotton, were subsequently woven into real textiles that were warmer and more comfortable. With the development of human civilization, a broad range of man-made fiber materials, such as nylon and Kevlar, gradually emerged and greatly improved life over the last century [2].

Textiles are typically used in clothes, apparel, and home furnishings. Recently, they have been integrated with smart devices that can sense environmental stimuli from mechanical, thermal, chemical, electrical, or magnetic sources and generate a response [3], enabling new functions. These functions include sensing, therapy, navigation, communication, and original fashion [4, 5]. Smart textiles have induced further powerful changes

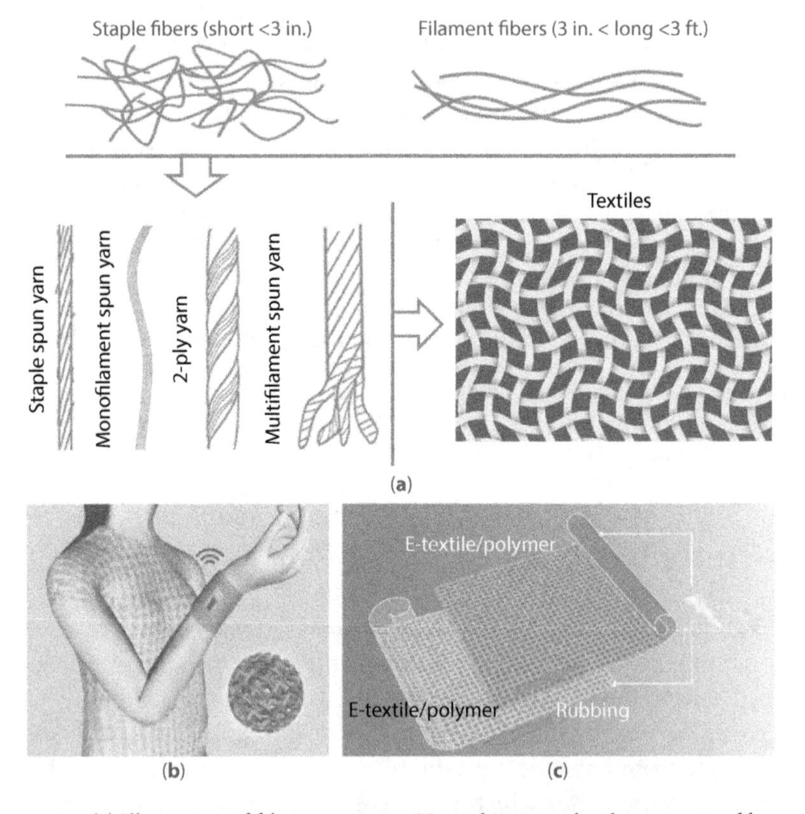

Figure 5.1 (a) Illustration of fabric structures. Note: this image has been prepared by the author. (b) Wireless body temperature sensor system triggered by the "power shirt." Reprinted with permission from [23]; Copyright 2014 American Chemical Society. (c) Schematic diagram of a generator based on smart textiles. Reprinted with permission from [24]; Copyright 2016 American Chemical Society.

across a broad field of industries and our lives (Figure 5.1b and c). Wearable self-powered devices can be easily manufactured when clothes convert energy sources, such as mechanical energy and solar energy, to electric energy. Various portable electronic facilities, such as mobile phones, can be more effectively and conveniently powered if our clothes can store electric energy. Meanwhile, the production of electronic devices from textiles for a more efficient use of electricity in many fields, such as displays and sensors, also exhibits potential. The dominance of electricity over our life and its ubiquity implies that the development of smart electronic textiles is important and has considerable potential for practical uses.

The variety of applications of smart textiles can be summarized into the following categories [2]. First, the conventional planar system cannot effectively meet the requirement of portable and wearable electronic products. Thus, the incorporation of solar cells as well as thermoelectric [6], piezoelectric [7, 8], and photoelectric [9–11] devices into textiles to solve the aforementioned problems has drawn significant interest. Taking the organic solar cell as an example, the output powers of a planar solar cell vary at different angles of incident lights, which is a problem for many applications; in contrast, solar textiles are developed to show a stable output power that was independent of the angle of the incident light. Such a property is particularly desired for applications for various outdoor activities and wireless electronics. Secondly, for better use of the electricity, electrochemically storage devices, including both lithium-ion batteries and supercapacitors, have been fabricated into textiles [12, 13]. These devices displayed high electrochemical performances, particularly as flexible powering systems. Thirdly, a variety of powered facilities such as chromatic and actuating devices have been incorporated into textiles and display high performances [14–16]. As we can see, the convergence of electronic components and advanced fibers with man-made textiles straddles the fields of materials science and digital electronics. However, such "smart" textiles above (also known as electronic or e-textiles) fall into the category of intelligent materials that sense and respond to environmental stimuli [17]. Within the field of wearable computing, smart textile applications range from medical monitoring of physiological signals [18], including heart rate [19], guided training, and rehabilitation of athletes [20], to assistance for emergency first responders [21], and to commercial applications where electronics including iPod controls, displays, and keyboards are integrated into daily clothing [22]. Smart textile research is still in its infancy, and there is little understanding of the mechanically demanding environment, e.g., electronic circuits face the change of strain during integration into textiles and subsequent wearing of the smart textile inside clothing.

To maintain essential textile properties (e.g., drapability), smart textiles are evolving to integrate increasing electronic functions at the fiber level [25]. Currently, most fibers are limited to a single functionality (e.g., electrical conductivity) due to intrinsic features of a single material. Fortunately, with the advancements in pervasive miniaturization in a wide range of technologies, the emergence of micro- and nano-electronics has created opportunities for the seamless integration of multifunction materials with textiles [26–28]. It is also clear that the multifunctional materials that are assembled in nano-electronics are usually grown or coated on the surface of fibers in a variety of nanostructures like nanorods (NRs) [29], nanowires (NWs) [30], nanoneedles (NNs) [31], nanoflowers (NFs) [32], nanobelts (NBs) [33], nanoflakes (NFKs) [34], nanotubes (NTs) [35], etc.

Among various nanostructures, NWs (e.g., a nanostructure with the diameter of the order of a nanometer (10^{-9} m), and the ratio of the length to the width being greater than 1000) have become the focus of intensive research owing to their unique applications in mesoscopic physics and fabrication of nanoscale devices [36]. In comparison with quantum dots and wells, NWs with the large length/diameter ratio provide a better system to improve the electrical and thermal transport and play an important role as interconnects and functional units in fabricating electronic, optoelectronic, electrochemical, and electromechanical devices. What's more, NWs with good toughness and mechanical flexibility are quite suitable for the realistic wearable equipment [37, 38].

In the chapter, recent progress in high-performance smart textiles based on NW materials is comprehensively described including metals, metal oxides, conducting polymers, sulfide, and other semiconductor materials. Furthermore, perspectives on future research are also discussed.

5.2 Advantages of Nanowires to Smart Textiles

The one-dimensional configuration and nanomaterial characteristics have rendered NW materials as promising building blocks to enhance the performance of smart textiles. Herein, a general perspective on the advantages of NW materials to smart textiles is presented.

5.2.1 Balance between Transparency and Conductivity

Transparent conductive electrodes are widely used in fabrication of various smart textiles, especially for optoelectronic textiles, where the transparency and electrical conductivity are two important and opposing factors that are crucial for the device performance. High transparency represents small losses of the incoming light, while low resistance means low internal resistance in optoelectronic smart textiles. The balance between transparency and resistance has always been a very tricky problem for high-performance devices. On one hand, a thinner transparent conductive layer will demonstrate higher transparency, but the sheet resistance will also be increased. On the other hand, a thicker transparent conductive layer shows lower sheet resistance, but high transparency has to be sacrificed. The imbalance between transparency and conductivity can be effectively solved. For example, NW materials, including silver (Ag) and copper (Cu) NW networks, not only provide fast charge transport but also offer enough interspace among the NWs for the penetration of incident light, which can strike an excellent balance between the two factors. To date, a high transmittance of over 90% and a low electrical resistance of $\sim 10\ \Omega\ sq^{-1}$ have been achieved based on Ag NWs, which are comparable to indium tin oxides (ITO) under the same conditions [39, 40]. Thus, NW materials have been widely investigated as transparent conductive electrodes to enhance the performance of optoelectronic textiles.

5.2.2 High Specific Surface Area

The interface between two phases has always been considered to be crucial for various applications, especially for optoelectronic textiles. Larger specific areas enable sufficient charge generation, separation, and collection, which lead to excellent device performance [41]. Compared with conventional bulk materials, materials in the nanoscale have demonstrated much larger specific surface areas and with higher interacting efficiencies, and thus are favorable for performance enhancement. Although the specific surface areas of NW materials are generally lower than their zero-dimensional counterparts, some strategies can be adopted to overcome this problem. One representative example is the roughened hierarchical structure of TiO_2/ZnO NWs, which takes the advantage of both dye adsorption and fast electron transportation to enhance the device performance of solar cells.

5.2.3 Direct Charge Transport Path

The electrical resistance of nanoparticles (NPs) is generally high due to many grain boundaries, which restrain the rapid charge transport [42, 43]. A NW configuration provides a direct charge transport path and decreases the electrical resistance, compared with NPs. For example, TiO_2 NWs have been widely investigated as one of the promising photoanode materials to replace conventional TiO_2 NPs for efficient dye-sensitized solar cells.

5.2.4 Oriented Assembly

The assembly method of nanomaterials is considered to be very important to extend their physical properties from the nanoscale to the macroscale. Oriented structures have been proven to efficiently improve the mechanical and electrical properties of nanomaterials on a certain direction and are favorable for improving device performance [44]. An anisotropic configuration is a necessary factor for the formation of an oriented structure. To this end, one-dimensional nanomaterials have demonstrated large potential to be assembled into highly oriented structures with much improved properties and device performance. To date, oriented metal oxide NWs, sulfide NWs, conducting polymer NWs, and other semiconductor NWs have all been prepared and utilized in high-performance smart textiles, and this will be discussed in the corresponding section.

5.3 Various Nanowires for Smart Textiles

5.3.1 Conductive Nanowires for Smart Textiles

Conductive textiles (fabric, thread, or yarn) are an important element in the realization of smart textiles. It can act as an interconnect between components to transfer signals or power such as health monitoring sensors [45] and solar cells [46]. Meanwhile, it can also be used in applications of electromagnetic interference (EMI) shielding [47], resistive fiber sensors, and wearable antennas [48] because of its own features. During the practical applications, these smart textiles still have to undergo great deformation, so electrical conductivity against flexibility and stretchability are critical for electrically conductive textiles. Commonly used textiles such as cotton, nylon, and polyester are all electrical insulators. Commercially available conductive textiles are typically either a solid metal wire, such as Cu and stainless steel, or a non-conductive thread coated with an ~1-μm-thick metal film (e.g., Ag). These options for conductive textiles are less than ideal as they tend to be stiff and brittle, which causes problems both during weaving and in practical applications. Furthermore, they can easily breakdown (resistance greatly increases) after repeated bends [49].

With the rise of composite materials, a variety of conductive materials including conductive polymers [50], graphene flakes [51], and carbon

NTs [52] are used as the conductive layers when commercial polymer fibers such as polyamide 6, Lycra, polyolefin, and polyurethane serve as the elastic substrates. Polypyrrole (PPy) and poly(3,4-ethylenedioxythiophene):poly(styrenesulfonate) (PEDOT:PSS) are the two mostly explored conducting polymers because of their acceptable electrical conductivity and piezoresistive properties. They can be coated onto commercial polymer fibers to form a conductive layer through various techniques such as solution dipping, *in situ* polymerization, and chemical vapor deposition [53–55]. However, the conductive layers are easily degenerated during stretching because of the smooth surfaces of polymer fibers [53, 54], and they are also unstable in air because of the absorption of oxygen and moisture [56]. Additionally, the conductivity of polymer-coated textiles is low [57] and therefore cannot be used in many applications. To solve these problems, conducting materials with one-dimensional nanostructure such as wires, rods, belts, and tubes are produced on the surface of the fiber substrates due to its excellent electrical conductivity and mechanical flexibility [58]. Unlike other nanostructure materials (e.g., NPs and nanosheets (NSs)), the elongated shape of NWs allows for a conductive film to be achieved at a far lower particle density and therefore there are fewer junctions. Then, the NWs connect with each other and form a conducting network that makes the conductive layer with high degree of bending and tensile properties. Meanwhile, the empty space between neighboring NWs provides transmittance for visible light [59]. Among various electrical conductive NWs, metal NWs and polymer NWs are widely used in smart textiles and are also the focus of this section.

5.3.1.1 Metal Nanowires for Smart Textiles

Since the successful syntheses of various metal NWs were reported [28, 60–62], metal NW networks have been widely used for conductors and have become one of the promising candidates to indium tin oxide transparent conductors [63–65]. Similarly, carbon NTs with the same one-dimensional nanostructures are also competitive due to its metallicity and extremely low resistance. However, the junction resistance between two overlapping NTs in the film is very high, which leads to resistances per unit length of thread on the order of 1 kΩ cm^{-1} [66]. Unlike carbon-based materials, metal NW junctions can be sintered to greatly reduce junction resistance. Because NW coatings are network state rather than continuous films, much fewer materials will be used to assemble films with the considerable optical and electrical properties. This can lead to lower material costs, lower weight, as well as greater mechanical flexibility. What's more, the NW coatings can be simply deposited without vacuum or complex processes. Currently, various metal NWs can be used as electrically conductive thread, including gold (Au), Ag, and Cu NWs.

5.3.1.1.1 Ag NWs

Ag is a typical conductor because it is more cost effective than Au or platinum, and is more stable in air than Cu. The solution process, such as the spray deposition of Ag NWs, is typically used to fabricate flexible electronics aiming at large-scale products (Figure 5.2a and b) [67, 68]. The first application of Ag NWs in smart textiles was reported by David T. Schoen *et al.* [69]. They chose the cheap, widely available, and chemically and mechanically robust cotton as the backbone. Cotton textiles can be coated easily with CNTs by submerging them in aqueous CNT ink. The cotton is then rinsed well in distilled water to remove excess surfactant. Ag NWs can be easily added to the conductive cotton by pipetting them directly from a methanol solution followed by drying on a hot plate at 95°C for 30 min. Then, copious rinsing was done to remove any excess solvent and surfactant, which was good for strong physical adhesion between NWs and cotton fibers. The final material is mechanically robust, with a low sheet resistance of ~1 Ω sq^{-1}. It can be mechanically manipulated for integration into the final filtering system, and this approach makes a gravity fed device operating at 100,000 L/(h m^2), which can inactivate >98% of bacteria with only several seconds of total incubation time. This excellent performance is attributed to the electrical properties rather than size exclusion, while the high surface area of the device coupled with large electric

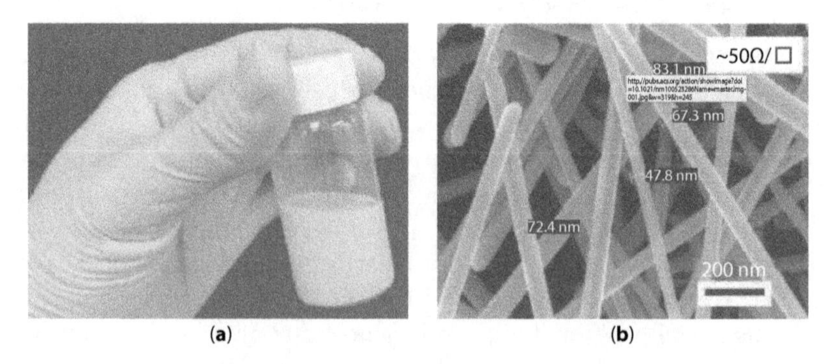

Figure 5.2 (a) Ag NW ink in ethanol solvent with concentration of 2.7 mg/mL. (b) The different densities of Ag NW films lead to different sheet resistances 50 Ω sq^{-1}. The diameters of the Ag NWs are in the range of 40–100 nm. Reprinted with permission from [40]; Copyright 2010 American Chemical Society.

field concentrations near the Ag NW tips allows for effective bacterial inactivation. Almost at the same moment, Madaria *et al.* [70] used spray coating to deposit Ag NWs on fabric to achieve high conductivity.

Different fabrics such as nylon, polyester, and cotton threads are also chosen to be coated with solution-synthesized Ag NWs via the dip-and-dry methods [71–75]. All types of textiles were cleaned in an ultrasonic bath using ethanol, alcohol, and deionized water for 5 min each. The NW solution adhered well to the surface of the cotton thread, but not to polyester and nylon. These latter two synthetic textiles required a chemical pretreatment. After cleaning, the polyester threads were submersed for 6 min in a solution consisting of 20 wt% NaOH and 80 wt% distilled water heated at 75°C, then dried in hot air. The nylon threads were submersed in a solution consisting of 91 wt% ethyl acetate and 9 wt% resorcinol for 1 min and then dried in air. As for cotton, polyester, and nylon, the density of the deposited NW film was varied through the concentration of the NWs in the coating solution and through the number of dipping steps. After deposition, the threads were annealed in air at 150°C for 30 min, and the Ag NW-fiber resistance per unit length of 0.8 Ω cm^{-1} could be achieved.

Furthermore, highly Ag NW-loaded cotton fabrics have been considered to be promising candidates for self-cleaning textiles [76] and heating applications due to their infrared (IR) reflection properties [59, 77]. Recently, a simple heating circuit fabricated from Ag NW-decorated cotton fabrics is operated with a 3-V battery, which clearly demonstrated the potential of heatable fabrics for mobile heating applications [59]. Besides, a highly stretchable and foldable conductive cord is developed for the first time by depositing Ag NWs on a polyurethane cord [72]. The step of prestrain to the cord during the deposition of Ag NWs resulted in a wrinkled Ag NW layer, thereby inducing structural stability towards stretching and bending. This conductive cord exhibited stable electric conduction under strain up to 167% and at a bending radius of 0.3 mm. Also, it could withstand a stretching test of 500 cycles at a strain of 50%. Owing to the convenience of solution process, Ag NWs have be successively assembled on the functional textile for external electrodes [78]. Ag NWs can also be directly coated onto polymer substrates to fabricate flexible solar cells as shown in Figure 5.3a, achieving a power conversion efficiency (PCE) of 5.02%, which is comparable to that of its glass substrate-based counterpart [39]. What's more, Ag NWs are also widely used in supercapacitors [79–83], because the high conductive nature of Ag NW promotes the rapid reaction kinetics during the energy storage measurements. Moreover, the high conductive nature of Ag NWs provides a smooth pathway for electron transportation during the energy storage process, and the binder-free nickel–cobalt

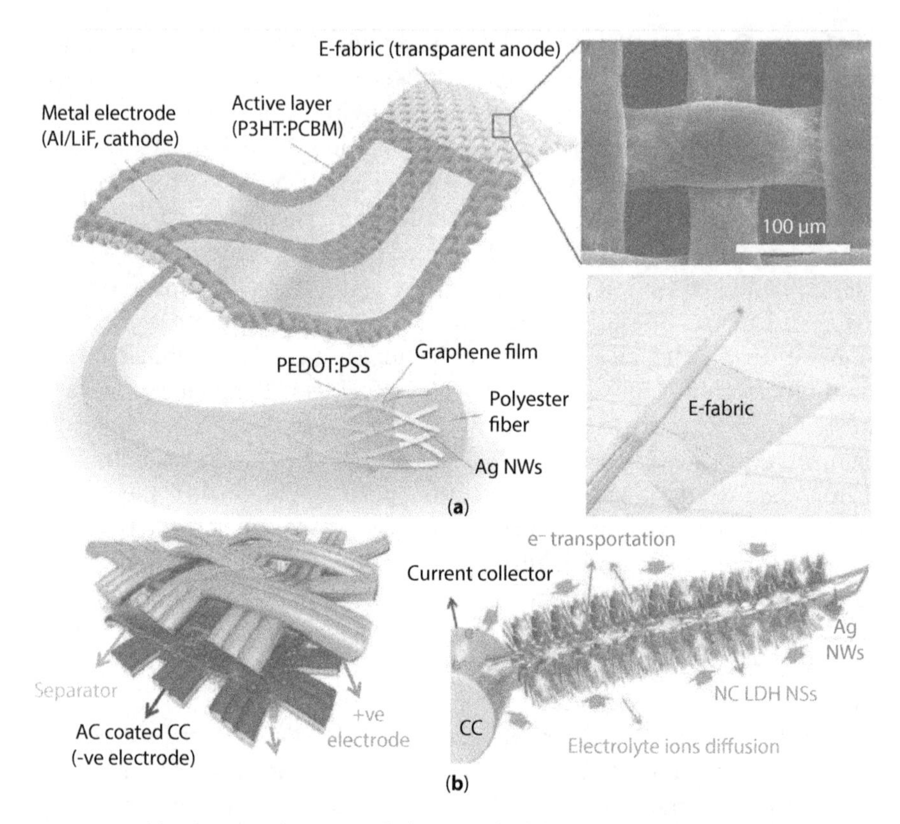

Figure 5.3 (a) Schematic diagram and photograph of the polymer solar textile with a polyester/Ag-NW film/graphene core-shell structure as a transparent anode. Reprinted with permission from [85]; Copyright 2016 Elsevier. (b) Schematic illustration of the fabricated asymmetrical supercapacitors with nickel–cobalt layered double hydroxides NSs@Ag NWs@CC and electrodes. Reprinted with permission from [84]; Copyright 2017 Elsevier.

layered double hydroxide NWs were deposited on Ag NW-fenced carbon cloth (CC) by a facile electrochemical deposition method with a chrono-amperometry voltage of −1.0 V for 120 s as shown in Figure 5.3b [84]. The electrically conductive and superhydrophilic nature of the hybrid nano-composite electrode led to relatively high areal capacitance (1133.3 mF cm^{-2} at 1 mA cm^{-2}) and good cycling stability (80.47% after 2000 cycles) compared to the electrode prepared without Ag NWs.

However, the solvents used in the existing techniques are ethanol and isopropyl alcohol (IPA), which leads to poor interaction and mismatch between the NW and textiles. Fortunately, the stretching mismatch can be effectively avoided through compositing processes instead of coating methods. Ag NWs with high electrical conductivities could be introduced into the polymer matrices to show better performances. For example, a styrene–butadiene–styrene fiber covered by Ag NWs could be readily obtained by a wet-spinning process, followed by adsorption and reduction of the Ag precursor [86]. The resulting composite fiber exhibited a high electrical conductivity of 2450 S cm^{-1} and a breaking strain of 900%. The fibers could be woven into gloves for detecting gestures and human motions.

There are still several problems needed to be solved. Firstly, all fabrics that are tested are natural and hydrophilic fabrics, like cotton. Ag NWs are hydrophilic due to the presence of the PVP polymer on them. Therefore, it is easier to coat natural fabrics with the NWs compared to synthetic fabrics. Coating synthetic fabrics like polyester with dyes or nanomaterials is known to be universal but problematic, because pretreatment is required for the textile before any deposition [87, 88]. Secondly, high concentration of NWs is needed to make the natural fabrics conductive. The final conductive fabric is thus unable to retain its original color.

5.3.1.1.2 Cu NWs

Recently, it was reported that Cu NWs also possessed electrical and chemical stability by coating with Cu–Ni alloying shells to protect Cu NWs from oxidation [28]. Then, they deposited the Cu NW inks on a variety of substrates such as fibers, PET, PDMS, and glass. The facile Cu NW electrodes were used as segments of external circuits for light-emitting diodes (LEDs) [89, 90]. What's more, Cu NWs are known to be lightweight, flexible, and stretchable, and thus could be integrated into commercial cloths and fabrics. For example, a straightforward solvothermal method was used to prepare a flexible conductive material that contains reduced graphene oxide (RGO) NSs bridging oriented Cu NWs [91] as shown in Figure 5.4. The composite film exhibits a high electrical performance (0.808 Ω sq^{-1})

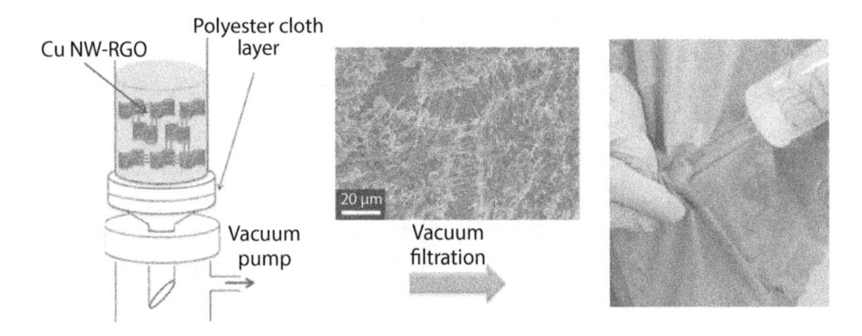

Figure 5.4 Fabrication of Cu NW-RGO composite wearable electrode. Reprinted with permission from [91]; Copyright 2015 American Chemical Society.

without considerable change over 30 days under ambient conditions. Moreover, the Cu NW-RGO composites can be deposited on polyester cloth as a lightweight wearable electrode with high durability and simple processability, which are very promising for a wide variety of electronic devices and more outstanding applications are expected to explore.

5.3.1.1.3 Au NWs

In addition to Ag NWs, Au NWs can also be applied as electrodes for smart textiles. Due to problems of gel drying of NWs that results in strong adhesives and noise signal, it can cause discomfort when worn for a long time. Dry electrodes offer a much more comfortable and durable alternative. An e-bra platform for sensors towards heart rate monitoring is reported and the electrodes in it composited by Au NWs [92]. However, the high cost of Au NWs limited their applications in the field of smart textiles to some extent.

5.3.1.2 Polymer Nanowires for Smart Textiles

Conductive polymers such as PPy, polyaniline (PANI), polythiophene (PTh), and poly-(p-phenylenevinylene) (PPV) are particularly appealing because of their flexibility, easy processability, and potential for low-cost fabrication combined with unique electrical, electronic, and optical properties similar to metals or semiconductors [93–95]. Moreover, it has been shown that one-dimensional materials such as conductive polymer NWs exhibit significantly improved physicochemical characteristics compared to their bulk counterparts. For instance, so many excellent properties, such as increased electrical conductivity, size-dependent excitation or emission, easier band-gap turn-ability, coulomb blockade, and so on, offer quite promising applications in a wide range of technological areas such as sensors [96–98], energy storage [99, 100], photodetectors [101], solar cells [102], electrolyte-gated transistors [103], and conversion devices [104, 105]. Besides, the flexible NW structure is quite suitable to be assembled on common textiles such as cotton, polyester, nylon, polyethylene terephthalate (PET), carbon fibers, and flax, which are usually used for energy storage and solar cells [106].

Among conductive polymer NWs, PANI NWs are used as electrodes because of outstanding energy storage capability and mechanical flexibility. For the first time, PANI NW arrays are successfully deposited onto the surface of single wall carbon NT (SWCNT)/cloth composite through dilute polymerization to obtain the PANI/SWCNT/cloth composite electrode, which is used to assemble the flexible supercapacitors directly [107]. A nonwoven cloth made from wood fiber was chosen as the flexible substrate, and conductive cloth was acquired by dipping the cloth into SWCNT ink (SWCNT/cloth). PANI NW arrays were then deposited on the substrates (pure cloth and SWCNT/cloth) by an in-suit dilute polymerization method in order to improve the capacitance of the electrodes.

The sandwich-structured flexible supercapacitors with the electrodes of SWCNT/cloth, PANI/cloth, and PANI/SWCNT/cloth were produced. High capacitance of 410 F g^{-1} was obtained in PANI/SWCNT/cloth-electrode-based supercapacitors, which was much higher than that of SWCNT/cloth (60 F g^{-1}) and PANI/cloth (290 F g^{-1}) based supercapacitors. The improved capacitance was ascribed to the synergistic effect of PANI NWs and SWCNT films.

Then, hierarchical PANI-coated $NiCo_2S_4$ NWs grown on carbon fibers ($NiCo_2S_4$@PANI/CF) were fabricated through hydrothermal and electrodeposition method [108] as shown in Figure 5.5. The core/shell heterostructure endowed the $NiCo_2S_4$@PANI/CF composite materials with high electron diffusion efficiency and abundant accessible electro-active sites. The PANI shell improved the structural stability of the core $NiCo_2S_4$ NWs. When employed as a free-standing electrode, the $NiCo_2S_4$@PANI/CF exhibited impressive electrochemical performance with a high specific areal capacitance of 4.74 F/cm^2 (1823 F/g) at 2 mA/cm^2 and an excellent cycling stability with capacitance retention of 86.2% after 5000 cycles. Furthermore, an asymmetric supercapacitor device was assembled using $NiCo_2S_4$@PANI/CF as anode and graphene/CF as cathode. The resultant

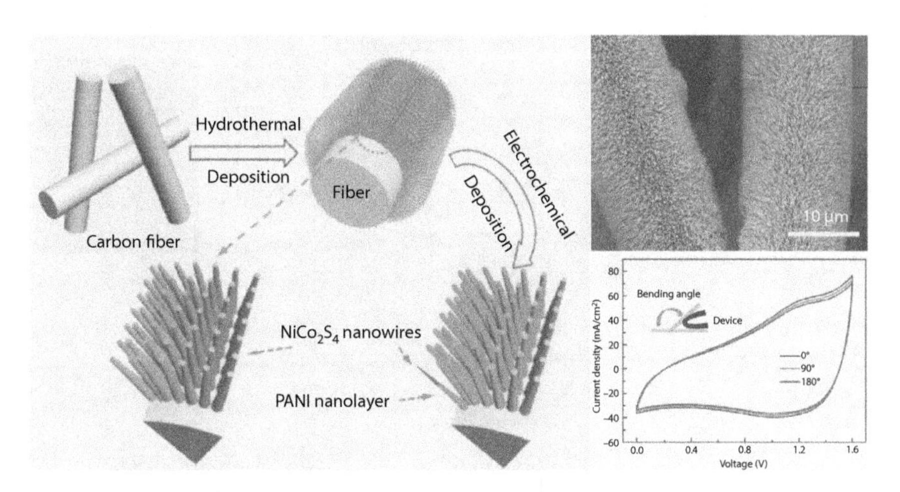

Figure 5.5 Schematic illustration and conceptual characterization of the synthetic procedure for $NiCo_2S_4$@PANI/CF composites. Reprinted with permission from [108]; Copyright 2017 Elsevier.

device delivers a high energy density of 64.92 W/kg at a power density of 276.23 W/kg, as well as considerable flexibility. The core/shell heterostructure design is expected to realize high-performance flexible supercapacitors for smart textiles.

More recently, *in situ* polymerization has been also used to grow PANI NWs on CNT yarns [109, 110]. PANI NW arrays were *in situ* polymerized on the surface of the CNT yarn to improve the energy storage capacity of the device. A layer of PVA gel was coated on the yarn surface to act as gel-like electrodes. Two coated yarns were twisted together to form a two-ply yarn supercapacitor. The as-prepared CNT@PANI yarn supercapacitor was characterized by electrochemical experiments and showed high capacitance and excellent superior cyclic charge–discharge stability. Later, Wang *et al.* [111] report high-performance wearable yarn solid-state supercapacitors that are formed from two composite yarns, each comprising of a platinum lament and a CNT yarn decorated with PANI NWs by *in situ* polymerization. The two-ply composite yarn supercapacitor delivers very high energy density and cycling durability.

Besides PANI NWs, PPy NWs are also deposited on cotton fabrics via *in situ* polymerization of pyrrole in the presence of the filtrate complex of $FeCl_3$ and methyl orange as a reactive self-degraded template (Figure 5.6) [112]. The obtained fabrics could be directly used as supercapacitor electrodes,

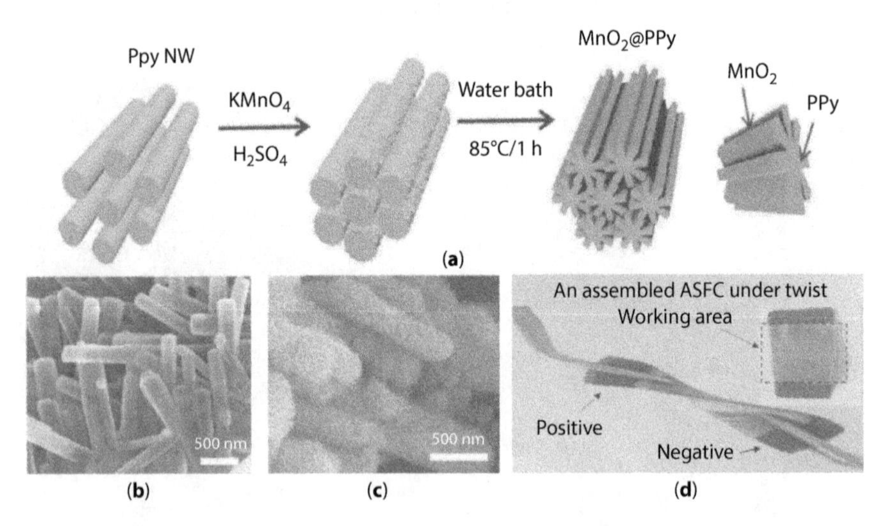

Figure 5.6 (a) Synthesis and characterization of MnO_2 NFs@PPy NWs core/shell nanostructures. (b) SEM images of PPy NWs. (c) SEM images of MnO_2@PPy core/shell nanostructures. (d) Digital images for asymmetrical flexible supercapacitor device. Reprinted with permission from [112]; Copyright 2017 Elsevier.

with a maximum specific capacitance of 325 F g^{-1} and an energy density of 24.7 Wh kg^{-1} at a current density of 0.6 mA cm^{-2}. The capacitance remained higher than 200 F g^{-1} after 500 cycles.

Depositing conductive polymer NWs on textiles not only can increase their electrical conductivity but also can enhance their binding force, resulting in better mechanical flexibility. Some polymer NWs may also provide additional pseudocapacitance to increase energy storage density. However, polymer NWs may limit the access of microspores in textile fibers by electrolyte ions, causing lower electrical double layer capacitor (EDLC).

5.3.2 Semiconducting Nanowires for Smart Textiles

Semiconductor NWs have been studied extensively for over 20 years due to their novel electronic, photonic, thermal, electrochemical, and mechanical properties [113]. They can enable the read of diagnostic assays with consumer electronic devices such as cell phones, smart phones, wearable technology, scanners, optical drives/disc players, and strip readers. Their outstanding features, such as good integration with electronic devices, novel sub-wavelength optical phenomena, large tolerance for mechanical deformations, compatibility with other microscopic and nanoscopic systems in nature, the decoupling of length scales associated with different physical phenomena in radial and axial directions, high surface-to-volume ratio, and so on, have brought a wide range of applications.

In order to achieve the multifunction of smart textiles, various semiconductor NWs have been also assembled on appropriate fabrics by microfabrication and dip-coating techniques. Furthermore, concerning various active materials, current smart textiles can be roughly categorized into three different types such as oxide NWs, sulfide NWs, and other NWs.

5.3.2.1 Oxide Nanowires for Smart Textiles

Among the inorganic semiconductor nanomaterials, metal oxide NWs are the focus of research efforts in nanotechnology since they exhibit excellent chemical and physical properties. They have been widely used in many areas, such as transparent electronics, piezoelectric transducers, ceramics, sensors, and electro-optical and electrochromic devices [114, 115]. Doubtlessly, it is quite significant to assemble metal oxide NWs into cotton/textiles to form wearable devices. Meanwhile, metal oxide NWs such as ZnO, SnO_2, In_2O_3, Cu_2O, Fe_2O_3, and V_2O_5, can be successfully synthesized by facile electrodeposition [12, 116] or *in situ* hydrothermal synthesis [117, 118].

ZnO NWs are perhaps the most widely studied materials among the oxides, and their conductivity can range from insulating to highly conductive with the use of external dopants. Meanwhile, intrinsic defects such as oxygen vacancies and zinc interstitials act as donors, which make the electrical conductivity of ZnO easy to control [119].

More interestingly, the piezoresistive functions of the individual ZnO NWs are critically linked to their geometrical features due to the difference in their deflection behaviors [120–123]. For example, a fiber-shaped piezoelectric nanogenerator was fabricated by radially growing piezoelectric ZnO NWs around Kevlar fibers (Figure 5.7a) [124]. The tips of NWs were separated from each other as a consequence of their small tilting angles, but their bottom ends were tightly connected. The space between the NWs was a magnitude of hundreds of nanometers, which was large enough for them to be bent to generate a piezoelectric potential. Two hybrid fibers were twisted to form the piezoelectric nanogenerator, which generated electricity on pulling/releasing the string. An output power density of 4–16 mW per square meter was expected if it was woven into a fabric. To improve the performance, a hybrid fiber-shaped piezoelectric nanogenerator was fabricated by coating a PVDF layer onto the ZnO NWs [125]. Then, self-powered piezoelectric textiles were realized by using fibers coated with ZnO NWs [126]. They were composed of two kinds of fibers that were interwoven with each other, one fiber grown with ZnO NWs and the other sequentially grown with ZnO NWs and coated with palladium. Schottky junctions formed between the metallic tip and NW allowed for charge separation, and an electric current could be extracted upon bending. The performance of the piezoelectric textile was relatively poor, with an open circuit voltage of 3 mV and a short-circuit current of 17 pA. Later, a new type of fully flexible, foldable nanopatterned wearable triboelectric nanogenerator with high power-generating performance and mechanical robustness was demonstrated for the first time based on ZnO NW (Figure 5.7b) [127]. Although ZnO has a lower piezoelectric constant when compared to other piezoelectric ceramic materials such as $PbTiO_3$ and $PbZrTiO_3$, it is much safer. Furthermore, ZnO can simultaneously possess piezoelectric and semiconducting properties, thereby resulting in a direct current/voltage output such as a ZnO NW-based pH sensor, which connected with the integrated nanogenerator [128]. The voltage across the nanosensor was monitored while changing the degree of acidity versus alkalinity in the target solution. An obvious and stable voltage variation corresponding to a local pH change was observed, indicating the high sensitivity of the "self-powered" pH meter.

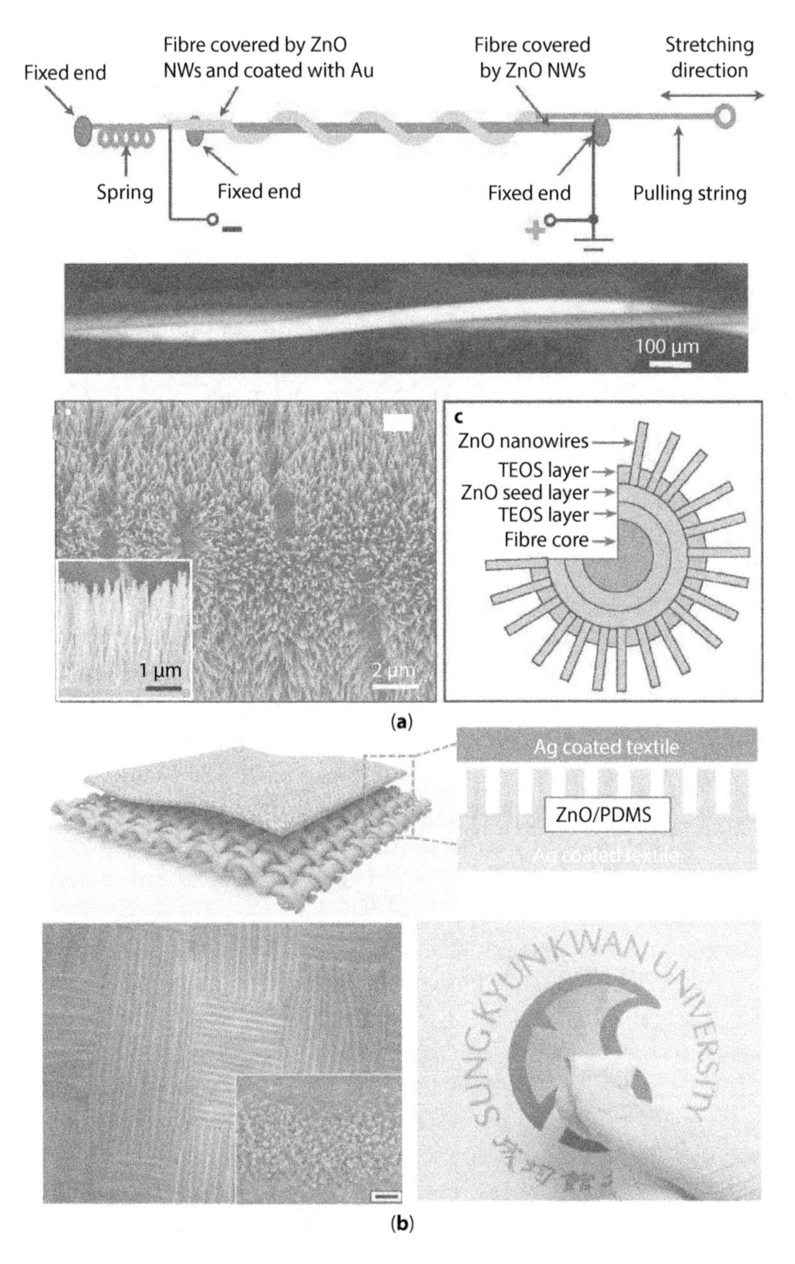

Figure 5.7 (a) Fiber-shaped piezoelectric nanogenerator based on two fibers coated with ZnO NWs. Reprinted with permission from [124]; Copyright 2008 Nature Publishing Group. (b) Schematic illustration of the flexible, foldable wearable triboelectric nanogenerator based on the ZnO NWs. Scale bar: 500 nm. Reprinted with permission from [127]; Copyright 2015 American Chemical Society.

Furthermore, based on the excellent photoelectric properties and extensive research [129], ZnO NWs have shown intriguing applications for various wearable optoelectronic devices. ZnO NW arrays were grown on Kevlar fiber and used as ultraviolet sensors with fast response and good repeatability, even when bended [130]. Lim *et al.* [29] have integrated ZnO NWs into a multifunctional wearable sensor working at room temperature as shown in Figure 5.8a. A zinc oxide seed layer was deposited by sputtering on the fabric, and ZnO NWs were grown by hydrothermal technique at 90°C. Then, Dong *et al.* reported that the ZnO NW arrays were uniformly grown on a Zn wire with high crystallinity, and poly(9-vinylcarbazole (PVK) and PEDOT:PSS layers were tightly contacting ZnO. It exhibits responsivity of 9.96 mA/W under zero bias, which is probably caused by the p–n heterojunction between NW arrays and PVK. However, the bending and poor interface contact of fiber-shaped devices typically lead to ineluctable degradation of performance, which is still a great challenge yet to be solved.

Zhu *et al.* [131] proposed an effective strategy, constructing inorganic–organic–graphene hybrid interfaces on a single fiber, to greatly improve the performances of fiber-shaped device. In the proposed structure, the ZnO NW arrays were grown vertically on the surface of a Zn wire (center core) and then wrapped up by PVK and graphene (outmost layer) as the two outer layers as shown in Figure 5.8b. These "soft" interfaces successfully built compact contacts between various functional layers even on curved interfaces, which markedly reduced the contact resistance. However, fiber-shaped photodetectors (PDs) made from bulky wire-based inorganic semiconductors are prone to damage under excessive bending. Therefore, a directly constructed photodetector textile (PDT) was introduced, and a large-area, organized, and dense weaving of fiber-shaped PDs was realized for the first time. To form the structure, ZnO NW arrays are grown uniformly on the surface of Ni wire textile to fabricate the Ni-based ZnO NW arrays textile, then Ag NWs and graphene film (outmost layer) are successively assembled on the functional textile. The precise energy level alignment of the structure is in favor of the separation and transportation of photon-generated carriers. The photo-responsivity at the bias of 1 V is 0.27 A W^{-1}, and the I_{light}/I_{dark} (I_{light} means light current and I_{dark} means dark current) ratio calculated form the current–voltage (I–V) curves reaches $\approx 10^2$, about two-orders-of-magnitude larger than that of the reported fiber-shaped PDs. Meanwhile, the whole structure also exhibits excellent durability under bending operations, which is ideal for wearable applications.

Recently, both built-in electric field and piezoelectric field are introduced into the photocatalytic process of CuS/ZnO NWs, and high

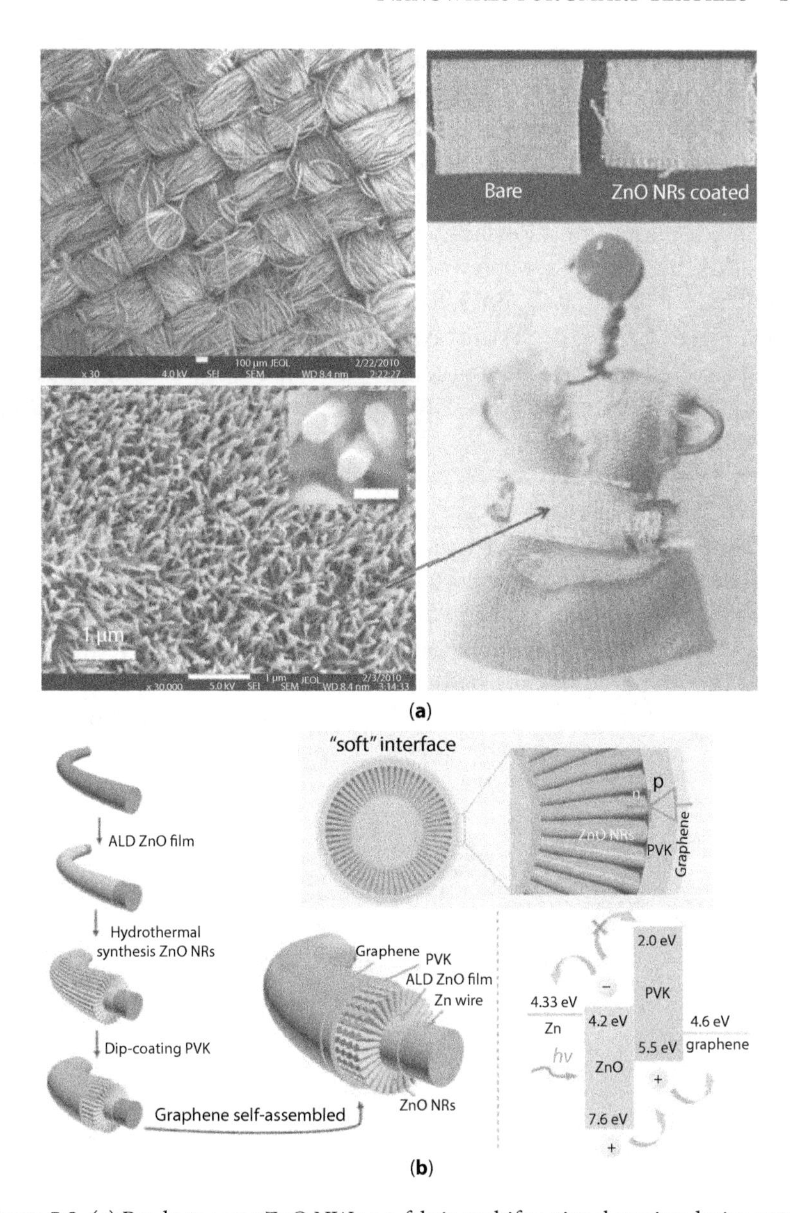

Figure 5.8 (a) Ready-to-wear ZnO NWs-on-fabric multifunctional sensing device sewn on a toy dress. Reprinted with permission from [29]; Copyright 2010 Elsevier.
(b) Structure and morphology characterization of the PD based on ZnO NWs. Reprinted with permission from [131]; Copyright 2017 American Chemical Society.

piezo-photocatalytic efficiency from the use of both solar and mechanical energy has been realized [132]. Besides, Wang's group at Georgia Institute of Technology reported the first fiber supercapacitors with a twisted fiber configuration made from ZnO NW electrodes [133]. The device was fabricated by entangling a flexible poly(methyl methacrylate) (PMMA) plastic wire covered with ZnO NWs around a Kevlar fiber covered with Au-coated ZnO NWs.

Besides ZnO NWs, a variety of active materials, such as $NiCo_2O_4$/PPy coaxial NWs [134], WO_3/SnO_2 core-shell NW arrays [135], Fe_2N NPs [136], aligned $Ca_2Ge_7O_{16}$ NW arrays [137], and $ZnCo_2O_4$ NW arrays [138], were deposited on these functional textiles. The electrochemical performance of these battery textiles could have been greatly increased. A high rate capability of 103 mAh g^{-1} at 90 C (Coulomb) and a good cyclic stability with 5.3% loss in specific capacity after 200 cycles at a rate of 10 C was achieved with the carbon textile as electrodes [139]. Hierarchical three-dimensional $ZnCo_2O_4$ NW arrays were hydrothermally synthesized on a CC as the anode, which was paired with a $LiCoO_2$/Al cathode and sealed in a flexible plastic package as shown in Figure 5.9 [138]. The $ZnCo_2O_4$ NW showed typical diameters in the range 80–100 nm with a length of about 5 mm. The prepared flexible battery presented an output voltage of 3.4 V with a reversible capacity of 1300 mAh g^{-1} for 40 cycles. Even though some

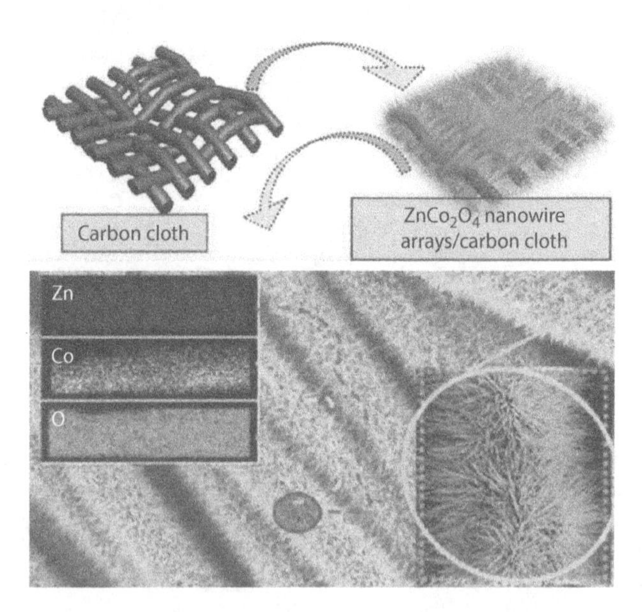

Figure 5.9 Schematic illustration of the synthesis of flexible three-dimensional $ZnCo_2O_4$ NW arrays/CC. Reprinted with permission from [138]; Copyright 2012 American Chemical Society.

interesting results regarding flexible batteries have been obtained, one of the main drawbacks about this technology is the use of a liquid electrolyte.

What's more, CuO NWs as functional materials for smart devices also emerged. Yu and Thomas [140] at the University of Central Florida in the United States designed a coaxial fiber electrode by depositing MnO_2 onto Au/Pd-coated CuO NW arrays on a copper cable. The CuO–AuPd–MnO_2 core-shell NW electrode showed a specific capacitance up to 1376 F g^{-1} (calculated based on the mass of MnO_2) in the KOH aqueous electrolyte in three-electrode tests. Co-axial fiber supercapacitors were assembled by placing an outer tubular electrode on an inner electrode (both precoated with the PVA/KOH electrolyte) with a porous separator. It showed an energy density of 0.55 mW h cm^3 and a power density of 413 mW cm^3. The high performance can be attributed to the unique architecture of the fiber, in which Au/Pd-coated CuO NW arrays with a large area and a high aspect ratio rendered more open space for the accessibility of the electrolyte and enhanced electrochemical-active sites for the reversible redox reactions while sharply reducing the internal resistance between Au, Pd, and MnO_2. Moreover, such coaxial fiber configuration helps to decrease the internal resistance compared to the twisted fiber configuration. More importantly, this work provides a promising approach to transmit electricity and store energy at the same time by integration of an electrical cable and an energy storage device into one unit.

5.3.2.2 Sulfide Nanowires for Smart Textiles

Similar to the oxides mentioned above, semiconducting metal sulfide NWs possess novel optical and electrical properties and are considered as building blocks for photovoltaic devices including dye sensitized cells, all-inorganic NP solar cells, and hybrid nanocrystal-polymer composite solar cells in addition to lasers and waveguides. The types of metal sulfide NWs are quite a lot, but few of them have been combined with textiles for electrochemical devices, including $Ni_{2.3\%}$–CoS_2 [141], $Ni_{3-x}Co_xS_4$ [142], Cu_7S_4 [143], and CoS_2 [144].

In the early stage of sulfide NW smart textiles, Li *et al.* [145] synthesized well-aligned Ni–Co sulfide NW arrays with a Ni–Co molar ratio of 1:1 on three-dimensional nickel foam by a facile two-step hydrothermal method. Owing to the low electronegativity of sulfur, Ni–Co sulfide NW arrays exhibit a more flexible structure and much higher conductivity compared with Ni–Co oxide NW arrays when used as active materials in supercapacitors. More importantly, the asymmetric supercapacitor, composed of Ni–Co sulfide NW arrays as the positive electrode and activated carbon as the negative electrode, reaches up to an energy density of 25 W h kg^{-1} and a power density of 3.57 kW kg^{-1} under a cell voltage of 1.8 V. The superior electrochemistry capacity demonstrates that the self-standing sulfide NW arrays are promising for high-performance supercapacitor applications. Then, $Ni_{3-x}Co_xS_4$ (x = 1.5, 2, 2.25, 2.5) NT arrays were prepared on CC by a two-step hydrothermal route, and their electrochemical performance was studied [142]. The aspect ratios and electrochemical performance of the $Ni_{3-x}Co_xS_4$ NT arrays can be controlled by manipulating the Co/Ni molar ratio. With the cooperation of the appropriate morphology and compositions, the specific capacitance of $Ni_{0.75}Co_{2.25}S_4$ can reach 1856 F g^{-1} at 1 A g^{-1}. The rate capability is 88% at a high current density of 50 A g^{-1}. A few months later, Hao *et al.* [146] found that hollow $NiCo_2S_4$ NT arrays can also be successfully grown on carbon textiles with robust adhesion through a two-step synthesis. Using $NiCo_2S_4$ NT arrays as the positive electrode, a high-performance asymmetric supercapacitor with a maximum voltage of 1.6 V has been fabricated.

Besides, Cu_7S_4 NWs, a kind of Faradic redox active materials, when coated on carbon fiber fabrics, can be directly used as binder-free electrodes for high-performance flexible solid-state supercapacitors [143]. The Cu_7S_4 NW-based supercapacitors exhibit excellent electrochemical performance such as high specific capacitance of 400 F g^{-1} at the scan rate of 10 mV s^{-1} and high energy density of 35 W h kg^{-1} at a power density of 200 W kg^{-1}. What's more, the supercapacitors also show many advantages of light weight, high flexibility, and long-term cycling stability by retaining 95% after 5000 charge–discharge cycles at a constant current of 10 mA. The high Faradic redox activity and high conductivity behavior of the Cu_7S_4 NWs result in a high pseudocapacitive performance with a relatively high specific energy and specific power.

At the same time, Fang *et al.* [141] reported the development of Ni promoted cobalt disulfide NW arrays supported on CC ($Ni_{2.3\%}$–CoS_2/CC) as an efficient bifunctional electrocatalyst for water splitting with superior activity and good durability in basic media, implying the great potential for water splitting (water splitting means the general term for a chemical reaction in which water is separated into oxygen and hydrogen) applications. Then, the design and engineering of low-cost and high-efficiency electrocatalysts for the hydrogen evolution reaction (HER) attracted growing interest in renewable energy research, and MoS_2 NS coated CoS_2 NW arrays supported on CC (MoS_2/CoS_2/CC) were prepared by a two-step procedure that entailed the hydrothermal growth of $Co(OH)_2$ NW arrays on CC followed by reaction with $(NH_4)_2MoS_4$ to grow an over-layer of MoS_2 NWs [144]. Electrochemical studies showed that the obtained three-dimensional electrode exhibited excellent HER activity with an over potential of ~87 mV at 10 mA cm^{-2}, a small Tafel slope of 73.4 mV dec^{-1}, and prominent electrochemical durability. The results above-presented may offer a new methodology for the design and engineering of effective multilevel structured catalysts for HER based on earth-abundant components.

However, rational design of cost-effective and non-precious-metal electrocatalysts for HER still remains a great challenge for future applications in sustainable energy storage and conversion systems [147] as shown in Figure 5.10a. Recently, a simple nitrogen-anion decoration

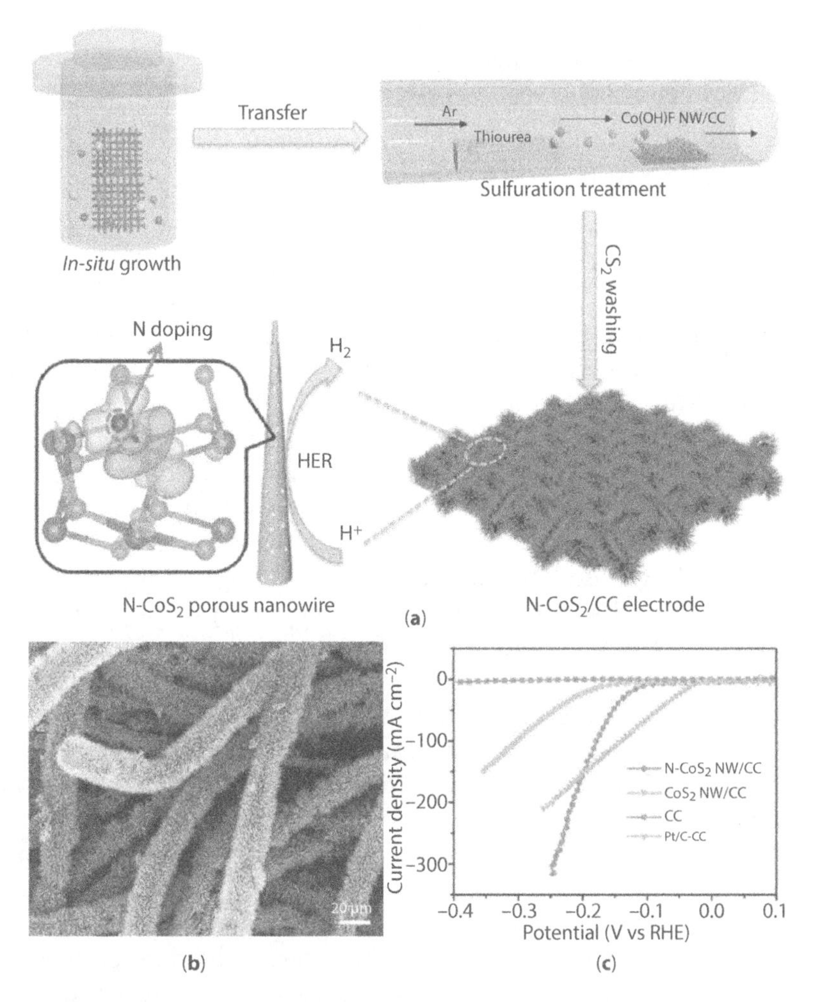

Figure 5.10 (a) Representation of reaction mechanism for N–CoS_2 NW/CC electrode material. (b) SEM images of N–CoS_2 NW/CC. (c) IR-corrected HER polarization curves and corresponding Tafel plots of blank CC, Pt/C–CC, N–CoS_2 NW/CC, and CoS_2 NW/CC electrodes. Reprinted with permission from [147]; Copyright 2017 American Chemical Society.

strategy was developed to realize the synergistic regulation of the catalytically active sites, electronic structure, and reaction dynamics in metallic CoS_2 porous NW arrays (Figure 5.10b). Specifically, the introduction of nitrogen-anion in the CoS_2 system not only modified the morphology, offering additional active sites, but also enhanced the electrical conductivity to promote rapid charge transfer for the HER process. The N–CoS_2 NW/CC electrode showed significantly enhanced HER performance with a lower overpotential and a larger exchange current density than the pristine one (Figure 5.10c). However, the application of sulfide NWs in smart textiles is still limited, and greater efforts are needed to explore for industrialization.

5.3.2.3 *Other Nanowires for Smart Textiles*

In addition to oxides and sulfides, there are still other semiconductor NWs, such as ZnSe [148], CdSe [149], CdTe [150], InP [151], GaAs [151], PbS, PbSe, $PbSe_xS_{1-x}$ [152], Si [153], and potassium vanadate (KVO) [154]. However, only several kinds of NWs (CdSe, CdTe, Si, KVO NWs) have been assembled on textiles or fibers, and few works have been done about NW-based smart textiles or fibers.

First, Si NWs are ideal materials for fundamental research due to their rich history of use in the microelectronics industry. In 2007, the Si NWs were firstly grown on CC via the vapor–liquid–solid reaction using silane gas as the Si source and Au as the catalyst for the decomposition of hydrogen Au tetrachloride [155]. A low operating electric field has been achieved with the emission current density of 1 mA/cm^2 at an operating electric field of 0.7 V/μm. The small operating electric field is resulted from a high field enhancement factor of 6.1×10^4 due to the combined effects of the high intrinsic aspect ratio of Si NWs and the woven geometry of CC. Such results may lead Si NWs field emitters to practical applications in vacuum microelectronic devices including microwave devices.

In addition to the excellent field emission properties, Si with extremely high theoretical charge capacity (4200 mAh g^{-1}) is considered as a promising candidate for substitution of the conventional graphite using as Li-ion anode materials [156]. For example, the complete lithium-ion batteries based on Si NWs and commercial $LiCoO_2$ materials were reported with a novel scaffold of hierarchical Si NW–carbon textile anodes (Figure 5.11a) [157]. The batteries showed enhanced specific capacity (2950 mAh g^{-1} at 0.2 C), good repeatability/rate capability (even 900 mAh g^{-1} at a high rate

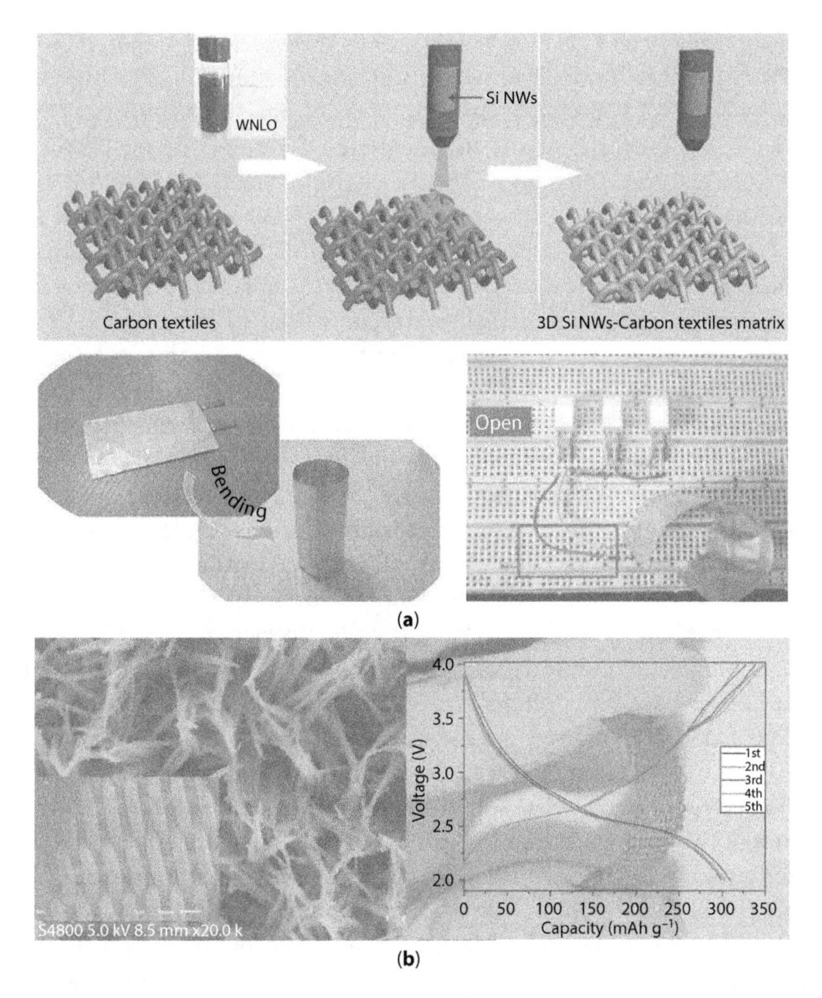

Figure 5.11 (a) The hierarchical Si NW–carbon textile matrix for high-performance advanced lithium-ion batteries. Reprinted with permission from [157]; Copyright 2013 Nature Publishing Group. (b) Flexible potassium vanadate NWs on Ti fabric as a binder-free cathode for high-performance advanced lithium-ion battery. Reprinted with permission from [154]; Copyright 2016 Elsevier.

of 5 C), long cycling life, and excellent stability in various external conditions (curvature, temperature, and humidity). Then, to achieve higher areal capacity, three-dimensional hierarchical SnO_2@Si NW arrays on CC are synthesized by a combination of the chemical vapor deposition (CVD) method for SnO_2 NWs and a subsequent Si thin film coating with a plasma-enhanced CVD (PE-CVD) route [158]. The as-prepared SnO_2@Si NW arrays are binder-free anode for Li ion batteries with a high areal capacity (e.g., 2.13 mAh cm^{-2} at a current density of 0.38 mA cm^{-2}), and good cycling performance is demonstrated for the core-shell SnO_2@Si NW/CC electrode. The high performance could be attributed to the three-dimensional hierarchical structures, in which NWs provide efficient electron transport and large interfacial area, improving the kinetic properties of the reaction. Above works light the way to replacing graphite anodes with Si-based electrodes, which were confirmed to have better performances, and more excellent works about Si NW-based smart textiles are expected in the future.

Second, the solution-synthesized CdSe and CdTe NWs have been also coated onto individual cotton fibers through a general approach [159]. To improve the NW-to-cotton affinity, cotton samples were chemically modified through cationic functionalization. Briefly, under alkaline conditions, ammonium epoxides react with the OH groups of cellulose to synthetize cotton with positively charged surfaces. Further, the cotton fibers with CdSe and CdTe NWs are optically active and exhibit size-dependent photoconductivity. The photoconductivity can be exploited to make cotton-based photodetectors and solar cells. Then, Zhang *et al.* [58] applied CdSe NW-grafted primary electrodes in combination with carbon NT yarns as counter electrodes for making fiber and fabric-shaped photoelectrochemical cells with different PCE from 1% to 2.9%. A unique feature of this process is that instead of making individual fiber cells, they directly weave single or multiple NT yarns with primary electrodes into a functional fabric. Their results demonstrate potential applications of semiconducting NWs and carbon NTs in woven photovoltaics. Besides, a facile and scalable strategy, based on a hydrothermal route, has been applied to fabricate KVO NWs on Ti fabrics [154]. The morphology, crystal structure, and chemical composition of the prepared sample are tested and presented in detail as shown in Figure 5.11b. When tested as cathode materials for lithium-ion batteries, the flexible KVO electrode has a high reversible capacity of 270 mAh g^{-1} at a current density of 100 mA g^{-1} after 300 cycles.

5.4 Perspectives on Future Research

Wearable electronic products including smart clothes, interfacing computers, and stretchable circuits are playing great roles in modern society and might shape fascinating lifestyles of human beings in the future. Now emerging smart textiles based on different functionalities, namely, electricity generation, electricity storage, electricity utilization, and their integration, have attracted a wide range of interest. These devices are generally composed of active materials sandwiched between two electrodes, and some of them need electrolytes. The main focus in recent years has been to optimize the materials and structures, because an upgrade in performance is accompanied by innovation in materials. NW materials, benefiting from their geometry configurations, stand out as promising candidates. Metals, metal oxides, conducting polymers, sulfide, and other semiconductor materials can be prepared as NWs, which have been carefully discussed in this chapter for their potential smart textile applications. Inheriting its intrinsic properties from these materials, the NW configuration introduces several advantages, such as efficient charge transport and collection, which can contribute to a better photovoltaic performance. The specific surface area makes NWs useful for constructing hierarchical structures for controlling and fine-tuning the performance of textiles.

NW materials have been widely utilized in applications in smart textiles, and typical devices are summarized in Table 5.1. Looking towards the future, there are still substantial opportunities in many areas, such as (1) synthesizing new NW materials or modifying existing species to further improve the electronic properties for better performance of smart devices; (2) developing and enriching the synthetic method of NW materials, and clarifying the synthesis parameters to accurately control the morphology and properties of the resulting materials, in order to efficiently and reliably fabricate high-performance smart devices; and (3) simplifying the fabrication process and reducing the fabrication cost. Current methods for the synthesis of NW materials are generally based on chemical vapor deposition, solution processing, a hydrothermal method, or anodic oxidation, which requires rigorous conditions or specific equipment. Considering the practical applications of smart devices, it is critically important to simplify the fabrication process and to reduce the fabrication cost of NW materials, which calls for deep investigations and exploration in the future. NW materials have been identified as promising candidates for fiber electrodes, which play a key role in achieving both high performance and stability, due to their combined high flexibility and electronic property based on the one-dimensional nanostructures. Increasing effort needs to be directed toward pushing smart textiles forward.

Table 5.1 The NW-based devices for smart textiles. Note: the table has been arranged by the author.

Type	Device Structure Based on Nanowires (NW)	Preparation Method	Electrical Property	Reference	Year
Solar cells	Ti thread/TiO_2 NWs/Pt counter electrode/transparent capillary tube	Alkali hydrothermal synthesis	PCE: 5.38%	[160]	2013
	Ti wire/TiO_2 NTs/P_3HT/$PC_{70}BM$/PEDOT:PSS/CNT fiber	Electrochemical anodic oxidation method	PCE: 0.15%	[161]	2012
	Carbon fiber/TiO_2 NWs/TiO_2 NP	Solvothermal process	PCE: 2.48%	[162]	2014
	Ti wire/smooth TiO_2 NT/hierarchical TiO_2 NT/electrolyte/capillary	Two-step anodization process	PCE: 8.6%	[163]	2014
	Ti wire/TiO_2 NT/dye molecule/electrolyte/CNT fiber	Electrochemical anodic oxidation method	PCE: 4.6%	[164]	2012
	Polyester/Ag NW/graphene/PEDOT:PSS/P_3HT/PCBM/Al/LiF	Spin-coating	PCE: 2.27%	[85]	2016

(*Continued*)

Table 5.1 The NW-based devices for smart textiles. Note: the table has been arranged by the author. (*Continued*)

Type	Device Structure Based on Nanowires (NW)	Preparation Method	Electrical Property	Reference	Year
Solar cells	Fe wire/ZnO NW arrays/ Pt wire	*In situ* hydrothermal synthesis	PCE: 0.96%	[165]	2012
	Si NW/Si shells	Chemical vapor deposition	PCE: 3.4 ± 0.2%	[166]	2007
	Ti wire/CdSe NWs	Chemical vapor deposition	PCE: 2.9%	[58]	2012
Supercapacitors	CC/Ag NW/nickel-cobalt/layered double hydroxides	Dip-coating	Chronoamperometry voltage of 1.0 V for 120 s. Energy density: 78.8 μWh cm^{-2} Specific capacitance: 1133.3 mF cm^{-2}	[84]	2017
	Au film/PMMA wire/ZnO NW arrays/Au film/ZnO NWs/Kevlar fiber	*In situ* hydrothermal synthesis	Specific capacitance: 2.4 mF cm^{-2}	[133]	2011
	Carbon fiber fabric/Cu_7S_4 NWs	*In situ* hydrothermal synthesis	Specific capacitance: 400 F g^{-1} Power density: 200 W kg^{-1} Energy density: 35 Wh kg^{-1}	[143]	2015

(*Continued*)

Table 5.1 The NW-based devices for smart textiles. Note: the table has been arranged by the author. (*Continued*)

Type	Device Structure Based on Nanowires (NW)	Preparation Method	Electrical Property	Reference	Year
Supercapacitors	CC/$Ni_{0.75}Co_{2.25}S_4$	*In situ* hydrothermal synthesis	Specific capacitance: 1856 F g^{-1} at 1 A g^{-1} Current density: 50 A $^{g-1}$	[142]	2015
	Carbon textiles/$NiCo_2O_4$@ PPy hybrid NW arrays	*In situ* hydrothermal synthesis	Specific capacitance: ~2244 F g^{-1}	[134]	2015
	CuO@AuPd@MnO_2 core-shell NWs	Heat treatment	Specific capacitance: 1376 F g^{-1} Power density: 0.55 mWh cm^{-3} Energy density: 413 mW cm^{-3}	[140]	2014
	Carbon fiber/ZnO NWs/ SnO_2 NWs/carbon papers	*In situ* hydrothermal synthesis	Specific capacitance: 12 mF cm^{-2} Energy density: 405 nWh cm^{-2}	[167]	2015
	Carbon textiles/$NiCo_2S_4$ NT arrays	*In situ* hydrothermal synthesis	Specific capacitance: 476 F g^{-1} at 20 A g^{-1} Pseudocapacitance: 1279 F g^{-1} at 20 A g^{-1}	[146]	2016
	Ni–Co sulfide NW arrays/ Ni foam	Two-step hydrothermal method	Energy density: 25 Wh kg^{-1} specific capacitance: 2415 F g^{-1}	[145]	2014

(*Continued*)

Table 5.1 The NW-based devices for smart textiles. Note: the table has been arranged by the author. (*Continued*)

Type	Device Structure Based on Nanowires (NW)	Preparation Method	Electrical Property	Reference	Year
Supercapacitors	Activated carbon fiber cloth/MnO_2 NFKs@PPy NWs	*In situ* hydrothermal synthesis	Specific capacitance: 276 F g^{-1} at 2 A g^{-1} Power density: 9000 W kg^{-1}	[112]	2017
	Cotton fabrics/PPy NWs	*In situ* polymerization	Specific capacitance: 325 F g^{-1} Energy density: 24.7 Wh kg^{-1} at 0.6 mA cm^{-2}	[168]	2015
	CNTZ@PANI NWs/PVA Yarn	Chemical vapor deposition	Capacitance: 38 mF cm^{-2}	[109]	2013
	Carbon fiber/$NiCo_2S_4$@ PANI core/shell NWs	*In situ* hydrothermal synthesis	Specific areal capacitance: 1823 F g^{-1} at 2 mA cm^{-2}	[108]	2017
	Cloth/PANI NW arrays/ SWCNT	Dilute polymerization	Specific capacitance: 410 F g^{-1}	[107]	2011
	Carbon NT composite yarn/PANI NWs/Pt wire	*In situ* polymerized	Specific capacitance: 91.67 mF cm^{-2} Energy density: 12.68 μWh cm^{-2}	[111]	2016

(*Continued*)

Table 5.1 The NW-based devices for smart textiles. Note: the table has been arranged by the author. (*Continued*)

Type	Device Structure Based on Nanowires (NW)	Preparation Method	Electrical Property	Reference	Year
Photodetectors	Zn wire/ZnO NWs/PVK/PEDOT/CNT/PDMS	*In situ* hydrothermal synthesis	I_{light}/I_{dark}: 2.18 Responsivity (A/W): 0.03 Rotation angle deg (°): 0–360	[169]	2016
	Ni wire/P-NiO/n-ZnO NWs/Au wire	*In situ* hydrothermal synthesis	I_{light}/I_{dark}: 4.9 Responsivity (A/W): 7.39 Response time (s): 10/18.1	[170]	2015
	Pt/ZnO NW arrays/Pt wire	*In situ* hydrothermal synthesis	I_{light}/I_{dark}: 18.5 Response time (s): 7.5/8.6 Rotation angle deg (°): 0–270	[171]	2014
	Zn wire/ZnO NW arrays/PVK/graphene	*In situ* hydrothermal synthesis	I_{light}/I_{dark}: 7.2 Response time (s): 0.28/2.2	[131]	2017
	Zn wire/ZnO NW arrays/PVK/PEDOT:PSS/Ag wire	*In situ* hydrothermal synthesis	I_{light}/I_{dark}: 1.5 Response time (s): 6/7	[131]	2017
	Carbon fiber/ITO/ZnO NW arrays/CdS NWs/ITO	*In situ* hydrothermal synthesis	I_{light}/I_{dark}: 1.3 Responsivity (A/W): 1.94 × 10^5 with light of 372 nm Response time (s): 10/	[172]	2013
	Cotton fibers/ZnO NW arrays	*In situ* hydrothermal synthesis	-	[173]	2013
	CdSe and CdTe NWs/cotton textiles/	Dip-coating	-	[159]	2014

(*Continued*)

Table 5.1 The NW-based devices for smart textiles. Note: the table has been arranged by the author. (*Continued*)

Type	Device Structure Based on Nanowires (NW)	Preparation Method	Electrical Property	Reference	Year
Electrical pH-responsive devices	Ag NWs/PANI	Vertical spinning	Switching ratio: 9.1×10^8	[174]	2015
Lithium batteries	Carbon fiber/WO_3@SnO_2 NW/	*In situ* hydrothermal synthesis	Reversible capacity (mAh g^{-1}): 1000	[135]	2014
	Carbon textile/$Ca_2Ge_7O_{16}$ NW arrays	*In situ* hydrothermal synthesis	Reversible capacity (mAh g^{-1}): 900–1100	[137]	2013
	$ZnCo_2O_4$ NW arrays/CC	*In situ* hydrothermal synthesis	Reversible capacity (mAh g^{-1}): 1300–1400	[138]	2012
	Carbon textiles/$NiCo_2O_4$ NW arrays	*In situ* hydrothermal synthesis	Reversible capacity (mAh g^{-1}): 1012	[175]	2014
	Ti fabric/potassium vanadate (KVO) NWs	*In situ* hydrothermal synthesis	Reversible capacity (mAh g^{-1}): 270	[154]	2017
	SnO_2@Si core-shell NW arrays	Chemical vapor deposition	Areal capacity: 2.13 mAh cm^{-2} at 0.38 mA cm^{-2}	[158]	2013
	Carbon textiles/Si NWs	Chemical vapor deposition	Specific capacity 2950 mAh g^{-1} at 0.2 C	[157]	2013

(*Continued*)

Table 5.1 The NW-based devices for smart textiles. Note: the table has been arranged by the author. (*Continued*)

Type	Device Structure Based on Nanowires (NW)	Preparation Method	Electrical Property	Reference	Year
Strain sensors	Nylon fibers/G-PEDOT:PSS/Ag NW/G-PEDOT:PSS	Dip-coating	Gauge factor: ~4 Strain range: 5%	[176]	2016
	Conductive fabric/Au NWs	Dip-coating	-	[92]	2011
	PET textiles/ZnO NWs/PDMS	*In situ* hydrothermal synthesis	Gauge factor: 0.96	[177]	2016
	Cotton fabric/ZnO NW arrays	*In situ* hydrothermal synthesis	Multifunctional sensors.	[29]	2010
Heatable textiles	Cotton fabrics/Ag NWs	Dip-coating	Heating range: 50°C	[59]	2016
Nanogenerators	Polyester/Ag NW/graphene	Blade coating	Effective output power (nW/cm^2): 7	[24]	2016
	Carbon fabric/aramid tow/ZnO NW arrays/carbon fabric	*In situ* hydrothermal synthesis	Voltage generation: 35 mV	[178]	2017
	Core fiber/Inner electrode/ZnO NW arrays/PVDF/outer electrode	*In situ* hydrothermal synthesis	Output voltage (V): 0.1 Current density (nA/cm^2): 10 Power density ($\mu W/cm^3$): 16	[125]	2012

(*Continued*)

Table 5.1 The NW-based devices for smart textiles. Note: the table has been arranged by the author. (*Continued*)

Type	Device Structure Based on Nanowires (NW)	Preparation Method	Electrical Property	Reference	Year
Nanogenerators	Kevlar 129 fibers/two layers of tetraethoxysilane/ZnO seed/two layers of tetraethoxysilane/ZnO NWs	*In situ* hydrothermal synthesis	Output power density (mW/cm^2): 20–80	[124]	2008
	Cr/ZnO NWs/Au	*In situ* hydrothermal synthesis	Output voltage (V): 1.26	[128]	2010
	Ag-coated woven textiles/ZnO NWs/PDMS	*In situ* hydrothermal synthesis	Output voltage (V): 120 Output current (μA): 65	[127]	2015
	Conductive textile/multi-walled CNT film/ZnO NWs	*In situ* hydrothermal synthesis	-	[179]	2014
Water sterilization	Cotton/Ag NW/CNT	Dip-coating	Efficiency: 98% at 100000 L/($h\ m^2$)	[69]	2010
Functionalized fabrics	Cotton microfibers/Ag NW	Dip and dry	Square resistance: 27.4 Ω/sq Antimicrobial capacity: 100% Static water contact angle and shedding angle: 156.2° ±3.2° and 7°.	[76]	2014

(*Continued*)

Table 5.1 The NW-based devices for smart textiles. Note: the table has been arranged by the author. (*Continued*)

Type	Device Structure Based on Nanowires (NW)	Preparation Method	Electrical Property	Reference	Year
Conductive fibers	PU/Ag NW	Dip-coating	The rate of conductivity degradation: $(R/Ro) = 5.8$ at $\varepsilon = 133\%$.	[72]	2017
	Styrene–butadiene–styrene/ Ag NW-Ag NPs	Wet spinning	Electrical conductivity: σ 0 = 2450 S cm^{-1} The rate of conductivity degradation, $\sigma/\sigma_0 = 4.4\%$ at 100% strain	[86]	2015
	Cotton/Ag NW	Dip-coating	Resistance: 0.8 Ω cm^{-1}	[71]	2015
	Friselina/Ag NW	Dip-coating	Square resistance: 30 Ω/sq	[180]	2015
	Fabric/Ag NW	Dip-coating	-	[70]	2011
	Cellulosic product/Ag NW	Dipping–drying	Resistance: 0.0047–0.0091 Ω in the strain range of 0–190%.	[181]	2015
	Polyester cloth/Cu NW–graphene	Dip-coating	Square resistance: 0.808 Ω/sq	[91]	2016

(*Continued*)

Table 5.1 The NW-based devices for smart textiles. Note: the table has been arranged by the author. (*Continued*)

Type	Device Structure Based on Nanowires (NW)	Preparation Method	Electrical Property	Reference	Year
Photocatalysis	Stainless-steel mesh/CuS/ZnO NWs	Two-step wet-chemical method	-	[132]	2016
Electrocatalysis	CC/$Ni_{2.3\%}$–CoS_2 NW arrays	*In situ* hydrothermal synthesis	-	[141]	2016
	MoS_2/CoS_2 NW arrays/CC	*In situ* hydrothermal synthesis	Tafel slope: 73.4 mV dec^{-1}	[144]	2015
	CoS_2 NW arrays/CC	*In situ* hydrothermal synthesis	Tafel slope: 118 mV dec^{-1}	[147]	2017
Personal thermal management	Cotton cloth/Ag NW	Dip-coating	-	[77]	2015
Solid phase microextraction	Fiber/PPy NWs	Electrochemical polymerization	-	[182]	2016
Field emission	CC/Si NWs	Chemical vapor deposition	Field enhancement factor: 6.1×10^4	[155]	2006

Reference

1. Jost, K., Dion, G. and Gogotsi, Y., Textile energy storage in perspective. *J. Mater. Chem. A*, 2, 10776–10787, 2014.
2. Weng, W., Chen, P., He, S., Sun, X. and Peng, H., Smart electronic textiles. *Angew. Chem. Int. Edit.*, 55, 6140–6169, 2016.
3. Stuart, M. A. C., Huck, W. T. S., Genzer, J., Muller, M., Ober, C., Stamm, M., Sukhorukov, G. B., Szleifer, I., Tsukruk, V. V., Urban, M., Winnik, F., Zauscher, S., Luzinov, I. and Minko, S., Emerging applications of stimuli-responsive polymer materials. *Nat. Mater.*, 9, 101–113, 2010.
4. Coyle, S., Wu, Y., Lau, K.-T., De Rossi, D., Wallace, G. and Diamond, D., Smart nanotextiles: A review of materials and applications. *MRS Bull.*, 32, 434–442, 2011.
5. Stoppa, M. and Chiolerio, A., Wearable electronics and smart textiles: A critical review. *Sensors*, 14, 11957, 2014.
6. Lee, J. A., Aliev, A. E., Bykova, J. S., de Andrade, M. J., Kim, D., Sim, H. J., Lepró, X., Zakhidov, A. A., Lee, J.-B., Spinks, G. M., Roth, S., Kim, S. J. and Baughman, R. H., Woven-yarn thermoelectric textiles. *Adv. Mater.*, 28, 5038–5044, 2016.
7. Soin, N., Shah, T. H., Anand, S. C., Geng, J., Pornwannachai, W., Mandal, P., Reid, D., Sharma, S., Hadimani, R. L., Bayramol, D. V. and Siores, E., Novel "3-D spacer" all fibre piezoelectric textiles for energy harvesting applications. *Energ. Environ. Sci.*, 7, 1670–1679, 2014.
8. Xue, J., Xu, L., Song, J. and Zeng, H., Wearable and visual strain sensors based on Zn_2GeO_4@ polypyrrole core@shell nanowire aerogels. *J. Mater. Chem. C*, 5, 11018, 2017.
9. Chen, T., Qiu, L., Yang, Z., Cai, Z., Ren, J., Li, H., Lin, H., Sun, X. and Peng, H., An integrated "energy wire" for both photoelectric conversion and energy storage. *Angew. Chem. Int. Edit.*, 51, 11977–11980, 2012.
10. Sun, H., You, X., Deng, J., Chen, X., Yang, Z., Chen, P., Fang, X. and Peng, H., A twisted wire-shaped dual-function energy device for photoelectric conversion and electrochemical storage. *Angew. Chem. Int. Edit.*, 53, 6664–6668, 2014.
11. Song, J., Li, J., Li, X., Xu, L., Dong, Y. and Zeng, H., Quantum dot light-emitting diodes based on inorganic perovskite cesium lead halides ($CsPbX_3$). *Adv. Mater.*, 27, 7162, 2015.
12. Hu, L., Pasta, M., La Mantia, F., Cui, L., Jeong, S., Deshazer, H. D., Choi, J. W., Han, S. M. and Cui, Y., Stretchable, porous, and conductive energy textiles. *Nano Lett.*, 10, 708–714, 2010.
13. Jost, K., Perez, C. R., McDonough, J. K., Presser, V., Heon, M., Dion, G. and Gogotsi, Y., Carbon coated textiles for flexible energy storage. *Energ. Environ. Sci.*, 4, 5060–5067, 2011.
14. Đorđević, D., Javoršek, A. and Hladnik, A., Comparison of chromatic adaptation transforms used in textile printing sample preparation. *Coloration Technol.*, 126, 275–281, 2010.

15. Kuo, W.-G., Luo, M. R. and Bez, H. E., Various chromatic-adaptation transformations tested using new colour appearance data in textiles. *Color Res. & Appl.*, 20, 313–327, 1995.
16. Tang, S. L. P., Recent developments in flexible wearable electronics for monitoring applications. *T. Int. Meas. Control*, 29, 283–300, 2007.
17. Lieva Van, L. and Carla, H., Smart clothing: A new life. *Int. J. Cloth. Sci. Tech.*, 16, 63–72, 2004.
18. Paradiso, R., Loriga, G., Taccini, N., Gemignani, A. and Ghelarducci, B., WEALTHY—A wearable healthcare system: New frontier on e-textile. *J. Tel. Inf. Tech.*, 105–113, 2005.
19. Cheng, M.-H., Chen, L.-C., Hung, Y.-C. and Yang, C. M., *A Real-Time Maximum-Likelihood Heart-Rate Estimator for Wearable Textile Sensors*, Engineering in Medicine and Biology Society, 2008. EMBS 2008. 30th Annual International Conference of the IEEE, IEEE, pp. 254–257, 2008.
20. Mattmann, C., Amft, O., Harms, H., Troster, G. and Clemens, F., Recognizing upper body postures using textile strain sensors. In *Proceedings of the 2007 11th IEEE International Symposium on Wearable Computers*, IEEE Computer Society, pp. 1–8, 2007.
21. Axisa, F., Dittmar, A. and Delhomme, G., *Smart Clothes for the Monitoring in Real Time and Conditions of Physiological, Emotional and Sensorial Reactions of Human*, Engineering in Medicine and Biology Society, 2003. Proceedings of the 25th Annual International Conference of the IEEE, IEEE, pp. 3744–3747, 2003.
22. Jung, S., Lauterbach, C., Strasser, M. and Weber, W., *Enabling Technologies for Disappearing Electronics in Smart Textiles*, Solid-State Circuits Conference, 2003. Digest of Technical Papers. ISSCC. 2003 IEEE International, IEEE, pp. 386–387, 2003.
23. Zhong, J., Zhang, Y., Zhong, Q., Hu, Q., Hu, B., Wang, Z. L. and Zhou, J., Fiber-based generator for wearable electronics and mobile medication. *ACS Nano*, 8, 6273–6280, 2014.
24. Wu, C., Kim, T. W., Li, F. and Guo, T., Wearable electricity generators fabricated utilizing transparent electronic textiles based on polyester/Ag nanowires/graphene core-shell nanocomposites. *ACS Nano*, 10, 6449–6457, 2016.
25. Wagner, S., Bonderover, E., Jordan, W. and Sturm, J., Electrotextiles: Concepts and challenges. *Int. J. Hi. Spe. Ele. Syst.*, 12, 391–399, 2002.
26. Hu, J., Meng, H., Li, G. and Ibekwe, S. I., A review of stimuli-responsive polymers for smart textile applications. *Smart Mater. Struct.*, 21, 053001, 2012.
27. Castano, L. M. and Flatau, A. B., Smart fabric sensors and e-textile technologies: A review. *Smart Mater. Struct.*, 23, 053001, 2014.
28. Song, J., Li, J., Xu, J. and Zeng, H., Superstable transparent conductive Cu@Cu_4Ni nanowire elastomer composites against oxidation, bending, stretching, and twisting for flexible and stretchable optoelectronics. *Nano Lett.*, 14, 6298–6305, 2014.

29. Lim, Z., Chia, Z., Kevin, M., Wong, A. and Ho, G., A facile approach towards ZnO nanorods conductive textile for room temperature multifunctional sensors. *Sensors Actuat. B: Chem.*, 151, 121–126, 2010.
30. Shakir, I., Ali, Z., Bae, J., Park, J. and Kang, D. J., Conformal coating of ultrathin $Ni(OH)_2$ on ZnO nanowires grown on textile fiber for efficient flexible energy storage devices. *RSC Adv.*, 4, 6324–6329, 2014.
31. Periasamy, C. and Chakrabarti, P., Time-dependent degradation of Pt/ZnO nanoneedle rectifying contact based piezoelectric nanogenerator. *J. Appl. Phys.*, 109, 054306, 2011.
32. Bao, L. and Li, X., Towards textile energy storage from cotton T-shirts. *Adv. Mater.*, 24, 3246–3252, 2012.
33. Huang, S., Wu, H., Zhou, M., Zhao, C., Yu, Z., Ruan, Z. and Pan, W., A flexible and transparent ceramic nanobelt network for soft electronics. *NPG Asia Mater.*, 6, e86, 2014.
34. Jiang, Y., Ling, X., Jiao, Z., Li, L., Ma, Q., Wu, M., Chu, Y. and Zhao, B., Flexible of multiwalled carbon nanotubes/manganese dioxide nanoflake textiles for high-performance electrochemical capacitors. *Electrochim. Acta*, 153, 246–253, 2015.
35. Kar, A., Smith, Y. R. and Subramanian, V., Improved photocatalytic degradation of textile dye using titanium dioxide nanotubes formed over titanium wires. *Environ. Sci. Tech.*, 43, 3260–3265, 2009.
36. Wang, Z. L., Characterizing the structure and properties of individual wire-like nanoentities. *Adv. Mater.*, 12, 1295–1298, 2000.
37. Xia, Y., Yang, P., Sun, Y., Wu, Y., Mayers, B., Gates, B., Yin, Y., Kim, F. and Yan, H., One-dimensional nanostructures: Synthesis, characterization, and applications. *Adv. Mater.*, 15, 353–389, 2003.
38. Langley, D., Giusti, G., Mayousse, C., Celle, C., Bellet, D. and Simonato, J.-P., Flexible transparent conductive materials based on silver nanowire networks: A review. *Nanotechnology*, 24, 452001, 2013.
39. Song, M., You, D. S., Lim, K., Park, S., Jung, S., Kim, C. S., Kim, D. H., Kim, D. G., Kim, J. K. and Park, J., Highly efficient and bendable organic solar cells with solution-processed silver nanowire electrodes. *Adv. Funct. Mater.*, 23, 4177–4184, 2013.
40. Hu, L., Kim, H. S., Lee, J.-Y., Peumans, P. and Cui, Y., Scalable coating and properties of transparent, flexible, silver nanowire electrodes. *ACS Nano*, 4, 2955, 2010.
41. He, Z., Phan, H., Liu, J., Nguyen, T. Q. and Tan, T. T. Y., Understanding TiO_2 size-dependent electron transport properties of a graphene–TiO_2 photoanode in dye-sensitized solar cells using conducting atomic force microscopy. *Adv. Mater.*, 25, 6900–6904, 2013.
42. Snaith, H. J. and Schmidt-Mende, L., Advances in liquid-electrolyte and solid-state dye-sensitized solar cells. *Adv. Mater.*, 19, 3187–3200, 2007.
43. Yum, J. H., Chen, P., Grätzel, M. and Nazeeruddin, M. K., Recent developments in solid-state dye-sensitized solar cells. *ChemSusChem*, 1, 699–707, 2008.

44. Sun, X., Chen, T., Yang, Z. and Peng, H., The alignment of carbon nanotubes: An effective route to extend their excellent properties to macroscopic scale. *Acc. Chem. Res.*, 46, 539–549, 2012.
45. Rai, P., Oh, S., Shyamkumar, P., Ramasamy, M., Harbaugh, R. E. and Varadan, V. K., Nano-bio-textile sensors with mobile wireless platform for wearable health monitoring of neurological and cardiovascular disorders. *J. Electrochem. Soc.*, 161, B3116–B3150, 2014.
46. Chen, T., Qiu, L., Yang, Z. and Peng, H., Novel solar cells in a wire format. *Chem. Soc. Rev.*, 42, 5031–5041, 2013.
47. Das, A., Krishnasamy, J., Alagirusamy, R. and Basu, A., Electromagnetic interference shielding effectiveness of SS/PET hybrid yarn incorporated woven fabrics. *Fiber Polym.*, 15, 169–174, 2014.
48. Salvado, R., Loss, C., Gonçalves, R. and Pinho, P., Textile materials for the design of wearable antennas: A survey. *Sensors*, 12, 15841–15857, 2012.
49. Lam Po Tang, S. and Stylios, G., An overview of smart technologies for clothing design and engineering. *Int. J. Cloth. Sci. Tech.*, 18, 108–128, 2006.
50. Tsukada, S., Nakashima, H. and Torimitsu, K., Conductive polymer combined silk fiber bundle for bioelectrical signal recording. *PloS one*, 7, e33689, 2012.
51. Samad, Y. A., Li, Y., Alhassan, S. M. and Liao, K., Non-destroyable graphene cladding on a range of textile and other fibers and fiber mats. *RSC Adv.*, 4, 16935–16938, 2014.
52. Cheng, H., Dong, Z., Hu, C., Zhao, Y., Hu, Y., Qu, L., Chen, N. and Dai, L., Textile electrodes woven by carbon nanotube–graphene hybrid fibers for flexible electrochemical capacitors. *Nanoscale*, 5, 3428–3434, 2013.
53. Wu, J., Zhou, D., Too, C. O. and Wallace, G. G., Conducting polymer coated lycra. *Synthetic Met.*, 155, 698–701, 2005.
54. Xue, P., Tao, X. and Tsang, H., *In situ* SEM studies on strain sensing mechanisms of PPy-coated electrically conducting fabrics. *Appl. Surf. Sci.*, 253, 3387–3392, 2007.
55. Xue, P., Wang, J. and Tao, X., Flexible textile strain sensors from polypyrrole-coated XLA™ elastic fibers. *High Perform. Polym.*, 26, 364–370, 2014.
56. Kawano, K., Pacios, R., Poplavskyy, D., Nelson, J., Bradley, D. D. and Durrant, J. R., Degradation of organic solar cells due to air exposure. *Sol. Energ. Mat. Sol. C*, 90, 3520–3530, 2006.
57. Knittel, D. and Schollmeyer, E., Electrically high-conductive textiles. *Synthetic Met.*, 159, 1433–1437, 2009.
58. Zhang, L., Shi, E., Ji, C., Li, Z., Li, P., Shang, Y., Li, Y., Wei, J., Wang, K. and Zhu, H., Fiber and fabric solar cells by directly weaving carbon nanotube yarns with CdSe nanowire-based electrodes. *Nanoscale*, 4, 4954–4959, 2012.
59. Doganay, D., Coskun, S., Genlik, S. P. and Unalan, H. E., Silver nanowire decorated heatable textiles. *Nanotechnology*, 27, 435201, 2016.
60. Sun, Y., Gates, B., Mayers, B. and Xia, Y., Crystalline silver nanowires by soft solution processing. *Nano Lett.*, 2, 165–168, 2002.

61. Chang, Y., Lye, M. L. and Zeng, H. C., Large-scale synthesis of high-quality ultralong copper nanowires. *Langmuir*, 21, 3746–3748, 2005.
62. Rathmell, A. R., Nguyen, M., Chi, M. and Wiley, B. J., Synthesis of oxidation-resistant cupronickel nanowires for transparent conducting nanowire networks. *Nano Lett.*, 12, 3193–3199, 2012.
63. Hsu, P.-C., Wang, S., Wu, H., Narasimhan, V. K., Kong, D., Lee, H. R. and Cui, Y., Performance enhancement of metal nanowire transparent conducting electrodes by mesoscale metal wires. *Nat. Commun.*, 4, 2522, 2013.
64. Soltanian, S., Rahmanian, R., Gholamkhass, B., Kiasari, N. M., Ko, F. and Servati, P., Highly stretchable, sparse, metallized nanofiber webs as thin, transferrable transparent conductors. *Adv. Energy Mater.*, 3, 1332–1337, 2013.
65. Garnett, E. C., Cai, W., Cha, J. J., Mahmood, F., Connor, S. T., Christoforo, M. G., Cui, Y., McGehee, M. D. and Brongersma, M. L., Self-limited plasmonic welding of silver nanowire junctions. *Nat. Mater.*, 11, 241–249, 2012.
66. Xue, P., Park, K., Tao, X., Chen, W. and Cheng, X., Electrically conductive yarns based on PVA/carbon nanotubes. *Composite Struct.*, 78, 271–277, 2007.
67. Kim, A., Won, Y., Woo, K., Kim, C. and Moon, J. *ACS Nano*, 7, 1081–1091, 2013.
68. Park, S.-E., Kim, S., Lee, D.-Y., Kim, E. and Hwang, J., Fabrication of silver nanowire transparent electrodes using electrohydrodynamic spray deposition for flexible organic solar cells. *J. Mater. Chem. A*, 1, 14286–14293, 2013.
69. Schoen, D. T., Schoen, A. P., Hu, L., Kim, H. S., Heilshorn, S. C. and Cui, Y., High speed water sterilization using one-dimensional nanostructures. *Nano Lett.*, 10, 3628–3632, 2010.
70. Madaria, A. R., Kumar, A. and Zhou, C., Large scale, highly conductive and patterned transparent films of silver nanowires on arbitrary substrates and their application in touch screens. *Nanotechnology*, 22, 245201, 2011.
71. Atwa, Y., Maheshwari, N. and Goldthorpe, I. A., Silver nanowire coated threads for electrically conductive textiles. *J. Mater. Chem. C*, 3, 3908–3912, 2015.
72. Kim, A., Ahn, J., Hwang, H., Lee, E. and Moon, J. A., Pre-strain strategy for developing a highly stretchable and foldable one-dimensional conductive cord based on a Ag nanowire network. *Nanoscale*, 9, 5773–5778, 2017.
73. Kim, I., Lee, E. G., Jang, E. and Cho, G., Characteristics of polyurethane nanowebs treated with silver nanowire solutions as strain sensors. *Textile Res. J.*, https://doi.org/10.1177/0040517517697647, 2017.
74. Eom, J., Heo, J.-S., Kim, M., Lee, J. H., Park, S. K. and Kim, Y.-H., Highly sensitive textile-based strain sensors using poly(3,4-ethylenedioxythiophene):polystyrene sulfonate/silver nanowire-coated nylon threads with poly-l-lysine surface modification. *RSC Adv.*, 7, 53373–53378, 2017.
75. Cai, L., Song, A. Y., Wu, P., Hsu, P.-C., Peng, Y., Chen, J., Liu, C., Catrysse, P. B., Liu, Y., Yang, A., Zhou, C., Zhou, C., Fan, S. and Cui, Y., Warming up

human body by nanoporous metallized polyethylene textile. *Nat. Commun.*, 8, 496, 2017.

76. Nateghi, M. R. and Shateri-Khalilabad, M., Silver nanowire-functionalized cotton fabric. *Carbohyd. Polym.*, 117, 160–168, 2015.
77. Hsu, P.-C., Liu, X., Liu, C., Xie, X., Lee, H. R., Welch, A. J., Zhao, T. and Cui, Y., Personal thermal management by metallic nanowire-coated textile. *Nano Lett.*, 15, 365–371, 2014.
78. Zhu, Z., Gu, Y., Wang, S., Zou, Y. and Zeng, H., Improving wearable photodetector textiles via precise energy level alignment and plasmonic effect. *Adv. Electron. Mater.*, 3, 1700281, 2017.
79. Yuksel, R., Coskun, S. and Unalan, H. E., Coaxial silver nanowire network core molybdenum oxide shell supercapacitor electrodes. *Electrochim. Acta*, 193, 39–44, 2016.
80. Yu, Z., Li, C., Abbitt, D. and Thomas, J., Flexible, sandwich-like Ag-nanowire/PEDOT: PSS-nanopillar/MnO_2 high performance supercapacitors. *J. Mater. Chem. A*, 2, 10923–10929, 2014.
81. Yuksel, R., Coskun, S., Kalay, Y. E. and Unalan, H. E., Flexible, silver nanowire network nickel hydroxide core-shell electrodes for supercapacitors. *J. Power Sources*, 328, 167–173, 2016.
82. Wu, S., Hui, K. and Hui, K., One-dimensional core–shell architecture composed of silver nanowire@hierarchical nickel–aluminum layered double hydroxide nanosheet as advanced electrode materials for pseudocapacitor. *J. Phys. Chem. C*, 119, 23358–23365, 2015.
83. Qiao, Z., Yang, X., Yang, S., Zhang, L. and Cao, B., 3D hierarchical MnO 2 nanorod/welded Ag-nanowire-network composites for high-performance supercapacitor electrodes. *Chem. Commun.*, 52, 7998–8001, 2016.
84. Sekhar, S. C., Nagaraju, G. and Yu, J. S., Conductive silver nanowires-fenced carbon cloth fibers-supported layered double hydroxide nanosheets as a flexible and binder-free electrode for high-performance asymmetric supercapacitors. *Nano Energy*, 36, 58–67, 2017.
85. Wu, C., Kim, T. W., Guo, T. and Li, F., Wearable ultra-lightweight solar textiles based on transparent electronic fabrics. *Nano Energy*, 32, 367–373, 2017.
86. Lee, S., Shin, S., Lee, S., Seo, J., Lee, J., Son, S., Cho, H. J., Algadi, H., Al-Sayari, S. and Kim, D. E., Ag nanowire reinforced highly stretchable conductive fibers for wearable electronics. *Adv. Funct. Mater.*, 25, 3114–3121, 2015.
87. Mihailović, D., Šaponjić, Z., Radoičić, M., Radetić, T., Jovančić, P., Nedeljković, J. and Radetić, M., Functionalization of polyester fabrics with alginates and TiO_2 nanoparticles. *Carbohyd. Polym.*, 79, 526–532, 2010.
88. Lu, Y., Xue, L. and Li, F., Adhesion enhancement between electroless nickel and polyester fabric by a palladium-free process. *Appl. Surf. Sci.*, 257, 3135–3139, 2011.
89. Xue, J., Song, J., Dong, Y., Xu, L., Li, J. and Zeng, H., Nanowire-based transparent conductors for flexible electronics and optoelectronics. *Sci. Bull.*, 62, 143–156, 2017.

90. Xue, J., Song, J., Zou, Y., Huo, C., Dong, Y., Xu, L., Li, J. and Zeng, H., Nickel concentration-dependent opto-electrical performances and stability of Cu@CuNi nanowire transparent conductors. *RSC Adv.*, 6, 91394–91400, 2016.
91. Zhang, W., Yin, Z., Chun, A., Yoo, J., Kim, Y. S. and Piao, Y., Bridging oriented copper nanowire–graphene composites for solution-processable, annealing-free, and air-stable flexible electrodes. *ACS Appl. Mater. Interfaces*, 8, 1733–1741, 2016.
92. Varadana, V. K., Kumara, P. S., Oha, S., Mathura, G. N., Raib, P. and Kegleya, L., *E-bra with Nanosensors, Smart Electronics and Smart Phone Communication Network for Heart Rate Monitoring*, International Society for Optics and Photonics Bellingham, WA, pp. 79800–79807, 2011.
93. Burroughes, J., Bradley, D., Brown, A., Marks, R., Mackay, K., Friend, R., Burns, P. and Holmes, A., Light-emitting diodes based on conjugated polymers. *Nature*, 347, 539–541, 1990.
94. Zhao, F., Shi, Y., Pan, L. and Yu, G., Multifunctional nanostructured conductive polymer gels: Synthesis, properties, and applications. *Acc. Chem. Res.*, 50, 1734–1743, 2017.
95. MacDiarmid, A. G., "Synthetic metals": A novel role for organic polymers (Nobel lecture). *Angew. Chem. Int. Edit.*, 40, 2581–2590, 2001.
96. Jung, Y. S., Jung, W., Tuller, H. L. and Ross, C., Nanowire conductive polymer gas sensor patterned using self-assembled block copolymer lithography. *Nano Lett.*, 8, 3776–3780, 2008.
97. Liu, H., Kameoka, J., Czaplewski, D. A. and Craighead, H., Polymeric nanowire chemical sensor. *Nano Lett.*, 4, 671–675, 2004.
98. Kaiser, A. B., Electronic transport properties of conducting polymers and carbon nanotubes. *Rep. Progress Phys.*, 64, 1, 2001.
99. Xia, X., Tu, J., Zhang, Y., Wang, X., Gu, C., Zhao, X.-B. and Fan, H. J., High-quality metal oxide core/shell nanowire arrays on conductive substrates for electrochemical energy storage. *ACS Nano*, 6, 5531–5538, 2012.
100. Xu, J., Wang, K., Zu, S.-Z., Han, B.-H. and Wei, Z., Hierarchical nanocomposites of polyaniline nanowire arrays on graphene oxide sheets with synergistic effect for energy storage. *ACS Nano*, 4, 5019–5026, 2010.
101. O'Brien, G. A., Quinn, A. J., Tanner, D. A. and Redmond, G., A single polymer nanowire photodetector. *Adv. Mater.*, 18, 2379–2383, 2006.
102. Xin, H., Reid, O. G., Ren, G., Kim, F. S., Ginger, D. S. and Jenekhe, S. A., Polymer nanowire/fullerene bulk heterojunction solar cells: How nanostructure determines photovoltaic properties. *ACS Nano*, 4, 1861–1872, 2010.
103. Alam, M. M., Wang, J., Guo, Y., Lee, S. P. and Tseng, H.-R., Electrolyte-gated transistors based on conducting polymer nanowire junction arrays. *J. Phys. Chem. B*, 109, 12777–12784, 2005.
104. Li, D., Huang, J. and Kaner, R. B., Polyaniline nanofibers: A unique polymer nanostructure for versatile applications. *Acc. Chem. Res.*, 42, 135–145, 2008.
105. Hochbaum, A. I. and Yang, P., Semiconductor nanowires for energy conversion. *Chem. Rev.*, 110, 527–546, 2009.

106. Zhai, S., Karahan, H. E., Wei, L., Qian, Q., Harris, A. T., Minett, A. I., Ramakrishna, S., Ng, A. K. and Chen, Y., Textile energy storage: Structural design concepts, material selection and future perspectives. *Energy Storage Mater.*, 3, 123–139, 2016.
107. Wang, K., Zhao, P., Zhou, X., Wu, H. and Wei, Z., Flexible supercapacitors based on cloth-supported electrodes of conducting polymer nanowire array/SWCNT composites. *J. Mater. Chem.*, 21, 16373–16378, 2011.
108. Liu, X., Wu, Z. and Yin, Y., Hierarchical $NiCo_2S_4$@PANI core/shell nanowires grown on carbon fiber with enhanced electrochemical performance for hybrid supercapacitors. *Chem. Eng. J.*, 323, 330–339, 2017.
109. Wang, K., Meng, Q., Zhang, Y., Wei, Z. and Miao, M., High-performance two-ply yarn supercapacitors based on carbon nanotubes and polyaniline nanowire arrays. *Adv. Mater.*, 25, 1494–1498, 2013.
110. Jin, L.-N., Shao, F., Jin, C., Zhang, J.-N., Liu, P., Guo, M.-X. and Bian, S.-W., High-performance textile supercapacitor electrode materials enhanced with three-dimensional carbon nanotubes/graphene conductive network and *in situ* polymerized polyaniline. *Electrochim. Acta*, 249, 387–394, 2017.
111. Wang, Q., Wu, Y., Li, T., Zhang, D., Miao, M. and Zhang, A., High performance two-ply carbon nanocomposite yarn supercapacitors enhanced with a platinum filament and *in situ* polymerized polyaniline nanowires. *J. Mater. Chem. A*, 4, 3828–3834, 2016.
112. He, W., Wang, C., Zhuge, F., Deng, X., Xu, X. and Zhai, T., Flexible and high energy density asymmetrical supercapacitors based on core/shell conducting polymer nanowires/manganese dioxide nanoflakes. *Nano Energy*, 35, 242–250, 2017.
113. Dasgupta, N. P., Sun, J., Liu, C., Brittman, S., Andrews, S. C., Lim, J., Gao, H., Yan, R. and Yang, P., 25th anniversary article: Semiconductor nanowires-synthesis, characterization, and applications. *Adv. Mater.*, 26, 2137–2184, 2014.
114. Shen, G., Chen, P.-C., Ryu, K. and Zhou, C., Devices and chemical sensing applications of metal oxide nanowires. *J. Mater. Chem.*, 19, 828–839, 2009.
115. Kolmakov, A. and Moskovits, M., Chemical sensing and catalysis by one-dimensional metal-oxide nanostructures. *Annu. Rev. Mater. Res.*, 34, 151–180, 2004.
116. Hu, L., Chen, W., Xie, X., Liu, N., Yang, Y., Wu, H., Yao, Y., Pasta, M., Alshareef, H. N. and Cui, Y., Symmetrical MnO_2–carbon nanotube-textile nanostructures for wearable pseudocapacitors with high mass loading. *ACS Nano*, 5, 8904–8913, 2011.
117. Yuan, C., Hou, L., Li, D., Shen, L., Zhang, F. and Zhang, X., Synthesis of flexible and porous cobalt hydroxide/conductive cotton textile sheet and its application in electrochemical capacitors. *Electrochim. Acta*, 56, 6683–6687, 2011.
118. Zhai, T., Fang, X., Liao, M., Xu, X., Zeng, H., Yoshio, B. and Golberg, D., A comprehensive review of one-dimensional metal-oxide nanostructure photodetectors. *Sensors*, 9, 6504–6529, 2009.

119. Janotti, A. and Van de Walle, C. G., Native point defects in ZnO. *Phys. Rev. B*, 76, 165202, 2007.
120. Liu, K., Wu, W., Chen, B., Chen, X. and Zhang, N., Continuous growth and improved PL property of ZnO nanoarrays with assistance of polyethylenimine. *Nanoscale*, 5, 5986–5993, 2013.
121. Xu, C., Shin, P., Cao, L. and Gao, D., Preferential growth of long ZnO nanowire array and its application in dye-sensitized solar cells. *J. Phys. Chem. C*, 114, 125–129, 2010.
122. Wang, Z. L. and Song, J., Piezoelectric nanogenerators based on zinc oxide nanowire arrays. *Science*, 312, 242–246, 2006.
123. Prabhakar, P., Nanowire reinforcements for improving the interlaminar properties of textile composites. In *Blast Mitigation Strategies in Marine Composite and Sandwich Structures*, S. Gopalakrishnan and Y. Rajapakse (Eds.), Singapore, pp. 281–299, 2018.
124. Qin, Y., Wang, X. and Wang, Z. L., Microfibre-nanowire hybrid structure for energy scavenging. *Nature*, 451, 809–813, 2008.
125. Lee, M., Chen, C. Y., Wang, S., Cha, S. N., Park, Y. J., Kim, J. M., Chou, L. J. and Wang, Z. L., A hybrid piezoelectric structure for wearable nanogenerators. *Adv. Mater.*, 24, 1759–1764, 2012.
126. Bai, S., Zhang, L., Xu, Q., Zheng, Y., Qin, Y. and Wang, Z. L., Two dimensional woven nanogenerator. *Nano Energy*, 2, 749–753, 2013.
127. Seung, W., Gupta, M. K., Lee, K. Y., Shin, K.-S., Lee, J.-H., Kim, T. Y., Kim, S., Lin, J., Kim, J. H. and Kim, S.-W., Nanopatterned textile-based wearable triboelectric nanogenerator. *ACS Nano*, 9, 3501–3509, 2015.
128. Xu, S., Qin, Y., Xu, C., Wei, Y., Yang, R. and Wang, Z. L., Self-powered nanowire devices. *Nat. Nanotechnol.*, 5, 366–373, 2010.
129. Djurišić, A. B. and Leung, Y. H., Optical properties of ZnO nanostructures. *Small*, 2, 944–961, 2006.
130. Liu, J., Wu, W., Bai, S. and Qin, Y., Synthesis of high crystallinity ZnO nanowire array on polymer substrate and flexible fiber-based sensor. *ACS Appl. Mater. Interfaces*, 3, 4197–4200, 2011.
131. Zhu, Z., Ju, D., Zou, Y., Dong, Y., Luo, L., Zhang, T., Shan, D. and Zeng, H., Boosting fiber-shaped photodetectors via "soft" interfaces. *ACS Appl. Mater. Interfaces*, 9, 12092–12099, 2017.
132. Hong, D., Zang, W., Guo, X., Fu, Y., He, H., Sun, J., Xing, L., Liu, B. and Xue, X., High piezo-photocatalytic efficiency of CuS/ZnO nanowires using both solar and mechanical energy for degrading organic dye. *ACS Appl. Mater. Interfaces*, 8, 21302–21314, 2016.
133. Bae, J., Song, M. K., Park, Y. J., Kim, J. M., Liu, M. and Wang, Z. L., Fiber supercapacitors made of nanowire-fiber hybrid structures for wearable/flexible energy storage. *Angew. Chem. Int. Edit.*, 50, 1683–1687, 2011.
134. Kong, D., Ren, W., Cheng, C., Wang, Y., Huang, Z. and Yang, H. Y., Three-dimensional $NiCo_2O_4$@ polypyrrole coaxial nanowire arrays on carbon

textiles for high-performance flexible asymmetric solid-state supercapacitor. *ACS Appl. Mater. Interfaces*, 7, 21334–21346, 2015.

135. Gao, L., Qu, F. and Wu, X., Hierarchical WO_3@SnO_2 core-shell nanowire arrays on carbon cloth: A new class of anode for high-performance lithium-ion batteries. *J. Mater. Chem. A*, 2, 7367–7372, 2014.
136. Balogun, M.-S., Yu, M., Huang, Y., Li, C., Fang, P., Liu, Y., Lu, X. and Tong, Y., Binder-free Fe_2N nanoparticles on carbon textile with high power density as novel anode for high-performance flexible lithium ion batteries. *Nano Energy*, 11, 348–355, 2015.
137. Li, W., Wang, X., Liu, B., Luo, S., Liu, Z., Hou, X., Xiang, Q., Chen, D. and Shen, G., Highly reversible lithium storage in hierarchical $Ca_2Ge_7O_{16}$ nanowire arrays/carbon textile anodes. *Chem. Eur. J.*, 19, 8650–8656, 2013.
138. Liu, B., Zhang, J., Wang, X., Chen, G., Chen, D., Zhou, C. and Shen, G., Hierarchical three-dimensional $ZnCo_2O_4$ nanowire arrays/carbon cloth anodes for a novel class of high-performance flexible lithium-ion batteries. *Nano Lett.*, 12, 3005–3011, 2012.
139. Shen, L., Ding, B., Nie, P., Cao, G. and Zhang, X., Advanced energy-storage architectures composed of spinel lithium metal oxide nanocrystal on carbon textiles. *Adv. Energy Mater.*, 3, 1484–1489, 2013.
140. Yu, Z. and Thomas, J., Energy storing electrical cables: Integrating energy storage and electrical conduction. *Adv. Mater.*, 26, 4279–4285, 2014.
141. Fang, W., Liu, D., Lu, Q., Sun, X. and Asiri, A. M., Nickel promoted cobalt disulfide nanowire array supported on carbon cloth: An efficient and stable bifunctional electrocatalyst for full water splitting. *Electrochem. Commun.*, 63, 60–64, 2016.
142. Ding, R., Gao, H., Zhang, M., Zhang, J. and Zhang, X., Controllable synthesis of $Ni_{3-x}Co_xS_4$ nanotube arrays with different aspect ratios grown on carbon cloth for high-capacity supercapacitors. *RSC Adv.*, 5, 48631–48637, 2015.
143. Javed, M. S., Dai, S., Wang, M., Xi, Y., Lang, Q., Guo, D. and Hu, C., Faradic redox active material of Cu_7S_4 nanowires with a high conductance for flexible solid state supercapacitors. *Nanoscale*, 7, 13610–13618, 2015.
144. Huang, J., Hou, D., Zhou, Y., Zhou, W., Li, G., Tang, Z., Li, L. and Chen, S., MoS_2 nanosheet-coated CoS_2 nanowire arrays on carbon cloth as three-dimensional electrodes for efficient electrocatalytic hydrogen evolution. *J. Mater. Chem. A*, 3, 22886–22891, 2015.
145. Li, Y., Cao, L., Qiao, L., Zhou, M., Yang, Y., Xiao, P. and Zhang, Y., Ni–Co sulfide nanowires on nickel foam with ultrahigh capacitance for asymmetric supercapacitors. *J. Mater. Chem. A*, 2, 6540–6548, 2014.
146. Hao, L., Shen, L., Wang, J., Xu, Y. and Zhang, X., Hollow $NiCo_2S_4$ nanotube arrays grown on carbon textile as a self-supported electrode for asymmetric supercapacitors. *RSC Adv.*, 6, 9950–9957, 2016.
147. Chen, P., Zhou, T., Chen, M., Tong, Y., Zhang, N., Peng, X., Chu, W., Wu, X., Wu, C. and Xie, Y., Enhanced catalytic activity in nitrogen-anion modified

metallic cobalt disulfide porous nanowire arrays for hydrogen evolution. *ACS Catalysis*, 7405–7411, 2017.
148. Petchsang, N., Shapoval, L., Vietmeyer, F., Yu, Y., Hodak, J. H., Tang, I.-M., Kosel, T. H. and Kuno, M., Low temperature solution-phase growth of ZnSe and ZnSe/CdSe core/shell nanowires. *Nanoscale*, 3, 3145–3151, 2011.
149. Kuno, M., An overview of solution-based semiconductor nanowires: Synthesis and optical studies. *Phys. Chem. Chem. Phys.*, 10, 620–639, 2008.
150. Kuno, M., Ahmad, O., Protasenko, V., Bacinello, D. and Kosel, T. H., Solution-based straight and branched CdTe nanowires. *Chem. Mater.*, 18, 5722–5732, 2006.
151. Trentler, T. J., Hickman, K. M., Goel, S. C., Viano, A. M., Gibbons, P. C. and Buhro, W. E., Solution-liquid-solid growth of crystalline III-V semiconductors: An analogy to vapor-liquid-solid growth. *Science*, 1791–1794, 1995.
152. Onicha, A. C., Petchsang, N., Kosel, T. H. and Kuno, M., Controlled synthesis of compositionally tunable ternary $PbSe_xS_{1-x}$ as well as binary PbSe and PbS nanowires. *ACS Nano*, 6, 2833, 2012.
153. Cui, Y., Duan, X., Hu, J. and Lieber, C. M., Doping and electrical transport in silicon nanowires. *J. Phys. Chem. B*, 104, 5213–5216, 2000.
154. Wang, C., Cao, Y., Luo, Z., Li, G., Xu, W., Xiong, C., He, G., Wang, Y., Li, S. and Liu, H., Flexible potassium vanadate nanowires on Ti fabric as a binder-free cathode for high-performance advanced lithium-ion battery. *Chem. Eng. J.*, 307, 382–388, 2017.
155. Zeng, B., Xiong, G., Chen, S., Wang, W., Wang, D. and Ren, Z., Field emission of silicon nanowires grown on carbon cloth. *Appl. Phys. Lett.*, 90, 033112, 2007.
156. Huggins, R. A., Lithium alloy negative electrodes. *J. Power Sources*, 81, 13–19, 1999.
157. Liu, B., Wang, X., Chen, H., Wang, Z., Chen, D., Cheng, Y.-B., Zhou, C. and Shen, G., Hierarchical silicon nanowires-carbon textiles matrix as a binder-free anode for high-performance advanced lithium-ion batteries. *Sci. Rep.*, 3, 1622, 2013.
158. Ren, W., Wang, C., Lu, L., Li, D., Cheng, C. and Liu, J., SnO_2@Si core–shell nanowire arrays on carbon cloth as a flexible anode for Li ion batteries. *J. Mater. Chem. A*, 1, 13433–13438, 2013.
159. Zhukovskyi, M., Sanchez-Botero, L., McDonald, M. P., Hinestroza, J. and Kuno, M., Nanowire-functionalized cotton textiles. *ACS Appl. Mater. Interfaces*, 6, 2262–2269, 2014.
160. Chen, L., Zhou, Y., Dai, H., Li, Z., Yu, T., Liu, J. and Zou, Z., Fiber dye-sensitized solar cells consisting of TiO_2 nanowires arrays on Ti thread as photoanodes through a low-cost, scalable route. *J. Mater. Chem. A*, 1, 11790–11794, 2013.
161. Chen, T., Qiu, L., Li, H. and Peng, H., Polymer photovoltaic wires based on aligned carbon nanotube fibers. *J. Mater. Chem.*, 22, 23655–23658, 2012.

162. Cai, X., Wu, H., Hou, S., Peng, M., Yu, X. and Zou, D., Dye-sensitized solar cells with vertically aligned TiO_2 nanowire arrays grown on carbon fibers. *Chem. Sus. Chem.*, 7, 474–482, 2014.
163. Liang, J., Zhang, G., Yang, Y. and Zhang, J., Highly ordered hierarchical TiO_2 nanotube arrays for flexible fiber-type dye-sensitized solar cells. *J. Mater. Chem. A*, 2, 19841–19847, 2014.
164. Chen, T., Qiu, L., Kia, H. G., Yang, Z. and Peng, H., Designing aligned inorganic nanotubes at the electrode interface: Towards highly efficient photovoltaic wires. *Adv. Mater.*, 24, 4623–4628, 2012.
165. Wang, W., Zhao, Q., Li, H., Wu, H., Zou, D. and Yu, D., Transparent, double-sided, ITO-free, flexible dye-sensitized solar cells based on metal wire/ZnO nanowire arrays. *Adv. Funct. Mater.*, 22, 2775–2782, 2012.
166. Tian, B., Zheng, X., Kempa, T. J., Fang, Y., Yu, N., Yu, G., Huang, J. and Lieber, C. M., Coaxial silicon nanowires as solar cells and nanoelectronic power sources. *Nature*, 449, 885–889, 2007.
167. Bae, J., Park, Y. J., Yang, J. C., Kim, H. W. and Kim, D. Y., Toward wearable and stretchable fabric-based supercapacitors: Novel ZnO and SnO_2 nanowires—Carbon fibre and carbon paper hybrid structure. *J. Solid State Electrochem.*, 19, 211–219, 2015.
168. Xu, J., Wang, D., Fan, L., Yuan, Y., Wei, W., Liu, R., Gu, S. and Xu, W., Fabric electrodes coated with polypyrrole nanorods for flexible supercapacitor application prepared via a reactive self-degraded template. *Org. Electron.*, 26, 292–299, 2015.
169. Dong, Y., Zou, Y., Song, J., Zhu, Z., Li, J. and Zeng, H., Self-powered fiber-shaped wearable omnidirectional photodetectors. *Nano Energy*, 30, 173–179, 2016.
170. Ko, Y. H., Nagaraju, G. and Yu, J. S., Wire-shaped ultraviolet photodetectors based on a nanostructured NiO/ZnO coaxial p–n heterojunction via thermal oxidation and hydrothermal growth processes. *Nanoscale*, 7, 2735–2742, 2015.
171. Chen, J., Ding, L., Zhang, X., Chu, L., Liu, N. and Gao, Y., Strain-enhanced cable-type 3D UV photodetecting of ZnO nanowires on a Ni wire by coupling of piezotronics effect and pn junction. *Optics Express*, 22, 3661–3668, 2014.
172. Zhang, F., Niu, S., Guo, W., Zhu, G., Liu, Y., Zhang, X. and Wang, Z. L., Piezo-phototronic effect enhanced visible/UV photodetector of a carbon-fiber/ZnO-CdS double-shell microwire. *ACS Nano*, 7, 4537–4544, 2013.
173. Athauda, T. J., Hari, P. and Ozer, R. R., Tuning physical and optical properties of ZnO nanowire arrays grown on cotton fibers. *ACS Appl. Mater. Interfaces*, 5, 6237–6246, 2013.
174. Huang, G.-W., Xiao, H.-M. and Fu, S.-Y., Electrical switch for smart pH self-adjusting system based on silver nanowire/polyaniline nanocomposite film. *ACS Nano*, 9, 3234–3242, 2015.

175. Shen, L., Che, Q., Li, H. and Zhang, X., Mesoporous $NiCo_2O_4$ nanowire arrays grown on carbon textiles as binder-free flexible electrodes for energy storage. *Adv. Funct. Mater.*, 24, 2630–2637, 2014.
176. Eom, J., Lee, W. and Kim, Y.-H., Textile-Based Wearable Sensors Using Metal-Nanowire Embedded Conductive Fibers, *SENSORS, IEEE*, pp. 1–3, 2016.
177. Lee, T., Lee, W., Kim, S. W., Kim, J. J. and Kim, B. S., Flexible textile strain wireless sensor functionalized with hybrid carbon nanomaterials supported ZnO nanowires with controlled aspect ratio. *Adv. Funct. Mater.*, 26, 6206–6214, 2016.
178. Malakooti, M. H., Patterson, B. A., Hwang, H.-S. and Sodano, H. A., *Development of Multifunctional Fiber Reinforced Polymer Composites through ZnO Nanowire Arrays*, Behavior and Mechanics of Multifunctional Materials and Composites 2016, International Society for Optics and Photonics, p. 98000L, 2016.
179. Khan, A., Edberg, J., Nur, O. and Willander, M., A novel investigation on carbon nanotube/ZnO, Ag/ZnO and Ag/carbon nanotube/ZnO nanowires junctions for harvesting piezoelectric potential on textile. *J. Appl. Phys.*, 116, 034505, 2014.
180. Menéndez, L. J. A., Suárez, M. F. M. and Plaza, D. G., Multifunctional textile fabrics with silver nanowires: Electrical conductive devices, 2016.
181. Cui, H.-W., Suganuma, K. and Uchida, H., Highly stretchable, electrically conductive textiles fabricated from silver nanowires and cupro fabrics using a simple dipping-drying method. *Nano Res.*, 8, 1604–1614, 2015.
182. Kamalabadi, M., Mohammadi, A. and Alizadeh, N., Polypyrrole nanowire as an excellent solid phase microextraction fiber for bisphenol A analysis in food samples followed by ion mobility spectrometry. *Talanta*, 156, 147–153, 2016.

6

Nanogenerators for Smart Textiles

Xiong Pu[1], Weiguo Hu[1] and Zhong Lin Wang[1,2*]

[1]Beijing Institute of Nanoenergy and Nanosystems, Chinese Academy of Sciences, Beijing, China
[2]School of Materials Science and Engineering, Georgia Institute of Technology, Atlanta, Georgia, USA

Abstract
Smart textiles or wearable electronics typically integrate various types of sensors, illuminating/displaying devices, and actuators into yarns, garments, accessories, or wearables, which all require power sources. Meantime, state-of-the-art batteries, the most commonly used portable power sources, are rigid, bulky, not washable, and in need of frequent recharging, making the smart textile uncomfortable and inconvenient, not to mention their unaddressed safety issue. Therefore, intensive research endeavors have been dedicated in recent years to energy-generation devices that could be integrated into smart textiles/wearable electronics, while maintaining the softness or comfort of textiles/wearables. Nanogenerators, converting mechanical motion/vibration energies into electricity, have been demonstrated to be promising for this end. This chapter provides an overview, perspective, and prospect on nanogenerators for smart textiles, with the specific focus on the progresses of two types of nanogenerators, i.e. piezoelectric and triboelectric nanogenerators.

Keywords: Piezoelectric nanogenerators, triboelectric nanogenerators, smart textiles, energy harvesting, hybrid, energy storage

6.1 Introduction

Smart textiles or electronic textiles (E-textiles), which integrate multifunctional electronic/optoelectronic devices into fashionable/stylish clothing, hold great promise for the next growth of the market of wearable

**Corresponding author*: zhong.wang@mse.gatech.edu

Nazire D. Yilmaz (ed.) Smart Textiles, (177–210) © 2019 Scrivener Publishing LLC

electronics [1, 2]. Various components of electronic devices have been demonstrated in smart garments or fabrics, including textile circuits [3], light emitting diodes (LED) [4], and a variety of sensors for temperature [5], pressure [6], medical diagnosis [7], etc. Still, an energy device is required to supply electrical power to the E-textile. It remains to be a challenge to integrate appropriate power devices into smart textiles. Conventional batteries are widely applied for portable electronics, but they typically require frequent recharging or replacements. Meanwhile, they fall short of the required flexibility, comfort, and lightweight for smart textiles. Therefore, tremendous efforts have been made to design energy-storage devices (batteries or supercapacitors) into fibers or fabrics [8]. Despite the advancements, limited energy density, safety concerns, and frequent recharging are almost inevitable drawbacks for electrochemical energy-storage devices. As a result, intensive research has been devoted to energy-harvesting textiles, which could convert the environmental energies into electricity for *in situ* power supply to the smart textile.

Generally, there are several different types of energy resources that are available in the working environment of smart textiles, such as sunlight, body thermal energy, and mechanical energies of human motions. Thermoelectric nanogenerators can convert thermal energies into electricity when there is a temperature gradient. Thermoelectric nanogenerators are typically based on solid-state p- and n-type semiconductors, but the efficiency and output performances are generally low. Different types of solar cells have been widely applied to convert solar energy into electricity. Meanwhile, recent studies have showed several preliminary prototypes of flexible solar cells that can be integrated into textiles or fabrics. However, solar cells are highly dependent on the location, season, and weather, and are unable to provide sustainable power supply. Mechanical vibration or human motions are ideal energy resources that are universally available for smart textiles. Two types of nanogenerators, i.e. piezoelectric nanogenerators (PENG) and triboelectric nanogenerators (TENG), have been intensively studied to harvest various types of mechanical energies, such as vibration [9], wind [10], water wave [11], and human motions [12–14]. Conventional electromagnetic generators can also convert mechanical energy into electricity, but they are not suitable for smart textiles mainly for the following two reasons: (1) the use of heavy and rigid magnet, and (2) the poor output performances for mechanical motions at low frequencies (<5 Hz). Nanogenerators, especially triboelectric nanogenerators, are well known for their abundant materials choices, versatile design structures, and excellent performances at low frequencies, and therefore are ideal candidates as power devices for smart textiles [15]. Furthermore, nanogenerators are also the promising candidates as the *energy of the new era*, i.e. the forthcoming era of Internet of Things (IoT) [16, 17]. Because billions of sensors distribute widely and communicate wirelessly in IoT, batteries will not be appropriate power devices since their recharging or maintenance becomes impossible. Nanogenerators with *in situ* energy conversion will be the ideal choice for continuous power supply.

In this chapter, fundamental theoretical background of these two types of nanogenerators will be first discussed, followed by a summary of recent progresses and representative prototypes of piezoelectric and triboelectric nanogenerators for smart textiles. Strategies that hybridize nanogenerators with other energy-harvesting devices or energy-storage devices will be introduced as well.

6.2 Working Mechanisms of Nanogenerators

The piezoelectric nanogenerator (PENG), based on the piezoelectric and semiconducting properties of aligned ZnO nanowires (NWs), was proposed by Prof. Wang's group in 2006 [18]. Tiny mechanical energies can be converted into electricity with piezoelectric nanogenerators. In 2012, Prof. Wang's group further invented the triboelectric nanogenerator, which is based on two well-known effects, i.e. contact electrification and electrostatic induction [19]. Though PENG and TENG have different effects for generating polarized charges, i.e. piezoelectric effect and triboelectrification, respectively, their theoretical foundations can both be traced back to Maxwell's displacement current.

6.2.1 Piezoelectric Nanogenerators

The concept of PENG was first demonstrated with the voltage/current generation by sweeping an AFM tip across ZnO NW arrays, as shown in Figure 6.1 [18]. ZnO has a hexagonal wurtzite structure (space group *C6mc*) with lattice parameters a = 0.3296 and c = 0.52065 nm. The structure of ZnO can be simply described as a number of alternating planes composed of tetrahedrally coordinated O^{2-} and Zn^{2+} ions, stacked alternately along the c-axis. The tetrahedral coordination in ZnO results in noncentral symmetric structure, and consequently piezoelectricity and a mechanical stress/strain-to-electrical voltage conversion. The ZnO NWs were grown on a conductive substrate, and the AFM tip was a Si tip coated with a Pt film. During the sweeping, the normal force of 5 nN (nanonewton) was maintained to be constant between the tip and the NW. The contact between

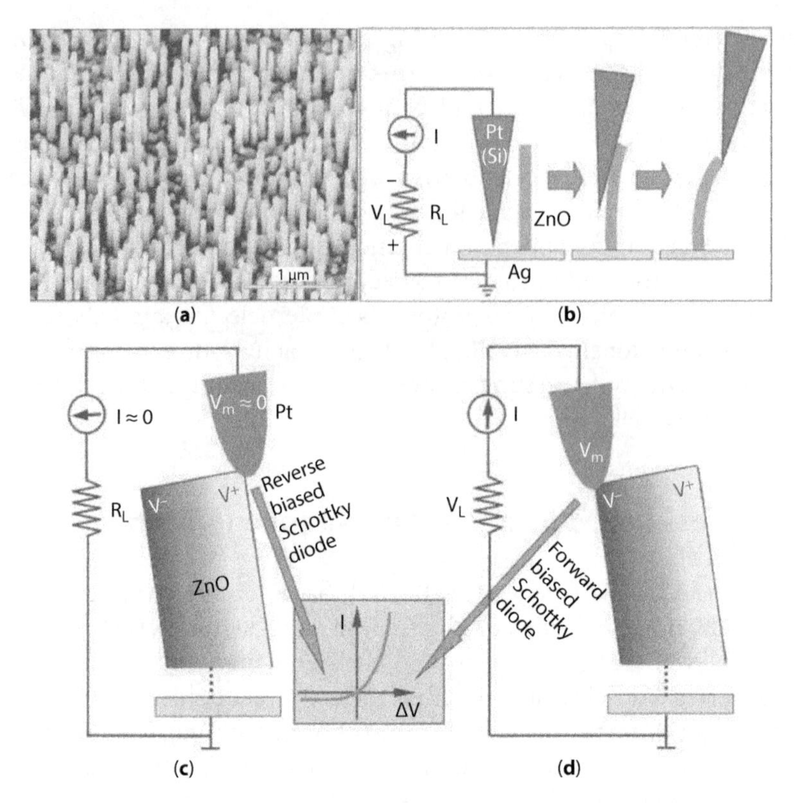

Figure 6.1 The working mechanism of a piezoelectric nanogenerator. (a) A scanning electron microscopy (SEM) image of aligned ZnO NWs. (b) Experimental setup and procedure for generating electricity by deforming a piezoelectric NW using a conductive AFM tip in contact mode. (c, d) Metal and semiconductor contacts between the AFM tip and the semiconductor ZnO NW at two reversed local contact potentials (positive and negative), showing reverse- and forward-biased Schottky rectifying behavior, respectively. Reproduced with permission [18]. Copyright 2006, The American Association for the Advancement of Science.

the AFM tip and NW is Schottky contact, while the contact between the NW and ground is ohmic. The metal-semiconductor Schottky contact has a large barrier height, which results in the rectifying properties of the junction. When the junction is forward biased, large amount of excited electrons in semiconductor can pass through the junction, while when the junction is reversely biased, electron flow is impeded by the Schottky barrier and only a very small leakage current can pass the junction, as schemed by the I–V curve in Figure 6.1c.

The deflection of the NW by the AFM tip creates a strain field, with the outer surface being tensile and the inner surface compressive. The asymmetric strain introduces an asymmetric potential at the two surfaces, with the compressed surface having negative potential (V^-) and the tensile surface positive potential (V^+). For the first step, the AFM conductive tip is in contact with the tensile surface with positive potential V^+ (Figure 6.1c). The Pt metal tip is ground, so the metal tip–ZnO semiconductor (M–S) interface is negatively biased for $\Delta V = (V_m - V^+) < 0$. As-synthesized ZnO NWs are n-type semiconductors, the M–S interface is a reverse-biased Schottky diode, and little current flows across the interface. In this step, the piezoelectric charges are created, separated, and accumulated at the surfaces. For the second step, when the AFM tip is in contact with the compressed side of the NW (Figure 6.1d), the M–S interface is positively biased for $\Delta V = (V_m - V^-) > 0$, so it produces a sudden increase in the output electric current.

Later, a PENG with horizontally aligned ZnO NWs on flexible substrate was also developed [20]. After that, substantial developments of PENG have been achieved [21–23]. PENGs were designed to simultaneously collect the electric current of numerous NWs together [24–27]; low-cost, flexible PENGs were also fabricated with new structure that ZnO NWs were sealed inside dielectric polymer film [28, 29]. The advantage of ZnO NW-based PENGs is that they can scavenge energies of tiny, random mechanical forces over a large range of frequencies.

6.2.2 Triboelectric Nanogenerators

For a TENG, the electricity generation is achieved through two processes. Firstly, surfaces of two different materials are brought into contact with each other, yielding the same amount of charges with opposite polarities at the two contacting surfaces, respectively [30, 31]. This process is so-called contact electrification or triboelectrification. The surface charges are typically immobile and can maintain for a long period of time, because materials with strong triboelectric effect are dielectric materials [32]. Secondly, when the two dielectric surfaces are separated, charges will be induced in the conductive electrodes coated on the backsides of the two dielectric materials, so that the local charge equilibrium can be achieved. Though the electrostatic charges are immobile, the induced charges can move between the two conductive electrodes through external circuits driven by the built-up electrostatic potentials [33]. If the contact-separation motion between the two dielectric materials is repeated, an alternative current will be generated.

As the triboelectrification is a universal effect, almost any materials can be utilized for electrification, such as metals, polymers, or natural leathers and woods. Therefore, the materials choices of the TENG are huge. In order to generate larger amount of triboelectric charges, the two materials used for the triboelectrification should be chosen according to the triboelectric series table, which was first proposed by John Carl Wilcke in 1757 (see Figure 6.2) [34]. A material toward the bottom of the series has stronger ability to gain electrons. When it touches a material near the top of the series, it will attain more negative charges. The further away two materials are from each other on the series table, the greater the charge transferred. The electrified tribo-charges can be further enhanced by introducing surface functionalization or nanostructure morphologies. The ability of a surface to gain or lose electrons can be manipulated by chemically introducing different functional groups or molecules, so as to improve the

	Aniline-formol resin	Polyvinyl alcohol	
	Polyformaldehyde 1.3-1.4	Polyester (Dacron) (PET)	
	Ethylcellulose	Polyisobutylene	
Positive	Polyamide 11	Polyuretane flexible sponge	
	Polyamide 6-6	Polyethylene terephthalate	
	Melanime formol	Polyvinyl butyral	
	Wool, knitted	Formo-phenolique, hardened	
	Silk, woven	Polychlorobutadiene	
	Polyethylene glycol succinate	Butadiene-acrylonitrile copolymer	
	Cellulose	Nature rubber	
	Cellulose acetate	Polyacrilonitrile	
	Polyethylene glycol adipate	Acrylonitrile-vinyl chloride	
	Polydiallyl phthalate	Polybisphenol carbonate	
	Cellulose (regenerated) sponge	Polychloroether	
	Cotton, woven	Polyvinylidine chloride (Saran)	
	Polyurethane elastomer	Poly(2,6-dimethyl polyphenyleneoxide)	
	Styrene-acrylonitrile copolymer	Polystyrene	
	Styrene-butadiene copolymer	Polyethylene	
	Wood	Polypropylene	
	Hard rubber	Polydiphenyl propane carbonate	
	Acetate, Rayon	Polyimide (Kapton)	
	Polymethyl methacrylate (Lucite)	Polyethylene terephtalate	Negative
	Polyvinyl alcohol	Polyvinyl chloride (PVC)	
	(continued)	Polytrifluorochloroethylene	
		Polytetrafluoroethylene (Teflon)	

Figure 6.2 Triboelectric series table of some common materials. Reproduced with permission [34]. Copyright 2013, American Chemical Society.

triboelectrification. The surface nanostructures can enlarge the surface areas, so that the effective contacting area for triboelectrification will be improved and more tribo-charges can be generated.

Rapid progresses of the TENG have been achieved after its invention, which are summarized in several recent review papers [16, 34–37]. Though TENGs with various structures have been demonstrated to harvest different types of mechanical energies, there are four basic working modes, as illustrated in Figure 6.3 [38]. (1) Vertical contact-separation mode (see Figure 6.3a). Two dissimilar dielectric films have relative contact-separation motion along the vertical direction repeatedly. Electrostatic charges generated when the two dielectric surfaces are in contact will induce free charges in the two electrodes deposited on the backsides. The free charges can flow back and forth through the external load when the two surfaces keep contact-separation motion repeatedly. (2) Lateral sliding mode (Figure 6.3b). The starting structure is similar to the vertical contact-separation mode, but

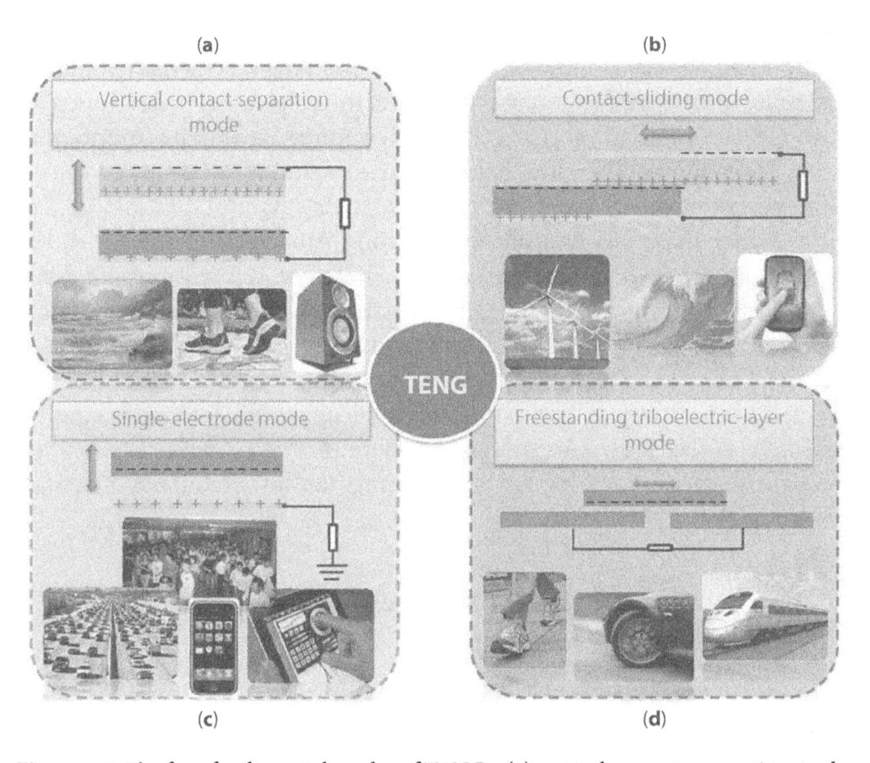

Figure 6.3 The four fundamental modes of TENGs: (a) vertical contact–separation mode; (b) in-plane contact–sliding mode; (c) single-electrode mode; and (d) freestanding mode. Reproduced with permission [38]. Copyright 2014, Royal Society of Chemistry.

the two dielectric films have lateral contact-sliding motion. A lateral polarization is created along the sliding direction, driving the flow of induced free electrons between the two electrodes. The lateral sliding can be planar sliding, cylinder rotation, or disc rotation. (3) Single-electrode mode (Figure 6.3c). The electrode attached on the bottom dielectric films is grounded. The electrified bottom dielectric film will induce free charges for local electric balance, which will flow back and forth through a load between the electrode and ground. The motion mode can be either vertical contact separation or lateral sliding. (4) Freestanding mode (Figure 6.3d). A pair of symmetric electrodes is placed underneath an electrified dielectric film, whose size and shape are the same as one electrode. The electrostatic charges in the surface of the dielectric film can maintain for a long period of time. The oscillation of the dielectric film between the pair of electrodes induces asymmetric charge distribution, which drives the back-and-forth electron flow between the two electrodes. The benefit of this mode is that the dielectric film does not have to contact the electrons, so that the possible wearing of materials can be avoided to achieve better mechanical durability.

The TENG has the following advantages: first, the mechanism or design is simple, and therefore a variety of structures have been developed to scavenge vast different mechanical energies: motional energies of rotation, sliding, pressing and bending, vibrational energies, water drops or waves, wind, acoustic sound, heart beating, to name a few [36]. Second, dielectric polymer materials typically have strong triboelectric effect, and the electrification only occurs at the surface. Therefore, it is feasible and facile to fabricate thin, lightweight, flexible, or stretchable TENGs for wearable electronics [39]. Last but most importantly, a recent study demonstrated that the TENG has much better output performance than that of electromagnetic generators for harvesting mechanical energies at low frequency, especially <5 Hz [15]. This unique attribute of TENG makes it the most competitive technology for harvesting irregular, low-frequency mechanical energies.

6.2.3 Theoretical Origin of Nanogenerators – Maxwell's Displacement Current

Mechanical energy is one important renewable energy source that is widely available. Various mechanisms, such as PENG, TENG, electromagnetic generator (EMG), etc., have been developed to convert different types of mechanical energies (including vibration [40], wind [41, 42], water waves [17, 43], sound [44], human motions [45], etc.) into electricity. Conventional electromagnetic generator has been widely applied for grid-scale electricity generation. However, the use of heavy and rigid magnet makes it inappropriate for small sensors and portable/wearable electronics. The TENG- and PENG-based energy devices are especially promising for sustainable power supply for huge amount of distributed sensors where battery replacement or recharging is impossible, making them the appropriate candidates for *the energy of new era* of IoT [16].

The theoretical foundations of the EMG and the TENG/PENG can all be traced back to Maxwell's equations, as shown in Figure 6.4 [17]. The output current of EMG is originated from the time variation of the magnetic field, and the current is related to $\frac{\partial \boldsymbol{B}}{\partial t}$, where $\boldsymbol{B}$ is the magnetic field and t is time. However, the current of the TENG and PENG is part of the Maxwell's displacement current, caused by the variation of the polarization

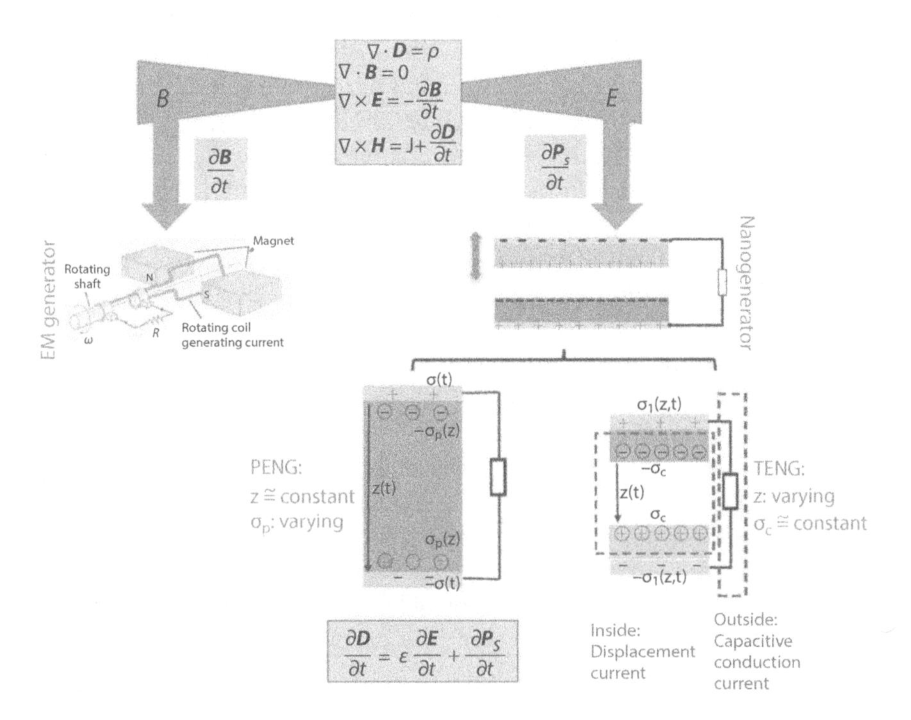

Figure 6.4 Fundamentals of the EMG and the PENG/TENG. The EMG is based on the time variation of magnetic field, while the PENG/TENG is based on the time variation of polarization field induced by surface polarization charges. The theoretical foundation of the PENG/TENG is Maxwell's displacement current. Reproduced with permission [17]. Copyright 2017, Elsevier.

field due to the surface polarized charges. The Maxwell's displacement current was first defined in 1861 and was described as [16, 17]

$$J_D = \frac{\partial \boldsymbol{D}}{\partial t} = \varepsilon_o \frac{\partial \boldsymbol{E}}{\partial t} + \frac{\partial \boldsymbol{P}}{\partial t} \tag{6.1}$$

where $\boldsymbol{J}_D$ is the displacement current, $\boldsymbol{D}$ is the electric displacement field, ε_o is the vacuum permittivity, $\boldsymbol{E}$ is the electric field, and $\boldsymbol{P}$ is the polarization field. For isotropic media, $J_D = \varepsilon \frac{\partial \boldsymbol{E}}{\partial t}$, where ε is the permittivity of the dielectrics. Whereas, when there are surface polarized charges, displace current is [16, 17]

$$J_D = \frac{\partial \boldsymbol{D}}{\partial t} = \varepsilon \frac{\partial \boldsymbol{E}}{\partial t} + \frac{\partial \boldsymbol{P}_s}{\partial t}, \tag{6.2}$$

where ε is the permittivity of the dielectrics, and $\boldsymbol{P}_s$ is the polarization field induced by surface polarized charges. The first term is the current due to the varying electric field, while the second term is induced by the varying polarization field of surface charges. For PENG, the surface polarization charges are generated at the two ends of a piezoelectric material, when it undergoes mechanical deformation. The time variation of this polarization field drives the electron flux through the external circuit, when two electrodes are attached on the piezoelectric material. For the TENG, the physical contact of two dielectric materials will generate electrostatic charges with opposite signs at the surfaces. The built electrostatic field drives the electron to flow through the external circuit. The internal circuit of the TENG is related to the displacement current, while the output current is the capacitive conduction current [17].

The first term of the Maxwell's displacement current leads to the discovery of electromagnetic wave theory and the development of various related technologies, such as radio, wireless communications, and Radar, to name a few. The second term, i.e. the time variation of surface polarized charges-induced polarization field, leads to the invention of PENG and TENG by Prof. Wang's group, and the proposed concepts of new energy technology, blue energy, self-powered sensors, and self-charging systems for the era of IoT [16].

6.3 Progresses of Nanogenerators for Smart Textiles

Intensive attentions have been recently paid to nanogenerators, and rapid progresses have been achieved. On one hand, improvements in the performances have been obtained for both TENGs and PENGs. On the other hand, studies have been conducted for specific potential applications, among which the smart textile is one of the most appealing areas. Though nanogenerators on rigid substrates can be possibly applied in smart textiles with certain degree of sacrifice of comfort, integrating nanogenerators in polymeric fibers/yarns and textiles is the ultimate goal for smart textiles. Therefore, this chapter will only focus on progresses on fiber-based and textile-based nanogenerators.

6.3.1 Piezoelectric Nanogenerators for Smart Textiles

6.3.1.1 Fiber-Based PENGs

Qin *et al.* [46] first reported a fiber-based PENG as shown in Figure 6.5. Kevlar 129 fibers were used as the 1D substrate. Radially aligned ZnO nanowires were grown on the fiber with a hydrothermal reaction. All of the ZnO nanowires are single crystalline and have a hexagonal cross-section with a diameter in the range of ~50–200 nm and a typical length of ~3.5 mm. Two layers of tetraethoxysilane (TEOS) were applied between the ZnO seed layer and Kevlar fiber, and between ZnO seed layer and ZnO nanowires. The Si–O bonds react with OH– on ZnO surface, and TEOS organic chains bind firmly to the aromatic polyamide Kevlar fiber. Therefore, ZnO wires and seed layers can be tightly bound on the Kevlar fiber.

The configuration of the PENG is as shown in Figure 6.5a. One ZnO nanowire-coated fiber was further covered with a layer of Au coating. Then, two fibers (one with top Au coating and the other without) were entangled into a two-ply fiber, the optical image of which is shown in Figure 6.5b. The gold-coated fiber was connected to the external circuit as the nanogenerator's output cathode. A scanning electron microscopy (SEM) image of the contacting interface of the two fibers is shown in Figure 6.5c. When the fiber is stretched as shown in Figure 6.5a, the Au-coated ZnO nanowires function as an array of brushing metal tips to deflect ZnO nanowires on the other fiber. Piezoelectric potential will be then generated across the bottom ZnO nanowire, with the stretched surface positive (V^+) and the compressed surface negative (V^-), as shown in Figure 6.5e. The negative-potential side has a forward-biased Schottky contact with the gold that allows the current to flow from the gold to the nanowire, as shown in Figure 6.5f.

One single fiber PENG can output an open-circuit voltage of 1 mV and short-circuit current of 5 pA, with a pulling force at a motor speed of

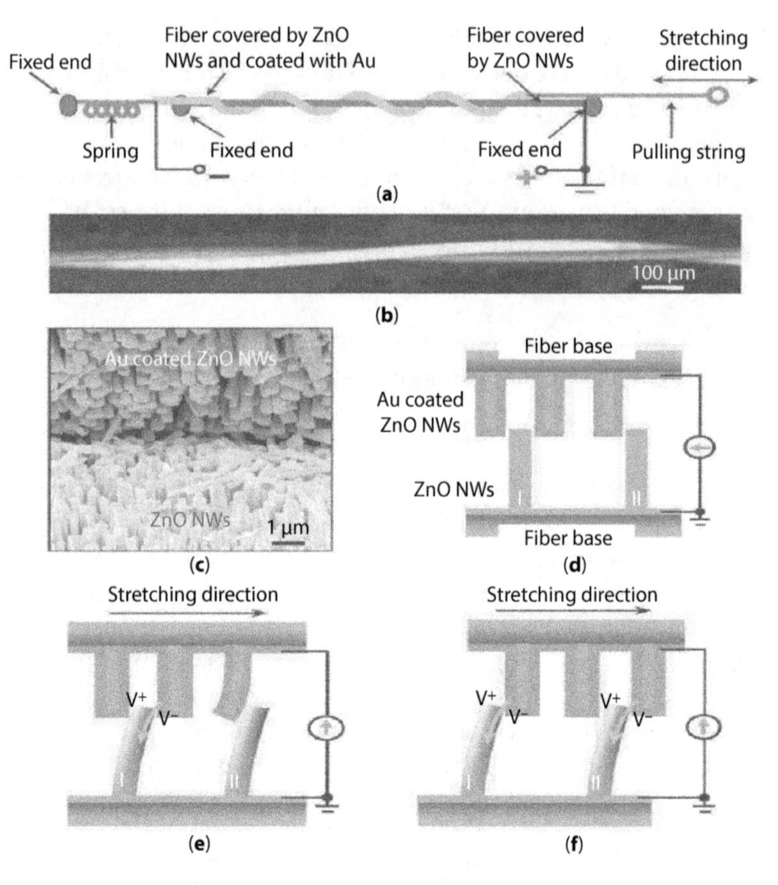

Figure 6.5 Design and electricity-generating mechanism of a fiber-based PENG. (a) Schematic illustration and (b) an optical image of the fiber nanogenerator. (c) SEM image of the interface of the two fibers. (d) Scheme of the contact of two fibers. (e) The piezoelectrical potential created across the ZnO nanowires. (f) Current generation through the external circuit. Reproduced with permission [46]. Copyright 2008, Nature Publishing Group.

80 rpm. A yarn of multiple fibers can be tested together to improve the output. A yarn consisting of six fibers can generate an average current of ~0.2 nA at the same pulling force, which is dramatically larger than that of the double-fiber PENG consisting of two fibers due to the significantly enhanced contacting areas. This fiber PENG is flexible and foldable, so it is possible to fabricate wearable and robust nanogenerators in the form of textile by weaving the fiber PENGs together.

Li *et al.* [47] proposed a core-shell design for fiber PENG, where carbon fiber electrode was covered by a textured ZnO thin film. The textured structure of the ZnO film resulted in a macroscopic piezopotential, which can produce a peak output voltage of 3.2 V and an average current density of 0.15 μA/cm^2. Air pressure, exhalation, and heartbeat pulse were demonstrated to drive the nanogenerator for alternating current outputs. The fiber PENG can both function as energy harvester and heartbeat sensor.

Chang *et al.* [48] utilized near-field electrospinning to produce and place poly(vinylidene fluoride) (PVDF) fibers to bridge two metal electrodes on substrates. Due to the strong electric field and stretching force during the electrospinning process, the dipoles of PVDF fibers were naturally aligned and β phase was obtained. When the substrate was stretched and released repeatedly, voltage and current outputs were recorded to be 5–30 mV and 0.5–3 nA, respectively, for PENG with a single PVDF fiber. Control experiments of nonpiezoelectric poly(ethylene oxide) fibers or randomly distributed PVDF fibers by conventional electrospinning showed no electrical outputs. The polymeric PVDF fiber-based PENG was claimed to be promising for self-powered textile with nanofibers produced on large area cloth.

6.3.1.2 *Textile-Based PENGs*

Bai *et al.* [49] developed a two-dimensional woven nanogenerator as schematically shown in Figures 6.6a and b. The fabric PENG was woven from two kinds of fibers, i.e. one was grown with radical ZnO nanowires and

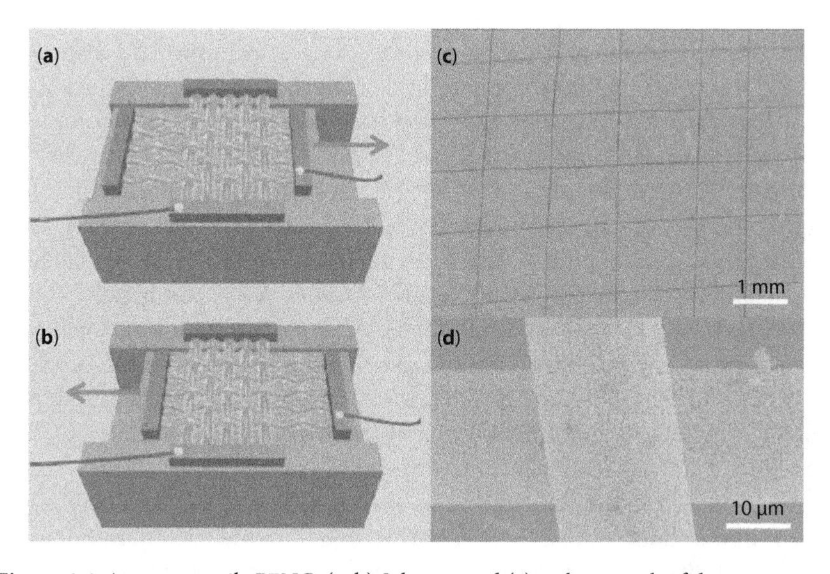

Figure 6.6 A woven textile PENG. (a, b) Schemes and (c) a photograph of the woven PENG. (d) A SEM image of one intersection point of two weft and warp fibers. Reproduced with permission [49]. Copyright 2013, Elsevier.

the other had an extra Pd coating on ZnO nanowires. Figure 6.6c shows an optical image of the woven PENG, and Figure 6.6d shows the intersection area of the two fibers. The working principle of the fabric PENG is similar to the fiber PENG discussed in Figure 6.5, only that the two contacting fibers are perpendicular to each other. This fabric PENG can generate electricity from the external tiny mechanical energies such as tiny wind and sound. The open-circuit voltage and short-circuit current of the woven PENG were 3 mV and 17 pA, respectively. Zhang *et al.* [50] also reported a woven fabric PENG. Cu wires were woven into a configuration of interdigitated electrodes. Polyvinylchloride fibers grown with $BaTiO_3$ nanowires were woven together with Cu wires but perpendicular to the Cu finger electrodes. When the fabric PENG was attached on an elbow pad, it can output 1.9 V voltage and 24 nA current, and power an LCD.

Khan *et al.* [51] reported a textile PENG with ZnO nanowire arrays grown on a commercial conductive fabric (ArgenMesh), which was made by weaving 55% silver and 45% nylon fibers with a final resistivity of <1 Ω/sq. An aqueous chemical growth at low temperature was used to prepare ZnO nanowires on textile substrates. The growth density of ZnO nanowires was measured to be 240±50 nanowires/μm^2 with a dominant *c*-axis direction. A nanoindentation technique was adopted for measuring mechanical and piezoelectric behaviors of ZnO nanowires. The generated output potential increased from 0 to 0.0048 V when the applied force increases from 0 to 300 mN.

Other than chemical growing methods, screen printing had also been utilized for applying piezoelectric ceramic materials on textiles. Almusallam *et al.* [52] reported a fabric PENG that used piezoelectric nano-composite film on flexible textile substrates. Lead zirconate titanate (PZT) particles, Ag nanoparticles, and polymers were mixed into a composite ink, which was then screen-printed on a conductive electrode-coated textile. Another top electrode was subsequently coated covering the piezoelectric composite film. The maximum energy density of the PENG under 800 N compressive forces was found to be 34 J/m^3 on a Kermel textile substrate, while for bending forces, the maximum energy density was found to be 14.3 J/m^3 on a cotton textile.

Polymeric PVDF is an ideal piezoelectric material for textile-based PENG, owing to its high flexibility and facile synthesis into fibers or fabrics. Soin *et al.* [53] designed three-dimensional piezoelectric fabrics based on "3D spacer" textile technology, as shown in Figure 6.7. Piezoelectrical PVDF fibers functioned as spacer yarns, which joined together and separated two knitted substrate fabrics as two electrodes, as shown by the SEM image of the fabric PENG in Figure 6.7b. Ag-coated polyamide 66 (PA66) yarns were used for fabricating the top and bottom conductive fabric

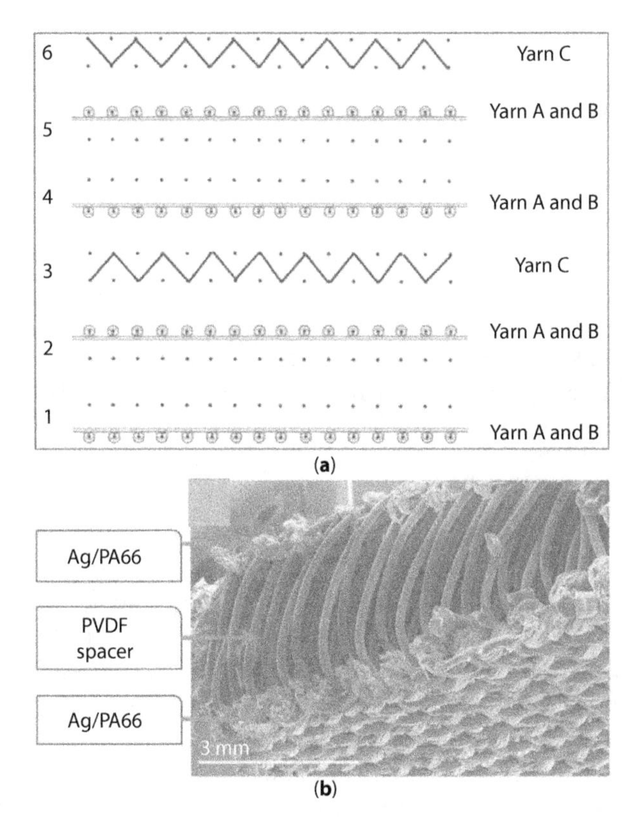

Figure 6.7 A 3D spacer all-textile PENG. (a) Scheme of the structure of the fabric PENG, and (b) a cross-sectional SEM image of the actual fabric clearly showing the position of piezoelectric and conductive yarns. Reproduced with permission [53]. Copyright 2014, Royal Society of Chemistry.

electrodes. The PVDF spacer yarns were manufactured by a continuous melt-spinning extrusion method. At an optimized drawing ratio, a high content of β phase (80%) was observed. In Figure 6.7a, yarns A, B, and C are conductive Ag-coated PA66, insulating polyester yarn, and piezoelectrical PVDF yarn, respectively. The polyester yarns have two functions: providing mechanical reinforcement to the 3D structure, and insulating the top and bottom conductive electrodes. The 3D spacer fabric PENG exhibited a power density in the range of 1.10 to 5.10 $\mu W/cm^2$ at applied impact pressures of 0.02 to 0.10 MPa.

Zeng *et al.* [54] reported a fabric PENG with simple stacked configuration. $NaNbO_3$ and PVDF composite nonwoven fabrics were synthesized by an electrospinning process. The mass ratio of $NaNO_3$/PVDF in the precursor mixture was 5:100. The piezoelectrical composite fabric was then sandwiched between two elastic conductive fabric electrodes, both of which were woven with segmented polyurethane multifilament yarns and Ag-coated polyamide multifilament yarns. The fabric PENG was then tested under a compressive pressure of 0.2 MPa. Stable and high peak values of V_{oc} and I_{sc} were measured to be 3.4 V and 4.4 mA, respectively.

In summary, fiber-based and textile-based PENGs have been demonstrated to be viable with various designs. Ceramic nanowires were typically utilized due to their better flexibility than bulk counterparts. Piezoelectric polymers, such as PVDF, are ideal for fiber/textile-PENGs due to their better flexibility and facile synthesis into yarns/textiles. Piezoelectric composite materials combining polymers and ceramics have also been used in textile-based PENG.

6.3.2 Triboelectric Nanogenerators for Smart Textiles

Due to the universality of triboelectrification effect, it is feasible to integrate a TENG directly into a fabric or cloth. Commonly used materials in textiles can be good candidates for triboelectrification with strong triboelectric effects, including the natural leather, silk, cotton, and synthetic polymers (polyester, nylon, etc.). Furthermore, yarns and textiles are composed of numerous microsized fibers, so the effective surface area for generating tribo-charges is large. Human body is also a rich source of various mechanical energies. Therefore, the fabric or cloth is an ideal vehicle for the TENG, so that the human motion energies can be harvested for sustainable power supply to various electronics.

6.3.2.1 Fiber-Based TENGs

The two electrodes of a TENG can be integrated in one single yarn or fiber. Zhong *et al.* [55] first reported a fiber-based TENG, as illustrated in Figure 6.8. Two modified cotton threads were entangled with each other into one composite yarn. One of the threads was coated with carbon nanotubes (CNTs); the other thread was coated successively by the CNT and a layer of polytetrafluoroethylene (PTFE). CNTs were first treated by ethanol flame and nitric acid solution to form a homemade CNT ink. Then, CNTs can be coated on the cotton threads through a "dipping and drying" process. PTFE was also coated by a dip-coating process, and an annealing process was conducted to enhance the adhesion. The CNT-coated cotton threads can achieve a conductivity of ~0.664 kΩ/cm. Finally, the two modified threads were entangled into a double-helical structure (Figure 6.8a). The composite yarn can then be woven into fabrics.

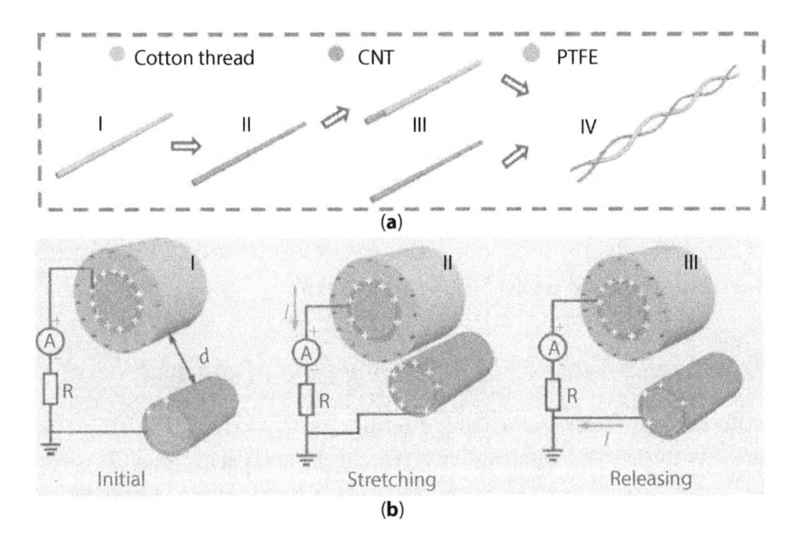

Figure 6.8 A fiber-based TENG. (a) Fabrication process and (b) working mechanism of the fiber-based TENG. Reproduced with permission [55]. Copyright 2014, American Chemical Society.

The working mechanism of fiber-based TENG is schematically illustrated in Figure 6.8b. PTFE layer of one thread and CNT coating of another thread are the two materials for triboelectrification. CNT coatings of two threads are connected through an external loading circuit. When the two modified cotton threads are stretched or pressed, the PTFE layer (green color) and CNT coating (red color) are in contact, yielding net negative charges on the PTFE surface and positive charges on the CNT coating of another thread. The repeat stretching–releasing of the composite yarn will lead to the relative contact–separation motions between the two threads, resulting in the alternating current through the external circuit. This process is similar to the common contact–separation mode of TENG.

The fiber-based TENG (~9.0 cm long and eight helix turns) can output a current with peak amplitude of 11.22 nA at 2.15% strain. When attached to a finger, the instantaneous power generated by small-scale finger motion reaches ~0.91 μW, which can power an electronic device such as a liquid crystalline display (LCD). A 2.2-μF commercial capacitor can be charged to 2.4 V in about 27 s. This proof-of-concept fiber-based TENG demonstrated the potential of power textile for self-powered smart garment.

Fiber-based TENG can also be realized by a configuration that the two electrodes are integrated in one yarn with core-shell structure. Sim *et al.* [56] reported a fiber-based TENG that the inner core electrode is a polyurethane fiber wrapped with silver-coated nylon yarn. Then, an electrospun

PVDF-TrFE layer and a CNT sheet were successively coated on the core fiber at pre-strained state, forming wrinkled coatings after the strain was released. The silver-coated nylon and PVDF-TrFE function as the positive and negative triboelectric materials, respectively. Due to the difference in the Poisson's ratio of the core fiber and shell coatings, contact–separation motion can be achieved during the repeated stretching–releasing of the fiber. Therefore, electricity generation was demonstrated, and the fiber can be reversibly stretched up to 50% tensile strain.

6.3.2.2 Textile-Based TENGs Starting from 1D Yarns/Fibers

A textile-based TENG can be generally designed by two approaches: (1) bottom-up process starting with treated fibers/yarns, and (2) direct use of 2D treated fabrics. For the first approach, Zhou *et al.* [57] first reported a textile TENG with woven structure. Strips of commercial silver fiber fabric were sandwiched in nylon and polyester strips, respectively. Subsequently, the two strips were woven into a fabric as wefts and warps. All weft silver fabric strips serve as one electrode, and all warp silver fabric strips serve as another. With contact–separation motion relative to another fabric, an alternating current can be generated through external circuit between the two electrodes in the textile TENG. A short-circuit current (I_{sc}) of 0.5 μA and open-circuit voltage (V_{oc}) of 27 V were observed when having relative motions to a common polyester fabric. Electricity generation has also been demonstrated when attaching the textile TENG to the moving parts of a human body, such as feet, legs, and arms. This TENG textile was composed of common textile materials; therefore, excellent flexibility and comfort can be expected.

Pu *et al.* [58] used electroless deposition method to apply a layer of conformal Ni coating on a wavy polyester textile so as to convert insulating polyester textile into conductive. Then, a thin layer of parylene was subsequently coated on the Ni-coated textile by a chemical vapor deposition (CVD) process. The TENG cloth was manufactured by weaving Ni-coated textile strips and parylene–Ni-coated strips as wefts and warps, respectively. The weft and warp Ni coatings were connected through an external circuit as shown in Figure 6.9a. Even though the TENG cloth was made of strips, all the processes were applicable to yarns, fibers, or threads. Comparing with typical conductive textiles dip-coated with CNTs or graphene, the conductivity of Ni cloth is higher. Meanwhile, the whole process is of low cost and suitable for scale-up synthesis. Furthermore, the woven TENG cloth retained the mechanical flexibility, air breathability, water washability, and thus comfortability of the original polyester cloth.

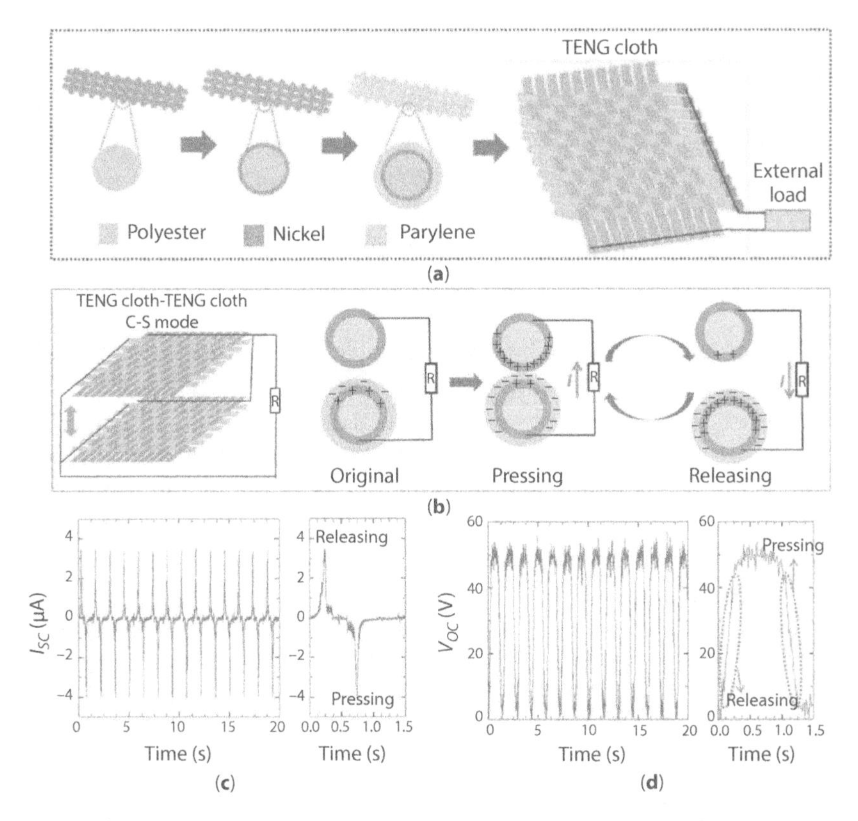

Figure 6.9 A woven textile-based TENG. (a) Fabrication process, (b) working mechanism, (c) current output, and (d) voltage output of the textile-based TENG. Reproduced with permission [58]. Copyright 2015, John Wiley & Sons.

Electricity generation had been demonstrated when rubbing two TENG cloths with each other, and having contact–separation motion between two TENG cloths, or between one TENG cloth and a common cloth, or between one TENG cloth and skin. So, the TENG cloth was versatile for harvesting various human motion energies. The energy generation process of the TENG cloth–TENG cloth motion mode was illustrated by Figure 6.9b. Ni coating is positive in the tribo-series and parylene is more negative, so static net negative charges will be produced on the surface of parylene at the contacting state. Induced free charges in the Ni coatings will then be driven by the electrostatic potential through the external circuit when pressing and releasing motions are repeated. Typical output of I_{sc} (~4 μA amplitude) and V_{oc} (~50 V amplitude) are shown in Figures 6.9c and d, respectively.

Later, Zhao *et al.* [59] fabricated a textile TENG with similar structure, but Cu-coated polyethylene terephthalate (PET) yarns and polyimide (PI)–Cu-coated PET yarns were used as the wefts and warps, respectively. They also conducted machine wash experiments to demonstrate the washability of the textile TENG. After 20 times of machine washes, the linear resistance of the Cu-PET yarn maintained <0.6 Ω/cm. The output current of the textile TENG had a significant reduction (from 13.78 to 3.52 mA/m^2) after the first three washes, but maintained 92% after the following wash cycles.

Lai *et al.* [60] reported a highly stretchable single-thread textile TENG. Silicone rubber-coated stainless-steel thread was sewn in an elastic textile substrate into a serpentine shape, so that the whole final textile can be stretched to a tensile strain up to ~100%. The radius of the fabricated thread is about 750 μm. The single energy-harvesting thread can harvest human-motion energy through contact with skin. The working mode was similar to a contact–separation mode, but the human skin served both as the positive material in the tribo-series and a reference electrode. The generating electric output reached peak values up to 200 V and 200 μA. The capability to harvest different kinds of mechanical energy from human body, including the movement of joints, walking, tapping, etc., was demonstrated.

The above textile TENGs are of 2D weaving structures. Dong *et al.* [61] reported a textile TENG with a 3D orthogonal woven (3DOW) structure for higher power output. Figure 6.10a schematically shows the manufacturing processes of the 3DOW TENG. The 3DOW textiles were fabricated from three types of yarns: 3-ply-twisted stainless steel (SS)/polyester fiber blended yarn (warp yarn), polydimethylsiloxane (PDMS)-coated energy-harvesting yarn (weft yarn), and binding yarns in thickness direction (Z-yarn). One layer of SS/polyester fibers was sandwiched in two layers of parallel PDMS-coated yarns. Z-yarns were woven to combine the top/bottom weft yarns and middle warp yarns. The 3D configuration of the design is as shown in Figure 6.10b. The authors tested a series of different connections of electrodes and found that double-electrode mode d32 (as shown in Figure 6.10c) showed the largest output. Single-electrode modes, including d1 (2D woven, one dielectric layer), d2 (2D woven, two dielectric layers), d31 (3DOW, weft-connection single-electrode mode), and d33 (3DOW, warp-connection single-electrode mode) showed much smaller I_{sc} (Figure 6.10d) and V_{oc} (Figure 6.10e) than double-electrode mode d32. Maximum peak power density can reach 263.36 mW/m^2 under a tapping frequency of 3 Hz. The textile 3DOW-TENG was then demonstrated for a broad range of applications, such as lighting up a warning indicator,

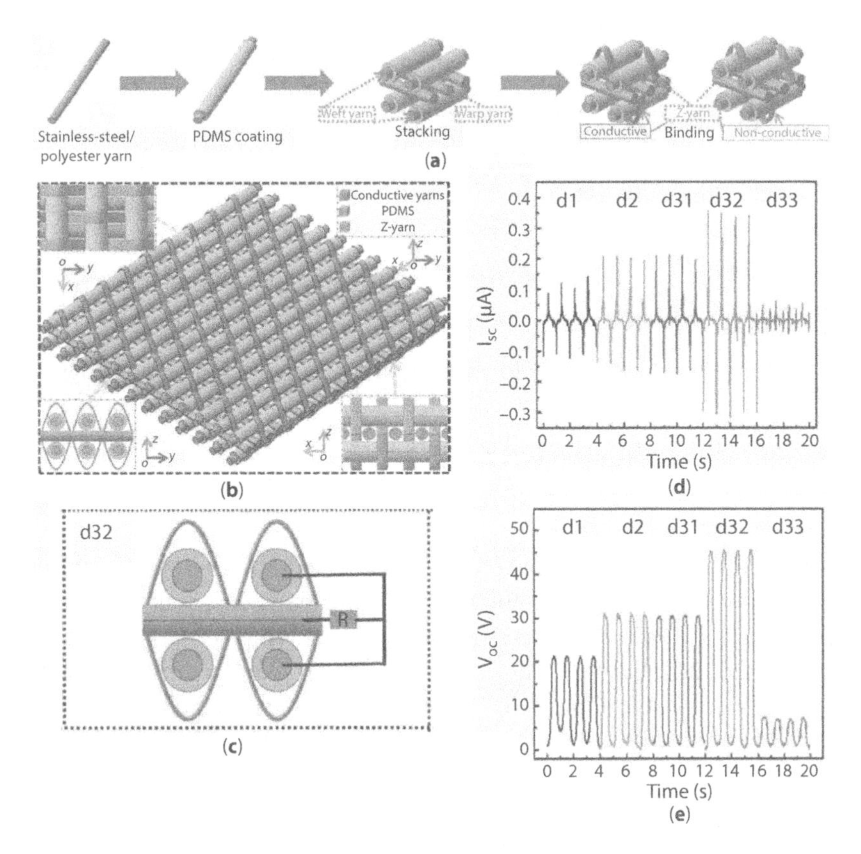

Figure 6.10 A 3D woven textile-based TENG. (a) Fabrication process and (b) structure illustration of the TENG textile. (c) The structure of a d32 type TENG textile. (d) Current and (e) voltage output of TENG textiles with different weaving structures. Reproduced with permission [61]. Copyright 2017, John Wiley & Sons.

charging a commercial capacitor, driving a digital watch, and tracking human motion signals. This study demonstrated that proper design of the weaving method was also an effective approach to improve the performances of textile TENG.

6.3.2.3 *Textile-Based TENGs Starting from 2D Fabrics*

Commercially available 2D fabrics can be treated with conductive coatings and dielectric films for TENGs. 2D treated fabrics can be directly used as electrodes without extra weaving or sewing processes [62–67]. Seung *et al.* [62] reported a textile TENG with a configuration as shown in Figure 6.11. Aligned ZnO nanorods were grown on a Ag-coated textile by a

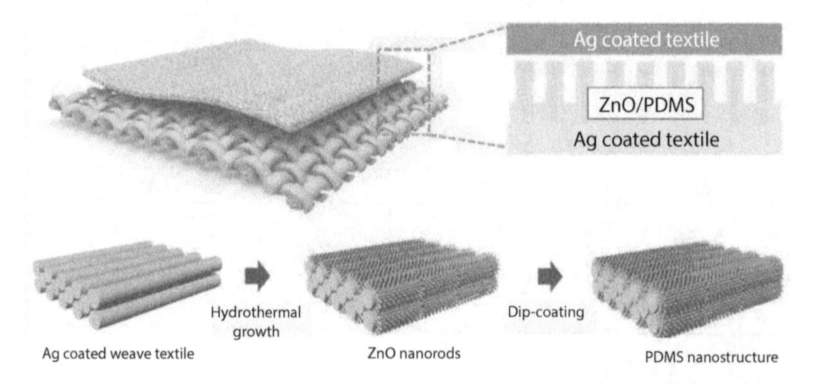

Figure 6.11 Structure illustration and fabrication process of a textile-based TENG with nanostructured patterns. Reproduced with permission [62]. Copyright 2015, American Chemical Society.

hydrothermal process, followed by dip-coating a layer of PDMS on the top. The nano-patterned morphology of the ZnO nanorods was maintained for the final PDMS coatings. Therefore, when having contact–separation motion relative to another Ag textile, the effective contacting area for triboelectrification was enhanced. The textile TENG featured a two-electrode contact–separation mode, with Ag as the tribo-positive material and PDMS as the tribo-negative one. According to the comparing experiment, a high output voltage and output current of about 120 V and 65 μA, respectively, were observed for a textile TENG with nano-patterns, while an output voltage and output current of 30 V and 20 μA, respectively, were obtained by a textile TENG with flat PDMS coating under the same mechanical compressive force. Finally, they demonstrated the self-powered operation of LEDs, an LCD, and a keyless vehicle entry system only with the output power of the textile TENG.

Some other studies reported textile TENGs with stacked structure by attaching conductive textiles with common nonconductive textiles, but the polymer films or sealing tapes will make the whole device not air-breathable [63, 65]. Zhang *et al.* [67] adopted a simple approach to sew conductive and dielectric textiles together with a stacked structure. Conductive fabrics were sewed at the backsides of a cotton and nylon fabric, respectively. A process of silanization was further conducted to introduce functional groups on the surface of the cotton fabric. Trichloro(*1H, 1H, 2H, 2H*-perfluorooctyl)silane was used to transform normally neutral cotton fabrics into a strong negative triboelectric, due to the grafted polymeric fluoroalkylated siloxane film on cotton. Considering nylon is a strong positive triboelectric material, the final device can have larger triboelectrification.

A fivefold V_{oc} increase, threefold J_{sc}, and threefold transferred charge increase were observed in sewn textile TENGs containing functionalized cotton cloths, as compared to that containing pristine cotton. From the two studies above, it is confirmed that surface nanoscale morphology and chemical functionalization are two effective approaches to improve the performances of textile TENGs.

Other than the contact–separation mode, freestanding-mode TENGs with interdigitated electrode configuration had also been realized in a textile TENG. Pu *et al.* [68] reported a textile TENG with the configuration as shown in Figure 6.12. A route of laser-scribing masking and electroless deposition of Ni was conducted for the synthesis of grating-structure electrodes on the textile (Figure 6.12a). On the top of the two electrodes, a layer of parylene was further coated by CVD method for triboelectrification. The final textile TENG was shown in Figure 6.12b. The textile TENG can be worn underneath the arm, as shown in Figure 6.12c. By swinging the arm, the electricity generated by the TENG can light LEDs and power LCD screens. The output performances can be improved by reducing the width of individual finger electrode. With optimized structure (finger electrode width is 1 mm), a peak power density of 3.2 W/m^2 was achieved at a sliding speed of 0.75 m/s. With the interdigitated grating structure, this textile TENG can convert low-frequency human motion energy into

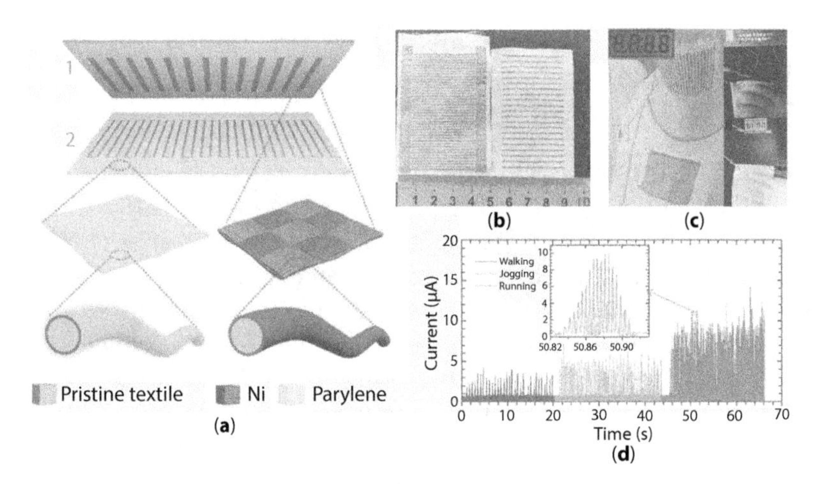

Figure 6.12 A textile-based freestanding mode TENG. (a) Structure illustration of the TENG with interdigitated electrodes. (b) A photo of the textile TENG. (c) A photo of the textile TENG wearing underneath the arm to power LEDs and LCD screens. (d) Current output of the textile TENG with different human motions. Reproduced with permission [68]. Copyright 2016, John Wiley & Sons.

high-frequency current outputs, as shown by rectified current in Figure 6.12d, where one single swing of arm output tens of current peaks. The current amplitude and output power are generally higher than previously discussed textile TENGs. This TENG fabric was also demonstrated to be flexible, washable, breathable, and compatible for integration into common textiles.

In summary, a variety of approaches have been demonstrated for textile TENGs to harvest human motion energies. These textile TENGs are generally flexible, wearable, washable, and compatible with common textiles. Nanostructured surface morphology, chemical functionalization, structure optimization, and textile weaving/sewing process can all possibly be optimized to improve the output performances.

6.3.3 Hybrid Nanogenerators for Smart Textiles

The mechanical energy of human body is generally irregular, so energy storage is needed to store the generated electricity by nanogenerators and provide stable power supply to smart textiles. Meanwhile, renewable energy resources (mechanical, solar, etc.) may not be always available in certain environment or during a certain period of time. Therefore, hybrid energy-harvesting devices that integrate two or more energy-harvesting technologies in one device could be an effective approach to combine the advantages of each individual unit and also to scavenge different types of energies in their working environment simultaneously.

6.3.3.1 Integrating with Energy-Storage Devices

Pu *et al.* [58] first proposed a self-charging power unit by integrating textile TENGs with a flexible belt-type lithium-ion battery (LIB). The prototype of the power unit is shown in Figure 6.13a. The textile TENG was placed underneath the arm to harvest human energy. A rectifier was used to convert the alternating current of TENG into direct current, so that the LIB belt can be charged to power a heartbeat monitor. The conductive Ni textiles were used as the current collector in the LIB, and $LiFePO_4$ and $Li_4Ti_5O_{12}$ were used as cathode and anode materials, respectively. The LIB belt can be folded completely for 180° without damage and performance degradation. A linear motor was utilized to mimic the TENG cloth–TENG cloth contact–separation motion of human at 0.7 Hz low frequency. The voltage of the LIB belt increases rapidly to the operational voltage (≈1.9 V) upon the charging by the TENG cloth. The charging times for the first, second, and third cycles are 4, 9, and 14 h, respectively. The corresponding

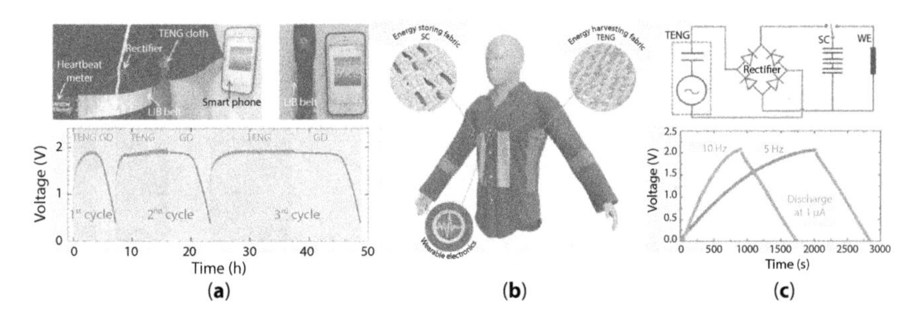

Figure 6.13 Textile TENG-based SCPSs. (a) Photos and charge/discharge profiles of an SCPS that integrates a textile TENG and a LIB belt. Reproduced with permission [58]. Copyright 2015, John Wiley & Sons. (b) Scheme, (c) equivalent circuit and charge/discharge profile of an all-textile SCPS that integrates a textile TENG and fiber supercapacitors. Reproduced with permission [69]. Copyright 2016, John Wiley & Sons.

discharge capacities are 1.3, 2.8, and 4.4 mA h/m^2, respectively, as shown in Figure 6.13b.

Later, Pu *et al.* [69] further constructed an all-textile self-charging power unit with yarn supercapacitors as energy-storage devices and TENG cloths as energy-harvesting devices, as schemed in Figure 6.13c. Reduced graphene oxides (rGOs), serving as the capacitive electrode materials, were coated on Ni-coated polyester yarns with a simple hydrothermal reaction and ascorbic acid reduction. Two rGO-coated yarns were placed in parallel with solid-state gel-type electrolyte. The resulting symmetric yarn supercapacitor achieved both high capacitance (13.0 mF/cm, 72.1 mF/cm^2) and stable cycling performances (96% for 10,000 cycles). Comparing with LIB belt, the yarn supercapacitors can also be woven or sewed into common textiles, so one single textile can possibly integrate both the energy-harvesting and energy-storage devices. With a rectifier, three supercapacitors connected in series can be charged by a textile TENG to 2.1 V, as shown in Figure 6.13d.

6.3.3.2 *Integrating with Energy-Harvesting Devices*

Piezoelectric and triboelectric nanogenerators can be integrated in one device for mechanical energy harvesting. Li *et al.* [70] reported a coaxial hybrid fiber nanogenerator with a single-electrode TENG (output 1) in the shell and a PENG (output 2) in the core, as shown in Figure 6.14a. ZnO nanorods were grown on a carbon fiber. Then, a copper layer was coated as one electrode of the PENG. An insulating nylon film was used to separate the PENG with the sheath TENG, which consists of a copper electrode covered by a PDMS film as triboelectrification layer.

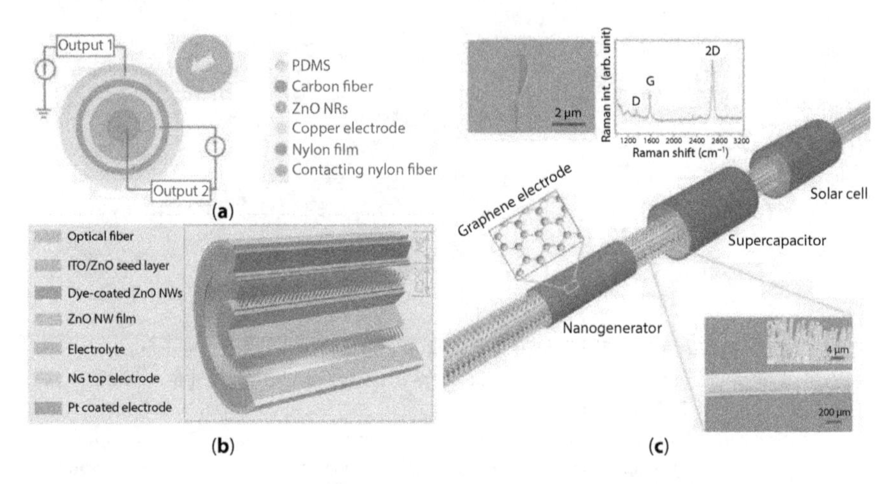

Figure 6.14 Fiber-based hybrid nanogenerators. (a) Scheme of a hybrid nanogenerator, where the inner core is a PENG and the sheath is a single-electrode TENG. Reproduced with permission [70]. Copyright 2014, American Chemical Society. (b) Scheme of a hybrid nanogenerator fabricated on an optical fiber, where the inner core DSSC and sheath PENG scavenge solar and mechanical energy, respectively. Reproduced with permission [71]. Copyright 2012, John Wiley & Sons. (c) A hybrid nanogenerator that integrates a PENG, a supercapacitor, and a solar cell on a single fiber. Reproduced with permission [72]. Copyright 2011, John Wiley & Sons.

Solar energy is also a suitable energy source for wearable electronics or smart textiles, but its time and weather dependence makes it hard to provide stable power alone. Therefore, integrating solar cells with nanogenerators is a promising approach. Pan *et al.* [71] integrated a dye-sensitized solar cell (DSSC) and PENG on a traditional optical fiber. Indium tin oxide (ITO) and dye-sensitized ZnO nanowire arrays were coated on the optical fiber as the photoelectrode for the inner DSSC, as shown in Figure 6.14b. Another ZnO coating at the outer shell served as the PENG. The high voltage of the PENG and high current of the DSSC were combined to charge capacitors for powering electronics.

Early in 2011, Bae *et al.* [72] constructed the first hybrid energy device with 1D fiber shape that a solar cell, a PENG, and a supercapacitor were all fabricated on different segments of a single fiber, as shown in Figure 6.14c. The radially grown ZnO NWs served as the photoelectrode in the solar cell, the piezoelectric layer in the PENG, and a capacitive electrode in the supercapacitor, respectively. Three segments of graphene layers were used as the second electrodes in all three devices. In this integration, solar energy and mechanical energy can all be harvested simultaneously, and the supercapacitor can be charged by the scavenged energy for stable power supply to electronics. These proof-of-concept studies showed the possibility of hybrid fiber energy devices that integrate different energy-harvesting technologies and energy-storage devices.

Considering that solar cells, TENGs, and supercapacitors can all be designed into 1D fibers/yarns, it is feasible to construct all-fiber hybrid energy textiles by weaving these components together. Chen *et al.* [73] reported a micro-cable power textile for simultaneously scavenging sunlight energy and mechanical motion energies, as shown in Figure 6.15a. Solid-state DSSC textile was woven with photoelectrode wires and Cu counter electrode wires as warps and wefts, respectively. Dye-sensitized ZnO coatings were grown on a Mn–Cu-coated polytetrafluoroethylene (PTFE) fiber. CuI was then coated on the top as the solid electrolyte, and the Cu counter electrode was perpendicular to the photoelectrode. The TENG textile was woven with Cu-coated PTFE stripes and Cu wires with a shuttle-flying process. The TENG and DSSC textiles can be woven in parallel or in series into a single cloth. The effect of different weaving textures (plain, twill, and satin) was studied. The as-woven hybrid power textile can charge a 2-mF capacitor in 1 min. The hybrid energy textile was demonstrated to charge a cellphone and power an electronic watch.

Pu *et al.* [68] integrated their reported textile TENG (described in Figure 6.12) with fiber DSSCs. Dye-sensitized TiO_2 layers were grown on

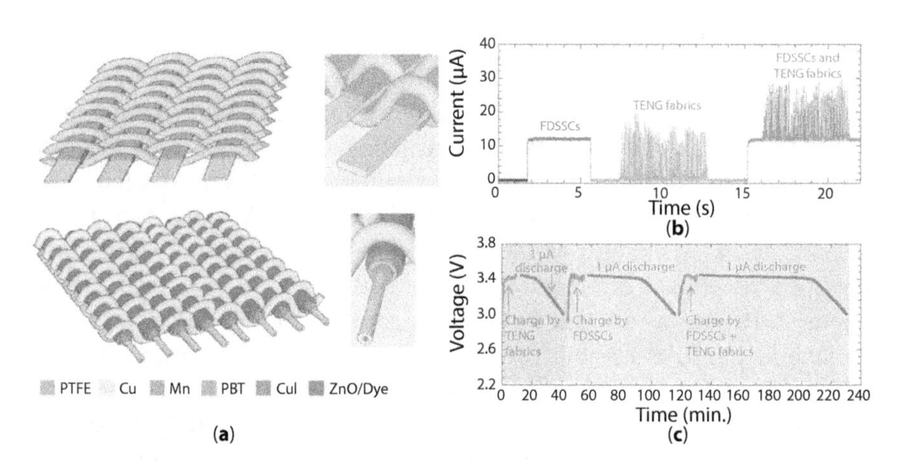

Figure 6.15 Textile-based hybrid nanogenerators. (a) Scheme of a power textile integrating a textile TENG and textile DSSCs. Reproduced with permission [73]. Copyright 2016, Nature Publishing Group. (b) Output current of a power textile integrating a textile TENG and textile DSSCs, and (c) voltage profiles of a lithium-ion battery charged by the power textile. Reproduced with permission [68]. Copyright 2016, John Wiley & Sons.

a Ti wire by a hydrothermal reaction and a dip-coating process. A Pt wire counter electrode was wrapped around the inner photoelectrode with a helix configuration. The fiber electrodes were inserted in a plastic tube, and liquid electrolyte was filled followed by sealing the tube with glue. The fiber DSSC achieved an average power efficiency of 6% with an open-circuit voltage of 0.68 V on average. The output current of the hybrid energy-harvesting textile was shown in Figure 6.15b. When combining a TENG textile and seven fiber DSSCs connected in series, a $LiFePO_4$–Li coin cell charged for 10 min can later be discharged at 1 μA for 98 min (Figure 6.15c).

Wen *et al.* [74] also proposed a self-charging power textile hybridizing all-fiber-shaped TENGs, solar cells, and supercapacitors. The fiber DSSC achieved 0.74 open-circuit voltage and efficiency of 5.64%. The TENG textile can output a current of ~0.91 μA when jogging. The fiber supercapacitor utilizing $RuO_2 \cdot xH_2O$ as the symmetric electrode materials exhibited a specific capacitance of 1.9 mF/cm. In 69 s, the supercapacitor can be charged to 1.8 V by the hybrid power textile.

In summary, all-fiber or all-textile hybrid energy textiles have been demonstrated to be viable. The hybrid energy textiles are compatible with existing textile engineering and show a certain degree of flexibility and wearability.

6.4 Conclusions and Prospects

In conclusion, rapid progresses have been reported in the field of textile nanogenerators. Both piezoelectric and triboelectric nanogenerators can be traced back to Maxwell's displacement current. Harvesting mechanical energies at low frequency is possibly the nanogenerators' killer application, considering their higher performances compared to conventional electromagnetic generators for human motion energies. Various fiber-based and textile-based nanogenerators have been designed. Meanwhile, hybrid energy textiles have also been demonstrated to be viable. Most of these energy textiles can function as common fabrics with excellent flexibility, washability, and breathability. Small electronics, ranging from LEDs to electronic watches, have been powered by these textile nanogenerators.

Despite these advancements, there are still several challenges. Further studies in terms of materials optimization, structure design, and performance improvement are still needed to render textile nanogenerators applicable for scale-up fabrication and feasible for practical applications. The fabrication of textile-based nanogenerators needs to be optimized to be compatible with the conventional textile engineering, and also consideration needs to be paid to their stylish design [69, 74]. For hybrid all-textile nanogenerators or nanogenerator-energy storage integrations, the optimization of the system efficiency is a challenging issue. The impedance/voltage mismatch between different energy-harvesting devices, or between energy-harvesting and storage devices, is a common problem leading to the low efficiency of the whole system. Even though several studies have shown promising improvements in the integration between TENG and energy-storage devices [75, 76], realization of power management in textile hybrid energy devices has not been reported yet. Nevertheless, in light of the growing advancements in this field and more attention received from the research community, textile nanogenerators are believed to be viable for practical applications in the near future.

References

1. Yetisen, A. K., Qu, H., Manbachi, A., Butt, H., Dokmeci, M. R., Hinestroza, J. P., Skorobogatiy, M., Khademhosseini, A. and Yun, S. H., Nanotechnology in textiles. *ACS Nano*, 10, 3042–3068, 2016.
2. Stoppa, M. and Chiolerio, A., Wearable electronics and smart textiles: A critical review. *Sensors*, 14, 11957, 2014.
3. Shim, B. S., Chen, W., Doty, C., Xu, C. and Kotov, N. A., Smart electronic yarns and wearable fabrics for human biomonitoring made by carbon nanotube coating with polyelectrolytes. *Nano Lett.*, 8, 4151, 2008.
4. Zhang, Z., Guo, K., Li, Y., Li, X., Guan, G., Li, H., Luo, Y., Zhao, F., Zhang, Q., Wei, B., Pei, Q. and Peng, H., A colour-tunable, weavable fibre-shaped polymer light-emitting electrochemical cell. *Nat. Photon.*, 9, 233, 2015.
5. El-Sherif, M. A., Yuan, J. and Macdiarmid, A., Fiber optic sensors and smart fabrics. *J. Intelligent Mater. Syst. Struct.*, 11, 407, 2000.
6. Rothmaier, M., Luong, M. and Clemens, F., Textile pressure sensor made of flexible plastic optical fibers. *Sensors*, 8, 4318, 2008.
7. Paul, G., Torah, R., Beeby, S. and Tudor, J., The development of screen printed conductive networks on textiles for biopotential monitoring applications. *Sens. Actuators A Phys.*, 206, 35, 2014.
8. Hu, L. and Cui, Y., Energy and environmental nanotechnology in conductive paper and textiles. *Energy Environ. Sci.*, 5, 6423, 2012.
9. Wang, S., Niu, S., Yang, J., Lin, L. and Wang, Z. L., Quantitative measurements of vibration amplitude using a contact-mode freestanding triboelectric nanogenerator. *ACS Nano*, 8, 12004, 2014.
10. Chen, B., Yang, Y., Wang, Z. L., Scavenging Wind Energy by Triboelectric Nanogenerators. *Adv. Energy Mater.*, 7, 1702649, 2017.
11. Zhu, G., Su, Y., Bai, P., Chen, J., Jing, Q., Yang, W. and Wang, Z. L., Harvesting water wave energy by asymmetric screening of electrostatic charges on a nanostructured hydrophobic thin-film surface. *ACS Nano*, 8, 6031, 2014.

12. Hou, T.-C., Yang, Y., Zhang, H., Chen, J., Chen, L.-J. and Wang, Z. L., Triboelectric nanogenerator built inside shoe insole for harvesting walking energy. *Nano Energy*, 2, 856, 2013.
13. Xie, Y., Wang, S., Niu, S., Lin, L., Jing, Q., Yang, J., Wu, Z. and Wang, Z. L., Grating-structured freestanding triboelectric-layer nanogenerator for harvesting mechanical energy at 85% total conversion efficiency. *Adv. Mater.*, 26, 6599, 2014.
14. Zhang, X.-S., Han, M.-D., Wang, R.-X., Meng, B., Zhu, F.-Y., Sun, X.-M., Hu, W., Wang, W., Li, Z.-H. and Zhang, H.-X., High-performance triboelectric nanogenerator with enhanced energy density based on single-step fluorocarbon plasma treatment. *Nano Energy*, 4, 123, 2014.
15. Zi, Y., Guo, H., Wen, Z., Yeh, M.-H., Hu, C. and Wang, Z. L., Harvesting low-frequency (<5 Hz) irregular mechanical energy: A possible killer application of triboelectric nanogenerator. *ACS Nano*, 10, 4797, 2016.
16. Wang, Z. L., On Maxwell's displacement current for energy and sensors: The origin of nanogenerators. *Mater. Today*, 20, 74, 2017.
17. Wang, Z. L., Jiang, T. and Xu, L., Toward the blue energy dream by triboelectric nanogenerator networks. *Nano Energy*, 39, 9, 2017.
18. Wang, Z. L. and Song, J., Piezoelectric nanogenerators based on zinc oxide nanowire arrays. *Science*, 312, 242, 2006.
19. Fan, F. R., Tang, W. and Wang, Z. L., Flexible nanogenerators for energy harvesting and self-powered electronics. *Adv. Mater.*, 28, 4283, 2016.
20. Yang, R., Qin, Y., Dai, L. and Wang, Z. L., Power generation with laterally packaged piezoelectric fine wires. *Nat. Nanotechnol.*, 4, 34, 2009.
21. Wang, Z. L., Zhu, G., Yang, Y., Wang, S. and Pan, C., Progress in nanogenerators for portable electronics. *Mater. Today*, 15, 532, 2012.
22. Hu, Y. and Wang, Z. L., Recent progress in piezoelectric nanogenerators as a sustainable power source in self-powered systems and active sensors. *Nano Energy*, 14, 3, 2015.
23. Kumar, B. and Kim, S.-W., Energy harvesting based on semiconducting piezoelectric ZnO nanostructures. *Nano Energy*, 1, 342, 2012.
24. Xu, S., Qin, Y., Xu, C., Wei, Y., Yang, R. and Wang, Z. L., Self-powered nanowire devices. *Nat. Nanotechnol.*, 5, 366, 2010.
25. Zhu, G., Yang, R., Wang, S. and Wang, Z. L., Flexible high-output nanogenerator based on lateral ZnO nanowire array. *Nano Lett.*, 10, 3151, 2010.
26. Xu, S., Wei, Y., Liu, J., Yang, R. and Wang, Z. L., Integrated multilayer nanogenerator fabricated using paired nanotip-to-nanowire brushes. *Nano Lett.*, 8, 4027, 2008.
27. Hu, Y., Zhang, Y., Xu, C., Lin, L., Snyder, R. L. and Wang, Z. L., Self-powered system with wireless data transmission. *Nano Lett.*, 11, 2572, 2011.
28. Zhu, G., Wang, A. C., Liu, Y., Zhou, Y. and Wang, Z. L., Functional electrical stimulation by nanogenerator with 58 V output voltage. *Nano Lett.*, 12, 3086, 2012.

29. Zhang, Y., Liu, C., Liu, J., Xiong, J., Liu, J., Zhang, K., Liu, Y., Peng, M., Yu, A., Zhang, A., Zhang, Y., Wang, Z., Zhai, J. and Wang, Z. L., Lattice strain induced remarkable enhancement in piezoelectric performance of ZnO-based flexible nanogenerators. *ACS Appl. Mater. Interfaces*, 8, 1381, 2016.
30. Lowell, J. and Rose-Innes, A., Contact electrification. *Adv. Phys.*, 29, 947, 1980.
31. Duke, C. B. and Fabish, T. J., Contact electrification of polymers: A quantitative model. *J. Appl. Phys.*, 49, 315, 1978.
32. Zhou, Y. S., Liu, Y., Zhu, G., Lin, Z.-H., Pan, C., Jing, Q. and Wang, Z. L., *In situ* quantitative study of nanoscale triboelectrification and patterning. *Nano Lett.*, 13, 2771, 2013.
33. Niu, S. and Wang, Z. L., Theoretical systems of triboelectric nanogenerators. *Nano Energy*, 14, 161, 2015.
34. Wang, Z. L., Triboelectric nanogenerators as new energy technology for self-powered systems and as active mechanical and chemical sensors. *ACS Nano*, 7, 9533, 2013.
35. Hinchet, R., Seung, W. and Kim, S.-W., Recent progress on flexible triboelectric nanogenerators for self-powered electronics. *ChemSusChem*, 8, 2327, 2015.
36. Wang, Z. L., Chen, J. and Lin, L., Progress in triboelectric nanogenerators as a new energy technology and self-powered sensors. *Energy Environ. Sci.*, 8, 2250, 2015.
37. Zhu, G., Peng, B., Chen, J., Jing, Q. and Wang, Z. L., Triboelectric nanogenerators as a new energy technology: From fundamentals, devices, to applications. *Nano Energy*, 14, 126, 2014.
38. Wang, Z. L., Triboelectric nanogenerators as new energy technology and self-powered sensors – principles, problems and perspectives. *Faraday Discuss.*, 176, 447, 2014.
39. Pu, X., Liu, M., Chen, X., Sun, J., Du, C., Zhang, Y., Zhai, J., Hu, W. and Wang, Z. L., Ultrastretchable, transparent triboelectric nanogenerator as electronic skin for biomechanical energy harvesting and tactile sensing. *Sci. Adv.*, 3, e1700015, 2017.
40. Zhang, H., Yang, Y., Su, Y., Chen, J., Adams, K., Lee, S., Hu, C. and Wang, Z. L., Triboelectric nanogenerator for harvesting vibration energy in full space and as self-powered acceleration sensor. *Adv. Funct. Mater.*, 24, 1401, 2014.
41. Zhao, Z., Pu, X., Du, C., Li, L., Jiang, C., Hu, W. and Wang, Z. L., Freestanding flag-type triboelectric nanogenerator for harvesting high-altitude wind energy from arbitrary directions. *ACS Nano*, 10, 1780, 2016.
42. Bae, J., Lee, J., Kim, S., Ha, J., Lee, B.-S., Park, Y., Choong, C., Kim, J.-B., Wang, Z. L., Kim, H.-Y., Park, J.-J. and Chung, U. I., Flutter-driven triboelectrification for harvesting wind energy. *Nat. Commun.*, 5, 4929, 2014.
43. Wang, Z. L., Catch wave power in floating nets. *Nature*, 542, 159, 2017.
44. Yu, A., Jiang, P. and Wang, Z. L., Nanogenerator as self-powered vibration sensor. *Nano Energy*, 1, 418, 2012.

45. Bai, P., Zhu, G., Lin, Z.-H., Jing, Q., Chen, J., Zhang, G., Ma, J. and Wang, Z. L., Integrated multilayered triboelectric nanogenerator for harvesting biomechanical energy from human motions. *ACS Nano*, 7, 3713, 2013.
46. Qin, Y., Wang, X. and Wang, Z. L., Microfibre-nanowire hybrid structure for energy scavenging. *Nature*, 451, 809, 2008.
47. Li, Z. and Wang, Z. L., Air/liquid-pressure and heartbeat-driven flexible fiber nanogenerators as a micro/nano-power source or diagnostic sensor. *Adv. Mater.*, 23, 84, 2011.
48. Chang, C., Tran, V. H., Wang, J., Fuh, Y.-K. and Lin, L., Direct-write piezoelectric polymeric nanogenerator with high energy conversion efficiency. *Nano Lett.*, 10, 726, 2010.
49. Bai, S., Zhang, L., Xu, Q., Zheng, Y., Qin, Y. and Wang, Z. L., Two dimensional woven nanogenerator. *Nano Energy*, 2, 749, 2013.
50. Zhang, M., Gao, T., Wang, J., Liao, J., Qiu, Y., Yang, Q., Xue, H., Shi, Z., Zhao, Y., Xiong, Z. and Chen, L., A hybrid fibers based wearable fabric piezoelectric nanogenerator for energy harvesting application. *Nano Energy*, 13, 298, 2015.
51. Khan, A., Hussain, M., Nur, O., Willander, M. and Broitman, E., Analysis of direct and converse piezoelectric responses from zinc oxide nanowires grown on a conductive fabric. *Phys. Status Solidi (a)*, 212, 579, 2015.
52. Almusallam, A., Luo, Z., Komolafe, A., Yang, K., Robinson, A., Torah, R. and Beeby, S., Flexible piezoelectric nano-composite films for kinetic energy harvesting from textiles. *Nano Energy*, 33, 146, 2017.
53. Soin, N., Shah, T. H., Anand, S. C., Geng, J., Pornwannachai, W., Mandal, P., Reid, D., Sharma, S., Hadimani, R. L., Bayramol, D. V. and Siores, E., Novel "3-D spacer" all fibre piezoelectric textiles for energy harvesting applications. *Energy Environ. Sci.*, 7, 1670, 2014.
54. Zeng, W., Tao, X.-M., Chen, S., Shang, S., Chan, H. L. W. and Choy, S. H., Highly durable all-fiber nanogenerator for mechanical energy harvesting. *Energy Environ. Sci.*, 6, 2631–2638, 2013.
55. Zhong, J., Zhang, Y., Zhong, Q., Hu, Q., Hu, B., Wang, Z. L. and Zhou, J., Fiber-based generator for wearable electronics and mobile medication. *ACS Nano*, 8, 6273, 2014.
56. Sim, H. J., Choi, C., Kim, S. H., Kim, K. M., Lee, C. J., Kim, Y. T., Lepró, X., Baughman, R. H. and Kim, S. J., Stretchable triboelectric fiber for self-powered kinematic sensing Textile. *Sci. Rep.*, 6, 35153, 2016.
57. Zhou, T., Zhang, C., Han, C. B., Fan, F. R., Tang, W. and Wang, Z. L., Woven structured triboelectric nanogenerator for wearable devices. *ACS Appl. Mater. Interfaces*, 6, 14695, 2014.
58. Pu, X., Li, L., Song, H., Du, C., Zhao, Z., Jiang, C., Cao, G., Hu, W. and Wang, Z. L., A self-charging power unit by integration of a textile triboelectric nanogenerator and a flexible lithium-ion battery for wearable electronics. *Adv. Mater.*, 27, 2472, 2015.

59. Zhao, Z., Yan, C., Liu, Z., Fu, X., Peng, L.-M., Hu, Y. and Zheng, Z., Machine-washable textile triboelectric nanogenerators for effective human respiratory monitoring through loom weaving of metallic yarns. *Adv. Mater.*, 28, 10267, 2016.
60. Lai, Y.-C., Deng, J., Zhang, S. L., Niu, S., Guo, H. and Wang, Z. L., Single-thread-based wearable and highly stretchable triboelectric nanogenerators and their applications in cloth-based self-powered human-interactive and biomedical sensing. *Adv. Funct. Mater.*, 27, 1604462, 2017.
61. Dong, K., Deng, J., Zi, Y., Wang, Y.-C., Xu, C., Zou, H., Ding, W., Dai, Y., Gu, B., Sun, B. and Wang, Z. L., 3D orthogonal woven triboelectric nanogenerator for effective biomechanical energy harvesting and as self-powered active motion sensors. *Adv. Mater.*, 29, 1702648, 2017.
62. Seung, W., Gupta, M. K., Lee, K. Y., Shin, K.-S., Lee, J.-H., Kim, T. Y., Kim, S., Lin, J., Kim, J. H. and Kim, S.-W., Nanopatterned textile-based wearable triboelectric nanogenerator. *ACS Nano*, 9, 3501, 2015.
63. Ko, Y. H., Nagaraju, G. and Yu, J. S., Multi-stacked PDMS-based triboelectric generators with conductive textile for efficient energy harvesting. *RSC Adv.*, 5, 6437, 2015.
64. Cui, N., Liu, J., Gu, L., Bai, S., Chen, X. and Qin, Y., Wearable triboelectric generator for powering the portable electronic devices. *ACS Appl. Mater. Interfaces*, 7, 18225, 2015.
65. Li, S., Zhong, Q., Zhong, J., Cheng, X., Wang, B., Hu, B. and Zhou, J., Cloth-based power shirt for wearable energy harvesting and clothes ornamentation. *ACS Appl. Mater. Interfaces*, 7, 14912, 2015.
66. Zhu, M., Huang, Y., Ng, W. S., Liu, J., Wang, Z., Wang, Z., Hu, H. and Zhi, C., 3D spacer fabric based multifunctional triboelectric nanogenerator with great feasibility for mechanized large-scale production. *Nano Energy*, 27, 439, 2016.
67. Zhang, L., Yu, Y., Eyer, G. P., Suo, G., Kozik, L. A., Fairbanks, M., Wang, X. and Andrew, T. L., All-textile triboelectric generator compatible with traditional textile process. *Adv. Mater. Technol.*, 1, 1600147, 2016.
68. Pu, X., Song, W., Liu, M., Sun, C., Du, C., Jiang, C., Huang, X., Zou, D., Hu, W. and Wang, Z. L., Wearable power-textiles by integrating fabric triboelectric nanogenerators and fiber-shaped dye-sensitized solar cells. *Adv. Energy Mater.*, 6, 1601048, 2016.
69. Pu, X., Li, L., Liu, M., Jiang, C., Du, C., Zhao, Z., Hu, W. and Wang, Z. L., Wearable self-charging power textile based on flexible yarn supercapacitors and fabric nanogenerators. *Adv. Mater.*, 28, 98, 2016.
70. Li, X., Lin, Z.-H., Cheng, G., Wen, X., Liu, Y., Niu, S. and Wang, Z. L., 3D fiber-based hybrid nanogenerator for energy harvesting and as a self-powered pressure sensor. *ACS Nano*, 8, 10674, 2014.
71. Pan, C., Guo, W., Dong, L., Zhu, G. and Wang, Z. L., Optical fiber-based core-shell coaxially structured hybrid cells for self-powered nanosystems. *Adv. Mater.*, 24, 3356, 2012.

72. Bae, J., Park, Y. J., Lee, M., Cha, S. N., Choi, Y. J., Lee, C. S., Kim, J. M. and Wang, Z. L., Single-fiber-based hybridization of energy converters and storage units using graphene as electrodes. *Adv. Mater.*, 23, 3446, 2011.
73. Chen, J., Huang, Y., Zhang, N., Zou, H., Liu, R., Tao, C., Fan, X. and Wang, Z. L., Micro-cable structured textile for simultaneously harvesting solar and mechanical energy. *Nat. Energy*, 1, 16138, 2016.
74. Wen, Z., Yeh, M.-H., Guo, H., Wang, J., Zi, Y., Xu, W., Deng, J., Zhu, L., Wang, X., Hu, C., Zhu, L., Sun, X. and Wang, Z. L., Self-powered textile for wearable electronics by hybridizing fiber-shaped nanogenerators, solar cells, and supercapacitors. *Sci. Adv.*, 2, e1600097, 2016.
75. Niu, S., Wang, X., Yi, F., Zhou, Y. S. and Wang, Z. L., A universal self-charging system driven by random biomechanical energy for sustainable operation of mobile electronics. *Nat. Commun.*, 6, 8975, 2015.
76. Pu, X., Liu, M., Li, L., Zhang, C., Pang, Y., Jiang, C., Shao, L., Hu, W. and Wang, Z. L., Efficient charging of Li-ion batteries with pulsed output current of triboelectric nanogenerators. *Adv. Sci.*, 3, 1500255, 2016.

7

Nanocomposites for Smart Textiles

Nazire Deniz Yilmaz

Textile Technologist Consultant, Denizli, Turkey

Abstract

Nanocomposites show promise for use in an array of different fields including transportation, aerospace, electronics, biomedicine, and protection applications. It is expected that nanocomposites will act in a positive way on our life quality in the near future. Nanocomposites exhibit design and property combinations, which are not possible for the case of conventional "micro" composites. Nanocomposites offer advanced multifunctions without interfering the comfort and aesthetics of textiles. As components of smart textiles, nanocomposites offer service for use in sensors, actuators, fire protection, defense applications, biosensing, self-cleaning, moisture management, thermoregulation, energy storing, and harvesting, among other sophisticated niche applications. While nanocomposites have already found use in various applications, there are numerous potential fields where nanocomposites can provide better service compared to conventional counterparts in the future.

Keywords: Nanocomposites, smart textiles, wearables, carbon nanotubes, nanocellulose, nanoparticles, nanoclay, conducting polymers

7.1 Introduction

Nanotechnology can be described as manufacturing and manipulating structures, at least one dimension of which falls within the range 1–100 nm. Nanotechnology deals with these nanoscale materials as well as materials containing nanoscale components, such as nanocomposites. As the dimension size falls into the nano range, the physical and chemical properties differ greatly compared to those of their bulk counterparts. Nanotechnology is an interdisciplinary area including chemistry, physics, materials science, and engineering [1, 2].

Email: naziredyilmaz@gmail.com

Nazire D. Yilmaz (ed.) Smart Textiles, (211–246) © 2019 Scrivener Publishing LLC

Extensive attention has been devoted to nanotechnology in the last two decades based on their outstanding properties [1]. Via nanotechnology, manufacturing of new structures exhibiting improved performance (due to ultrahigh surface area), lower handling cost (e.g. self-healing, low material content), and additional functionality (adaptive materials, electronic/opto-electronic/magnetic material properties) have become possible [3]. Nanocomposites show promise for use in an array of different fields including transportation, aerospace, electronics, biomedicine, and protection applications. It is expected that nanocomposites will affect our life quality positively in the near future [1, 2].

Nanocomposites are composite materials that include finely dispersed phases at least one of which is with dimensions in the nanoscale such as carbon nanotubes or lamellar clay [4]. Nanocomposites show promise especially for sophisticated, niche applications [5]. Nanocomposites can especially offer advanced multiple functions by overcoming the limitations related with conventional materials. By offering functionalities, nanocomposites can serve as components of smart textiles without interfering with the comfort and aesthetics [2, 6].

The nanoscale phase of these composites presents dimensions, one of which must be in the nanometric range that is less than 100 nm. There must be repeat distances at the nanoscale between the different phases of the composite to be rendered as a "nanocomposite." Interestingly, nanocomposites are not just materials of the advanced technology; they can be also found in natural structures [7].

Nanocomposites exhibit design and property combinations, which are not possible for the case of conventional "micro" composites. In other words, nanocomposites are free from some limitations that are valid for macro and micro counterparts. Material characteristics show changes when the dimensions approach a critical level. At the nanometric scale, interactions at the interfaces between different phases substantially increase. Nanometric dimensions lead to great surface area-to-volume ratios. And it is the interfaces that play a very important role in determining the properties of composite materials [2, 3]. Hence, with nanocomposites, it is possible to obtain characteristics that exhibit great differences in comparison to their components [1].

In order to obtain smart functionality, very complicated structures are needed. This necessitates rigid bulky systems when using conventional materials. If lightweight, flexible, smaller devices are needed, nanocomposites should be used, which offer integrated components in much smaller scales. This makes nanocomposites very important in terms of smart textiles. These give smart responses to external stimuli as designed in the development stage. Energy is needed in order to make response possible. This energy can be transferred from the stimuli to the smart component. The necessity of energy transfer renders the structure extra complicated. This results in integrated and sophisticated systems such as nanocomposites. Smart nanocomposites are capable of data processing, analysis, and response plus energy transfer/harvest [8]. Some are able to adapt themselves to the changes in the surrounding, that is, changes in temperature, pressure, light intensity, electrical field strength, etc. [9].

In terms of smart textiles, flexibility is a very important factor that provides comfort during use and longer service life. Multifunctional sensors are advantageous that they allow less material and energy consumption and provide more comfort. Multifunctionality, most of the time, necessitates a composite structure that comprises contents of different functions together [10]. Lightweight, flexibility, comfort, and multifunctionality are all possible only with nanocomposites.

Nanocomposites have also found use in various smart textile applications including sensors [10, 11], defense protection [3], self-cleaning [12], antibacterial [13], moisture management [14], fire protection [15], actuators [16–18], and energy harvesting [19].

This chapter investigates nanocomposites for use in smart textile applications. The following section presents the classification of nanocomposites. The third section studies structure and property relationships of nanocomposites. The fourth section refers to production methods of nanocomposites. The fifth section focuses on components of nanocomposites. The sixth section describes nanocomposite types. The seventh section reviews smart applications of nanocomposites. The last section concludes the chapter. Based on the innumerable functionalities that nanocomposites can offer, it is impossible to cover all of them in a single chapter. Thus, the chapter presents a summary of the recent advancements in nanocomposite research targeting different smart textile applications.

7.2 Classification of Nanocomposites

Similar to conventional composites, nanocomposites can be categorized into three groups in terms of their matrix materials:

- Ceramic matrix nanocomposites
- Metal matrix nanocomposites
- Polymer matrix nanocomposites [20].

It is projected that nanocomposites with metal and ceramic matrices will pose important impact on a number of fields such defense, electronics, and aerospace, whereas polymer nanocomposites show promise for use in sensors, microelectronics, nanogenerators, protection, smart textiles, and so on. Nanocomposites can act as components of sensors, catalysts, and electrodes, among other applications [2, 21, 22].

Other than the matrix type, nanocomposites can also be classified mainly based on the form of their reinforcements as follows:

- Fiber-reinforced composites
- Particulate composites
- Laminar composites.

Here, fiber-reinforced composites can further be grouped as discontinuous fiber- or continuous fiber-reinforced composites [23, 24].

According to another approach, nanocomposites can be classified into four groups:

- Zero-dimensional (core-shell)
- One-dimensional (nanotube-, nanofiber-reinforced)
- Two-dimensional (lamellar)
- Three-dimensional (nanoparticulate-reinforced) [20].

Nanocomposites can be also grouped in terms of the dimensions of their components. If all three dimensions fall in the nano range, they are referred to as isodimensional nanoparticles such as metal nanoparticles, semiconductor nanoclusters, and spherical silica. The second class is qualified by the two dimensions that fall into the nano range, such as cellulose whiskers, nanofibers, and carbon nanotubes, while the remaining dimension is in the micro range. These have found extensive use in the nanocomposites area. The last group has only one of its dimensions in the nanorange. This group includes sheets with thickness in nanometers. These include nanoclays and layered silicates [1, 2].

7.2.1 Nanocomposites Based on Matrix Types

Among different types of nanocomposites, ceramic matrix nanocomposites present a special group. Ceramic materials exhibit good level of resistance to wear, temperature, and chemicals. The downside of these materials is brittleness, which prevents wide use in various applications. To overcome this obstacle, nanofillers, which have energy-dissipating capability including nanofibers, whiskers, nanoparticles, and platelets, are included in ceramic matrices. These agents act as load carrying and/or bridging components. The reinforcements deflect the crack and/or provide load bridging effect, and protects nano- and microcracks from transforming into macrocracks [2, 25]. Ceramic matrix nanocomposites based on matrices including SiN, SiC, and Al_2O_3. Al_2O_3 are widely used for reinforcing ceramic-matrix nanocomposites to increase strength and toughness. Ceramic matrix materials can be produced via different methods including conventional powder technique, polymer precursor method, spray pyrolysis, chemical and physical vapor deposition techniques, sol–gel technique, and template synthesis techniques [2, 26, 27].

Metal matrix nanocomposites include metal or alloy matrices, which are inherently ductile, and nanofillers incorporated within them. Metal matrix composites comprise matrices of Al, Fe, Cr, Mg, and W. They show high toughness, strength, and modulus. Metal matrix nanocomposites can be prepared via spray pyrolysis, chemical and physical vapor deposition techniques, electro-deposition, and sol–gel technique. Metal matrix nanocomposites are durable to high temperatures. They are promising for various application areas such as aerospace, transportation, and structural applications [2, 20].

Compared to the other two nanocomposite types, polymer matrix nanocomposites form a special class. Polymer matrix composites are based on matrix materials such as polyolefins (polyethylene, polypropylene), condensation polymers (polyester, polyamid), vinyls, and other special polymers [2, 28]. Polymers generally present some advantages such as lightweight, easy processibility, and corrosion resistance. On the other hand, their shortcomings include low strength and low thermal resistance. Nevertheless, there are some properties that are desired for certain applications while being considered as weakness for others. An example of this is low modulus. It is a drawback for components in structural elements [2]. However, it is a must for textile applications that necessitate flexibility, elasticity, and resilience such as smart textile applications [29]. Whereas metal and ceramic matrix nanocomposites have the potential to cater various industries such as transportation, aerospace, and defense, polymer matrix nanocomposites show promise for use in sensors, batteries, microelectronics, and wearables [2, 30, 31]. The remaining of this chapter will focus mainly on polymer matrix nanocomposites.

7.3 Structure and Properties of Nanocomposites

As composites, nanocomposites exhibit heterogeneous structures. Hence, their characteristics are influenced by the properties of the components, the component composition, structure, and interactions at the interfaces as in the case of microcomposites. Nevertheless, they present more complicated structures compared to those conventional microcomposites [5].

In composites, it is the interface that has great influence on the resulting performance. At the interface region, the properties show changes in comparison to the rest of the component phases. At this area, mobility of polymer chain, curing level, and crystalline structure are different from the remaining parts. If the interface area can be increased, the influence on the final characteristics of composites becomes stronger. This is the case with nanocomposites, as well. The nanoscale particles exhibit very high surface areas per volume and mass, leading to great interface areas. Thus, targeted performance is attained at much lower loadings compared to conventional microcomposites due to the great interface areas [3].

Nanocomposites are not without difficulties. There are challenges in terms of controlling the elemental composition as well as stoichiometry in the nanophases [2]. Aggregation and orientation are among the major issues related with the nanocomposites [5]. It should be noted that we have not been able to fully understand yet nanocomposites [2, 3].

The formation of interphases in nanocomposites has not been fully understood yet. However, the characteristics of the interphases play a very important part in determining the properties of the resultant nanocomposites [5]. Furthermore, interactions at the interfaces have become more important in nanocomposites compared to the conventional ones, as the former ones present much greater interface areas with respect to the mass and volume, as mentioned before [5].

The reinforcement materials are exposed to surface modification methods in order to prevent agglomeration and to enhance adhesion to the matrix material. Prevention of agglomeration is necessary to obtain a uniform composite structure. In nanoscale dimensions, agglomeration comes out as a major challenge when compared to micro-counterparts in terms of great surface areas. Incorporation of surfactants as well as chemical functionalization are utilized for interfacial bond enhancement. In a different approach, silane coupling agents may be applied for producing repulsion of nanoparticles to enhance dispersion and introducing better compatibility with the matrix [12]. Physical blending, *in situ* polymerization, and ultrasonication are some other methods to establish uniform dispersion of nano-reinforcement elements in matrices [2, 27]. Besides other properties, uniformity also affects transparency positively [9], which may be advantageous in terms of aesthetics of wearables.

7.4 Production Methods of Nanocomposites

There are a number of methods for producing composites. These include (for thermosetting polymers) hand lay-up, spraying, compression, resin transfer molding, injection molding, pultrusion, and foam molding; and (for thermoplastic matrix polymers) extrusion, injection molding, thermoforming, compression, foam molding, and co-extrusion methods [32].

In nanocomposites, reinforcements represent nano-range dimensions. The reinforcers include carbon nanotubes, nanocellulose, conducting polymers, metal oxides, silica, and so on [11]. Nanoscale reinforcements are produced by different techniques such as chemical and mechanical milling, vapor deposition, and co-precipitation, among other methods [2, 20]. Means of production of nanofibers with nanocomposite structure can be listed as coaxial core-shell electrospinning, conjugate electrospinning, and island-in-the-sea methods [22].

Polymer matrix nanocomposites can be produced via various processes. These methods include melt homogenization [5], *in situ* polymerization [9, 27], sol–gel method [20], electrodeposition, solution dispersion [27, 33], template synthesis, and other advanced processes such as self-assembly and atomic layer deposition [2, 12]. Figure 7.1 depicts smart nanocomposite production via the spraying layer by layer method.

It is very important to find the optimum concentration of nanofillers to attain the best characteristics as possible. Concentrations higher than the

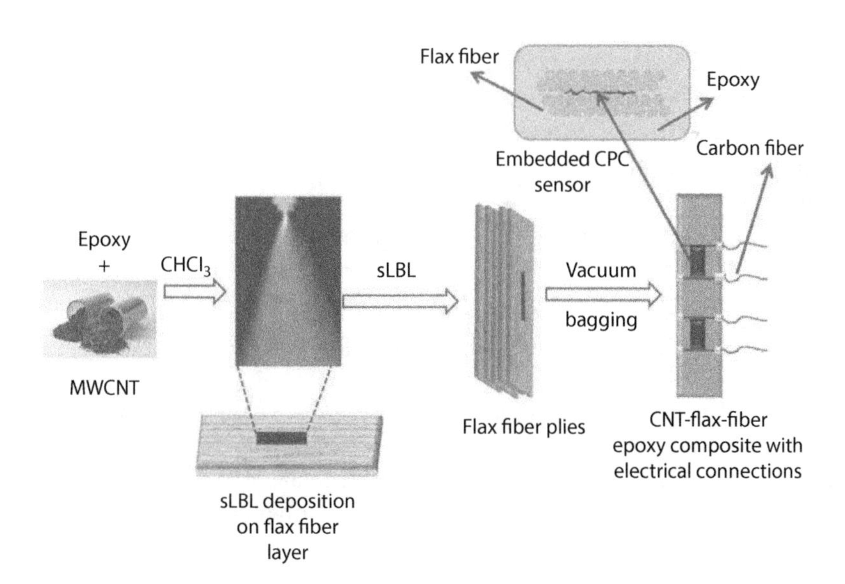

Figure 7.1 Production method of carbon nanotube-flax-epoxy nanocomposites. sLBL: spraying layer by layer [34]. Note: The source [34] is an Open Access article distributed under the terms of the Creative Commons Attribution License (http://creativecommons.org/licenses/by/4.0/), which permits unrestricted use, distribution, and reproduction in any medium, provided the original work is properly cited.

optimum point result in agglomeration and consequent decrease in strength parameters [2]. Efficient dispersion of the nanofiller in the polymer matrix leads to a higher interface area. This results in constriction of matrix motion and, thus, enhances mechanical and thermal characteristics [1].

Good dispersion of nanoparticles in the matrix is a must to achieve the designed functions. Good dispersion is attained when there is no aggregation. Aggregation tendency is determined by the type and the concentration of the nanomaterial, matrix types, the nanomaterial–matrix interaction, and the production method [1].

Even though considerable progress has been achieved in preparation of polymer nanocomposites, more understanding is still needed to determine the best method for specific polymer-reinforcement combinations, as well as the optimal matrix/reinforcement ratios to attain the best properties at a cost-efficient manner [2].

7.5 Nanocomposite Components

In order to enhance performance characteristics of polymers, various fillers, such as fibers, whiskers, platelets, or particles, both in the micro- and nanoscale, are used. These reinforcements impart thermal durability, resistance to impact, flame retardancy, high strength, and electrical resistance. Metallic and ceramic fillers are used for achieving further optical, magnetic, and electrical properties. Via use of nanosized fillers, these unique features can be attained without compromising capabilities of polymers like flexibility, film formability, and ease of processing. Furthermore, using nanofillers, these functionalities can be provided at contents much lower compared to their microsized counterparts. Nanoscale reinforcements result in improvements in not only smart functionalities but also basic mechanical properties such as strength, toughness, thermal stability, hardness, etc. [1, 2].

Nanoscale reinforcements may be conductive (metallic nanoparticles), semiconductive, or insulating [2]. The components of nanocomposites include carbon nanotubes, nanoparticles of metals, metal oxides and inorganic materials, [1], nanocellulose, conducting polymers, and nanoclay [11].

7.5.1 Carbon Nanotubes

Roger Bacon was the first to find multiwalled carbon nanotubes in the 1950s [35]. The first nanotubes to be observed were multiwalled carbon nanotubes as shown in Figure 7.2. Multiwalled carbon nanotubes are

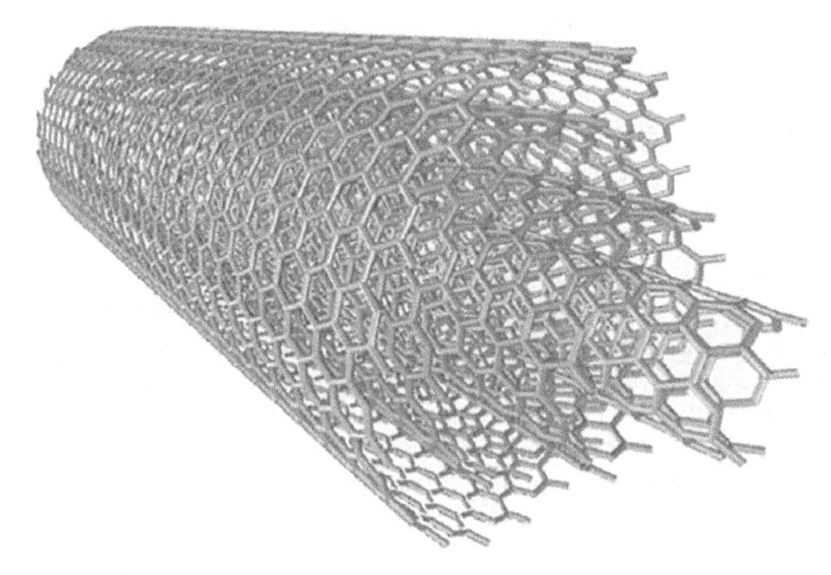

Figure 7.2 Structure of a multiwalled carbon nanotube [36]. (Reprinted from reference [36], with permission of Elsevier.)

composed of two or more coaxial cylindrical graphene sheets, while single-walled carbon nanotubes include only one graphene cylinder sheet. Both multi- and single-walled carbon nanotubes are microcrystals and exhibit solid physical properties, while the diameters fall in the range of molecular dimensions. Invention of carbon nanotubes carried nanocomposite research to the next level. Carbon nanotubes present outstanding mechanical, thermal, and electrical characteristics [2, 3].

As the carbon atoms show a helical arrangement and the dimensions are in the nano range, electrical properties are remarkably affected. Thus, even though graphite exhibits semimetal properties, carbon nanotubes can show metallic or semiconducting characteristics. The invention of carbon nanotubes has contributed to the attention on fullerenes. Despite their outstanding properties, fullerenes have found limited use in composites. The structure of carbon nanotubes shows difference in comparison to conventional carbon fibers used as reinforcement in the composite industry. However, in fact, carbon nanotubes exhibit the most perfect ordered carbon fiber structure investigated at the atomic level [3].

Single-walled carbon nanotubes are generally found in bundles. Here, individual single-walled carbon nanotubes present stiffness values substantially higher than those of the bundle they belong to. Incorporation of carbon nanotubes into polymers leads to improvements in electrical and thermal properties [3]. Single-walled carbon nanotubes exhibit a density level corresponding to 1/6 of steel, and that of multiwalled carbon nanotubes is around 1/2 of the density of aluminum. Tensile strength values of carbon nanotubes are higher than high-strength steel, whereas rigidity is close to diamond. They present substantial resilience; in other words, they can endure flexing in great angles and high strains before getting damaged. In this aspect, they are superior to carbon fibers, which suffer brittleness. Moreover, carbon nanotubes possess theoretical electrical and thermal conductivity levels, which are close to those of diamond. Their thermal expansion factor is nearly zero. Their thermal stabilities in vacuum and air conditions are higher than those of metal wires. Their specific surface areas are around 3000 $m^2\ g^{-1}$. Whereas the theoretical values are as mentioned, the measured values differ based on experimental conditions [2].

Whereas carbon nanotubes are generally produced via chemical vapor deposition methods, carbon nanotube-reinforced polymer nanocomposites can be produced via conventional blending, solution mixing, melt blending, and *in situ* polymerization [2]. Incorporation of carbon nanotubes in nanocomposites leads to remarkably enhanced mechanical properties such as toughness, strength, and stiffness [1, 2].

Carbon nanotubes also exhibit antibacterial activity. Direct contact results in killing *E. coli* bacteria, which may be due to puncture of the cells and leaking of their cellular material. Nevertheless, adverse effects on humans may also be present, which would be of concern to people involved in production processes of carbon nanotubes rather than the final consumers [1]. Nanocomposites of carbon nanotubes were used for pressure-, optical-, thermal-, and biosensors, self-cleaning, antibacterial, energy storing, and energy-harvesting applications, among others [22, 36–38]. Even though carbon nanotubes pose some challenges during handling, their outstanding characteristics show promise for a wide array of applications [2].

7.5.2 Carbon Nanofiber

Carbon nanofibers (CNF) form a unique structure falling between carbon fibers with diameters between 5 and 10 μm and carbon nanotubes of diameters 1–10 nm. Carbon nanofibers exhibit greater surface areas as compared to carbon fibers and offer advantages entailed with this. The diameter of carbon nanofibers is around 100–200 nm, whereas the length can vary from 100 μm to several centimeters. They exhibit aspect ratios typically greater than 100. They can exhibit different shapes including truncated cones, whole cones, and stacked coins [2, 3]. Carbon nanofiber-based nanocomposites have been utilized in nanocomposites for use in sensor applications [39].

7.5.3 Nanocellulose

Cellulose is one of the two most abundant materials available on earth, the other one being starch [40]. Cellulose is a biodegradable and biocompatible component of composites based on renewable resources [41]. Cellulose, in nano form, has found use in sensors (ion sensors, strain sensors, metal ion sensors), biosensors, energy harvesting [42], and actuator applications [11, 43].

Nanocellulose, besides common natural cellulose, exhibits high strength, low thermal expansion coefficient, transparency [17], hydrophilic nature, and dispersibility in aqua [43, 44]. Naturally, cellulosic fibers exhibit a composite structure including bundles of microfibrils showing highly ordered crystalline and irregular amorphous structures [43, 45]. After removal of the amorphous parts via acid hydrolysis processes, nanocrystals are left behind, which are long cellulose crystals exhibiting very high stiffness and strength [1]. Different self-assembled nanocrystal forms of cellulose include cellulose nanowhiskers and microfibrils [11].

The key factor for use of cellulose in smart applications is the interaction between the negative charge of cellulose and the charged analytes. This has strong influence on ion transfer as well as selective permeability. The possibility of modifying cellulose suggests flexibility to cater different demands [43]. A difficulty in working cellulose with conventional polymers is the incompatibility between the hydrophilic polar cellulose structure and hydrophobic apolar polymers. In order to solve this issue, chemical surface treatments have been developed [45].

In a relevant study, Thielemans *et al.* [46] developed membranes of cellulose nanowhiskers exhibiting selective permeability based on the charge of species. In the preparation stage, aqueous cellulose nanocrystal solutions were produced via esterification of surface hydroxyl groups, which render cellulose highly reactive, by using sulfuric acid solutions.

The surface charges of nanocellulose affect the mediator performance of sensors [47]. Nanocellulose was reported to show high permeability to cationic analytes, while low permeability to negatively charged ions [46]. Another nanocomposite was produced by polymerization of pyrrole on microfibrillated cellulose to give a polypyrrole-microfibrillated system presenting open porous hydrogel structure. This composite shows promise for use in ion exchange and energy-storing applications [48].

A nanocellulose component used in composites is bacterial (microbial) cellulose, that is, cellulose produced by bacteria rather than the widely known cellulose found in plant structures. Bacterial cellulose exhibits nanostructure [44]. A glucose biosensor comprising bacterial cellulose was reported by Lv *et al.* [42] exhibiting flexibility and self-powering as shown in Figure 7.3. Nanocomposites of bacterial cellulose and gold were developed for use in biosensors [44].

In a strain sensor application, nanocomposites of bacterial cellulose—double-walled carbon nanotubes—were produced. Higher content of carbon nanotubes resulted in higher strain sensitivity [49].

Nanocellulose have also found use in sensors detecting pH and heavy metal ions. Here, detector molecules or colorimetric reagents are immobilized on cellulose surfaces [43, 50].

Cellulose has also found use in actuator and strain sensor applications. In 2002, interestingly, an actuation mechanism in cellulose paper was discovered [16]. Then, this paper was referred to as electroactive paper. The electroactive paper underwent a bending motion as electric voltage was applied. Voltage, frequency and paper type are the major factors affecting the bending motion [17]. Ion migration and piezoelectric property impart electroactivity. Piezoelectric function is obtained when applied mechanical strain leads to electric generation or polarization. This effect can be observed in some non-conducting materials including ceramics and quartz crystals. Piezoelectric effect can be obtained from cellulose due to its dipolar orientation as well as monoclinic crystal form. This property can be further enhanced through application of electric and magnetic fields, and mechanical strain [43]. Yang *et al.* [18] reported improved piezoelectric properties and stiffness as a result of applied mechanical stretching on

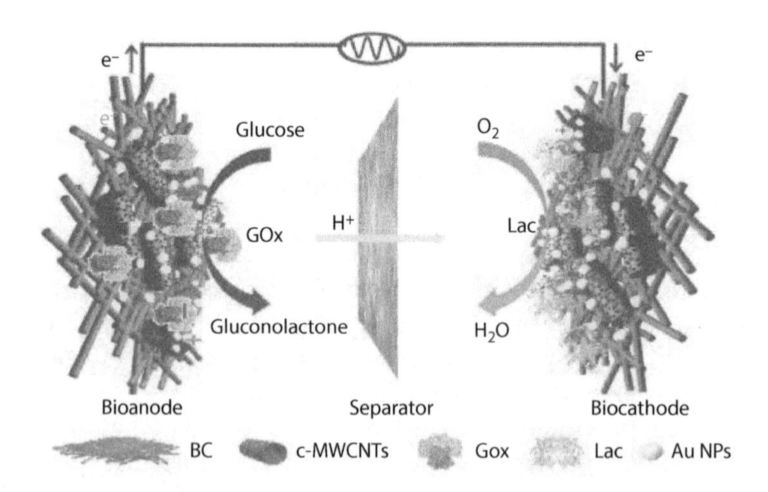

Figure 7.3 A glucose biosensor based on gold nanoparticle–bacterial cellulose nanocomposite. (Reprinted from reference [42], with permission of Elsevier.)

electroactive paper. Stretching leads to orientation in randomly oriented cellulose chains in amorphous regions.

Electroactive paper based on cellulose can be prepared as composite structures with carbon nanotubes, metal oxides, and polymers. Electroactive papers can be used in strain sensors, biosensors, chemical sensors and actuators, and energy-harvesting applications [17]. Applying polypyrrole on the electroactive paper improved actuating performance. Introduction of single-walled carbon nanotubes and multiwalled carbon nanotubes to cellulose electroactive paper leads to improved force and actuation frequency. Metal oxide incorporation in the cellulose improves chemical durability, mechanical strength, conducting property, and sensitivity to light. As seen from the mentioned studies, applicability of electroactive paper in smart applications can be greatly extended via introduction of nanofillers to overcome its drawbacks [17].

7.5.4 Conducting Polymers

An important subgroup of nanocomposites is conducting polymer-based composites. Via rational design and effective optimization of conductive polymer-based nanocomposites' parameters, advancement in research and feasibility of commercialization can be achieved [2, 3]. These conducting polymer nanocomposites may exhibit magnetic susceptibility, dielectric, piezoresistive, catalytic, or energy-storage and -harvesting capabilities [1, 2].

Conducting polymers including polypyrrole and polyaniline have found common use in smart applications. Conducting polymers generally exhibit low strength in comparison to conventional polymers. Moreover, they exhibit low solubility and dispersibility in common organic solvents. Thus, it is not very feasible to coat them via common methods. Hence, it has become advantageous to use them in a nanocomposite composition with additional components [43].

In a study, a conductive paper has been prepared from a nanocomposite of polypyrrole and cellulose. They obtained improved absorption for larger anions. The results show promise for use in sensors for biomedicine applications where biomarker, DNA, and other protein extraction are necessary. The nanocomposites were prepared by using one of two oxidizing agents: Fe (III) chloride or phosphomolybdic acid. Use of Fe (III) chloride led to better ion transport, surface area, conductivity, and higher current [51].

A nanocomposite of cellulose nanocrystal–polypyrrole–glucose oxidase enzyme was developed as a biosensor in diabetes treatment. Here, the cellulose nanocrystal–polypyrrole structure was used to immobilize the enzyme, whereas the nanosized pores facilitated electron transfer. Esmaeili *et al.* [52] obtained high glucose sensitivity and good repeatability. The limit of detection was 50 $\pm$ 10 μM, and exclusion of interfering species such as uric acid, ascorbic acid, and cholesterol was also achieved.

7.5.5 Nanoparticles

Nanoparticles are often defined by their diameters, which are less than 100 nm. Nanoparticles can be of different organic or inorganic materials. Different nanoparticles that have found use in nanocomposites can be given as [1, 3]:

- Metals (aluminum, iron, gold, silver, nickel, palladium, platinum,...)
- Metal oxides (titanium oxide, zinc oxide, Al_2O_3,...)
- Nonmetal oxide (silica—SiO_2)
- Other (silicon carbide SiC).

The type of nanoparticle included in the nanocomposite determines the resultant electrical, thermal, and mechanical characteristics. To give examples, aluminum nanoparticles impart conductivity, silicon carbide improves mechanical strength and resistance to corrosives, and silica enhances tensile and impact properties [3].

Silica, titanium, and aluminum nanoparticles have found use in forming barrier property against gases as well as to attain self-cleaning and antimicrobial effects [1].

Nanoparticles of metals can be used in catalysts, and inert metals can be used in biomedical applications such as tumor treatments. Wang *et al.* [53] produced polymer solar cells incorporating aluminum-doped ZnO nanoparticles with surfactant agents including ethanolamine, ethylenediamine, diethylenetriamine, and triethylenetetramine and reached super power conversion efficiency levels for similar systems by exceeding 10%.

Metal oxide nanoparticles impart electrical and thermal conductivity, mechanical strength, and barrier effect, besides other functions including UV protection, antibacterial, and photocatalytic (self-cleaning) activity. Metal oxide nanoparticles include MnO_2, SnO_2, WO_3, SiO_2, Fe_2O_3, ZrO_2, PtO_2, and TiO_2 [1, 2].

Titania or titanium dioxide (TiO_2) has common use for photocatalytic, UV protection, and antibacterial functions. Silver has found use for antibacterial effects [1]. Inclusion of silver nanoparticles at very low loadings as low as 1 wt% leads to increased mechanical strength and thermal durability [2]. As with other materials, high surface area resulted from nano dimensions increases effectivity of silver [1].

7.5.6 Nanoclays

Clay is generally defined as materials composed of very fine inorganic components including hydrated aluminum silicates. Phyllosilicate, also referred to as the clay mineral, includes hydrate silicates of aluminum and magnesium and shows layered structure. Each clay mineral includes two forms of sheets, which are tetrahedral and octahedral [3].

Among layered silicates, hectorite, saponite, and montmorillonite are the ones that enjoy the widest use in nanocomposites. Montmorillonite, the most popular one among them, exhibits high surface reactivity. Each layered sheet exhibits thickness around 10 Å (1 nm), whereas the remaining dimensions fall within the range between 30 nm and several microns. Their aspect ratio varies between 10 and 1000, while the specific surface area is approximately 750 m^2/g. The aspect ratio falls as the layers break during nanocomposite production. On the other hand, if good dispersion cannot be achieved, the effective aspect ratio falls as well [3].

Nanocomposites of montmorillonite were reported to show enhancements in ionic conductivity of PEO. This was attributed to the fact that the PEO cannot crystallize due to intercalation, as crystallites are naturally not conductive. The conductivity is slightly affected by the ambient temperature [2]. Another nanocomposite of montmorillonite exhibited decreased flammability [15].

Polymer matrix nanocomposites containing silicate layers can be prepared by intercalative methods. Via this technique, various nanocomposites presenting intercalated or exfoliated structures can be obtained. The structure is determined by the penetration level of polymer matrix into the galleries of the silicate. In case of the intercalated nanocomposites, slight entrance of polymer chains between the sheets is observed, where the stacked order is maintained. On the other hand, in terms of the exfoliated form, a high affinity between the polymer and clay leads to separation of sheets and loss of stacked structure [1, 2].

In their natural state, silicate layers may not form nanocomposites. Normally, silicate layers are not miscible with polymers as they include hydrate sodium or potassium ions. In order to increase affinity of layered silicates to polymers, the chemical structure should be changed from hydrophilic to organophilic. This can be achieved by organic modification. Such modified nanolayers of silicates are named as organosilanes or nanoclays [3]. Hence, the ability of forming nanocomposite structure depends on the components, production technique, and the affinity between the layered silicate and the matrix polymer; consequently, the type of the composite is determined: tactoid, intercalated, or exfoliated [2].

If efficient separation of sheets cannot be achieved, the tactoid form cannot be broken down and no nanocomposite structure can be achieved, as the resultant composite exhibits a phase-separated structure as in the case of conventional microcomposites. In this form, the nanocomposite structure is not formed as the polymer and the clay are immiscible. It is crucial issue to break the tactoid structure in order to achieve nanocomposite structure. In the other case, in the nanocomposites of intercalation, the polymer enters into the silicate layers in a crystallographical order. Generally, intercalation occurs as a few polymer molecule arrays in each layer. Ordered clay sheets result in very good properties, which are more pronounced in high aspect ratios [1–3].

Polymer-silicate layer nanocomposites have received intensive attention based on their outstanding characteristics such as high stiffness, strength, thermal durability, and resistivity against flammability. Silicate layers show flexibility as their properties can be modified via ion exchange reactions with cations. Very low loadings of layered silicates can lead to substantial increases in properties. Moreover, the possibility of melt-mixing polymers with silicate layers without the need for organic solvents is another positive aspect of intercalation chemistry. A very strong interface interaction can be achieved between polymers and the silicate layer, much higher compared to that observed in conventional composites. High aspect ratio results in high stiffness in the resulting nanocomposite [2].

7.5.7 Nanowires

Nanowires can also be incorporated in nanocomposites for use in smart textile applications including energy harvesting as in flexible solar cells. In a related example, utilization of silver nanowires in supercapacitors [19] leads to high power conversion efficiency related to low resistance due to high conductivity of silver nanowires, which promotes charge mobility [54]. Catenacci *et al.* [55] produced a core-shell copper–silver nanowire welded composite including a silicon elastomer, which presented a serpentine pattern. Based on the pattern and silicon matrix, the composite system showed 300% stretchable. As expected, the capability of stretching is very important for textile applications.

In another application of nanowires, copper nanowires were vertically aligned in polydimethylsiloxane matrix to act as thermal interface materials, which are of high importance for heat management of electronic devices [56]. A transparent electrode was produced via embedding copper nanowires in polymethyl methacrylate matrix. The matrix did not only provide transparency; it also imparted copper nanowire chemical stability. When the fashion aspect of textiles is considered, it becomes apparent that transparency is an important parameter of smart components to be used in smart textile applications [29].

A different nanowire example is that of polyaniline. Polyaniline is a conducting polymer. Nanocomposites of polyaniline nanowire–polyamide nanofiber–cellulose acetate film systems were developed. First, polyamide nanofibers were produced via electrospinning. The nanowires were then infiltrated with cellulose acetate to form a transparent film. Followingly, polyaniline nanowires were grown on the substrate through an *in situ* polymerization technique. The produced electrode presented good durability against cyclic bending [57].

7.5.8 Others

A number of other components have found use in nanocomposites for use in smart textile applications. These include common natural fibers including flax and wood fibers [34, 58], and synthetic polymers including but not limited to polyamide, polypropylene, polyurethane, polyethylene glycol, polyacrylate, polydimethylsiloxane, epoxy, polyvinyl alcohol, and polycaprolactone [27, 34, 56, 59].

On natural animal-based fibers (wool, silk), limited studies are present in terms of use in nanocomposite smart textile applications. Whereas wool has been found as a textile substrate rather than a nanocomposite component, few works are present for silk [60, 61]. Silk fibroin is a very interesting material. It is bio-based and biodegradable and shows properties similar to synthetic polymer fibers. It has applications in nanocomposites with target uses rather limited to biomedicine applications due to its cost [61]. In a study, nanocomposite of silk-gold was produced through electrospinning and tested *in vivo* for replacing nerves including nerve conducting velocity, compound muscle action potential, and motor unit potential [60]. In another study, nanofibrous webs of silk fibroin and silver nanoparticles were produced for antibacterial function [62]. It would be interesting to see applications of silk fibers, especially the fibroin section, in nanocomposites for smart textile applications. Some nanocomponents used in smart nanocomposites and the attained functionalities are listed in Table 7.1.

Table 7.1 Some nanocomponents used in smart nanocomposites and the attained functionalities. The table has been arranged by the author.

Nanocomponent group	Nanocomponent	Functionality	Reference
Carbon	Carbon nanotube	Strain sensitivity Optical sensitivity Thermal sensitivity Electrical field sensitivity Glucose biosensor Self-powering Energy harvesting	[10, 36–38, 42, 49, 63–70]
	Carbon nanofiber	Sensor conductivity	[39]
Nanocellulose	Cellulose nanocrystal	Actuator Biosensor Glucose sensor selective permeability Ion sensor pH sensor Ion exchange Energy storing Energy harvesting	[16, 18, 43, 46–48, 52, 70]
	Bacterial cellulose	Glucose biosensor Self-powering	[42]

Conducting polymers	Polypyrrole	Ion exchange biosensor Glucose sensor Energy storing Energy harvesting Actuator Electromagnetic shielding Microwave absorption	[17, 48, 52, 71]
	Polyaniline	Electromagnetic shielding Microwave absorption	[71, 72]
Metal nanoparticles	Copper nanoparticle	Antibacterial Energy harvesting	[13, 73]
	Silver nanoparticle	Antibacterial Conductive Energy harvesting	[62, 74, 75]
	Gold nanoparticle	Glucose biosensor Self-powering	[42, 76, 77]
	Aluminum nanoparticle	Energy harvesting	[53, 78]

(*Continued*)

Table 7.1 Some nanocomponents used in smart nanocomposites and the attained functionalities. The table has been arranged by the author. (*Continued*)

Nanocomponent group	Nanocomponent	Functionality	Reference
Metal oxide nanoparticles	Titania (TiO_2)	Self-cleaning Antibacterial UV absorption	[37, 79–81]
	ZnO	Energy harvesting	[53, 78]
	SiO_2	Moisture management	[14]
	Fe_3O_4	Electromagnetic shielding Microwave absorption	[71, 72]
Nanoclays	Montmorillonite	Flame retardancy	[3, 15]
Nanowires	Silver nanowire	Conductive Supercapacitor Energy storing Energy harvesting Antibacterial	[19, 55, 78, 82]
	Copper nanowire	Conductive Heat management	[55, 56]
	Polyaniline nanowire	Electrode	[57]
	Zinc oxide nanowire	Energy harvesting	[83]

7.6 Nanocomposite Forms

Textiles offering smart functionalities may be produced by different means. In a straight forward approach, yarns may be completely made of metals or metal alloys. In another way, yarns/fibers may be coated with conducting metals/polymers. Or conducting particles such as carbon or metal particles may be incorporated in fibers. Thus, one should not limit the use of nanocomposites to coatings on fabrics. On the contrary, by imparting nanocomposite structure to the fiber, the textile itself can be bestowed smart functionality without the need for smart coating [8]. Accordingly, nanocomposites for smart textile applications can be in different forms including laminated nanocomposites, nanocomposite fibers, nanocomposite membranes, nanocomposite coatings, and nanocomposite hydrogels.

7.6.1 Laminated Nanocomposites

Some composites exhibit laminated /layered composition. Some of these layered composites may contain nanostructured layers. A common example to this is composite filtration systems. Nanofibrous web layers commonly find place in these filters. Nanofibrous webs are generally produced via the electrospinning method. High porosity ratios and small pore dimensions exhibit advantages in terms of filtration efficiency [22].

Laminate composites have found use in moisture management applications. Moisture management is very critical in terms of ensuring body comfort, especially for sportswear. A dual-layer approach is generally adopted. In the first layer close to the body, hydrophobicity is preferred, whereas hydrophilic content is located in the outer layer. This results in the push–pull effect where the inner layer pushes moisture to the outer layer via capillary force and in turn to wicking action [22]. In a related study, the inner layer was a hydrophobic polydopamine-treated nonwoven layer, and the outer was a hydrophilic electrospun nanofiber membrane of polyacrylonitrile–SiO_2 [14]. Janus fabrics are another example of superamphiphilicity, which shows superhydrophobicity at one side and superhydrophilicity at the other [84].

7.6.2 Nanocomposite Fibers

A different form of nanocomposites is the fiber form. In this form, the fiber exhibits a nanocomposite structure itself. In this way, different functionalities can be incorporated into the fiber. An example is fire protection. Montmorillonite clay and intumescent flame-retardant agents can be used as nanofillers in a fiber. It was reported that addition of carbon nanotubes also enhanced fire protection capability [22].

In a nanofiber exhibiting nanocomposite structure, electrical conduction and magnetism have been simultaneously achieved. In the nanofiber, polyvinyl pyrrolidone was selected as the matrix, whereas varying amounts of polyaniline and Fe_3O_4 nanoparticles were incorporated to fine-tune conductivity and magnetism properties. The prepared nanofiber shows promise for use in electromagnetic shielding and microwave absorption applications [71].

7.6.3 Nanocomposite Membranes

Membranes of electrospun nanofibers are also used for different applications. They take place as lightweight breathable layers in protective clothing. They form breathable, flexible membranes with high filtration efficiency and low pressure drop. They may be used with other porous layers such as polyurethane foams with open cells, that is, interconnected pores. Electrospun nanofibrous webs are also used in rechargeable lithium ion batteries [33].

Various methods can be found for manufacturing nanofibers such as chemical vapor deposition, composite spinning, drawing, template synthesis, melt blowing, self-assembly, electrostatic spinning, and so on [22]. Among these methods, electrostatic spinning has been the most popular one depending on its versatility and ease of setup [85].

In its own, electrospun nanofiber webs present low durability against compression and other mechanical effects. Thus, electrospun nanofiber membranes are generally used in composite systems with other components or layers, such in laminated composites. In an application, electrospun nanofibrous webs are utilized as carrier matrices of phase change materials for thermoregulation applications [59]. Phase change materials are materials that can store heat energy when the ambient temperature is higher than a critical point and vice versa, while keeping their temperature nearly constant. Advantages of electrospun nanofibrous webs can be given as high porosity, high surface area, and ease of encapsulation in terms of acting as carriers or phase change materials [12, 22, 59].

Composite systems are preferred for filtration applications in order to achieve high efficiency levels. Nanocomposites are especially promising for filtering small structures while maintaining the necessary pressure drop, system integrity, and strength performances. Electrospun nanofiber membranes offer high porosity, low pore dimensions, which are necessary for filtration applications. So, electrospun nanofiber membranes are utilized as layers of composite filtration systems. In a close field, electrospun nanofiber webs may be utilized as a membrane of protective clothing. They present air and water vapor permeability, elasticity, and good filtration performance [33, 65].

7.6.4 Nanocomposite Coatings

Coating of textiles with nanoparticles does not result in durable coatings most of the time. One of the effective means of preparing nanoparticle coating with washing fastness is producing nanocomposites where the functional nanoparticle is embedded in a polymer matrix. Improvement of bonding results in durable coating effect with better human—and environmental and ecological—safety. Via use of such nanocomposites, functions including UV resistance, antimicrobial effect, conductivity, fire protection, and self-cleaning can be attained. The performance characteristics of nanocomposites may surpass those of each component. Appropriate selection of the optimum polymer system may lead to reduced agglomeration [12].

Via adoption of a nanocomposite system, chemical species showing affinity to textile substrates can be introduced, besides the main functional groups. This way, coatings durable to washing and other effects may be produced. The prepared nanocomposites may be coated onto the textiles via different methods such as dip–dry or blade coat–dry with subsequent cure processes. Nanocomposite systems offer durable thin coatings presenting transparency and multifunctionality. Nanocomposites used in textile coatings include metal nanoparticles, metal oxides, titania and zinc oxide, graphene, carbon nanotubes, as well as phase change materials [12, 33].

In order to improve bonding of nanoscale materials on substrates for achieving continuation of the designed functions, nanofibers and/or nanofillers may be treated with physical or chemical surface modification procedures. Another means of ensuring durable nano-effects is embedding functional nanomaterials in polymer matrices, which have the affinity to the textile substrate [6, 33].

In a nanocomposite coating application, silver nanowires and fluorosilane were coated on a cotton fabric. Silver nanowires contributed to conductivity as well as antibacterial effect of the fabric. Superhydrophobicity was obtained with a contact angle of 156° from the fluorosilane–silver nanowire-coated fabric as well [82].

7.6.5 Nanocomposite Hydrogels

It should not be considered that composite and nanocomposite materials are all rigid materials. Hydrogels constitute a very special interesting stimuli-responsive material group, which presents composite structures [9]. As hydrogels are covered in another chapter of this book [86], they are not investigated in this chapter.

7.7 Functions of Nanocomposites in Smart Textiles

7.7.1 Sensors

Sensors, a very important component of smart devices, are capable of measuring a physical quantity and converting it into a signal that can be detected by a human observer or an electronic device. Electrode takes a critical part in the sensor determining the sensitivity and efficiency. Mediators facilitate ion transfer between the analyte and the electrode. Mediators can be composed of different materials such as conducting polymers, metallocenes, and conductive nanocellulose composites [43]. Mediators are generally covered on surfaces of electrodes, taking place between the analyte and the electrode. The properties of the mediators heavily influence the response rate of the sensor. Low response rate is a very critical issue in terms of sensor applications. Response rates in bulk polymers are rather low due to the time necessary for target molecules to penetrate into the polymer. However, in the case where nanostructuring is achieved via nanofibers or nanotubes, response rate can be highly increased. Furthermore, as known, nanoscale materials exhibit a very high surface area per volume, which enhances sensitivity as well as the response rate. These render nanocomposites ideal materials for use in sensors [11, 43].

Sensors, as key elements of smart devices, can function based on electrical, mechanical, or optical mechanisms. They can be subgrouped as electrochemical, piezoelectric, acoustic, fluorescent, and colorimetric sensors, etc. [11]. Biosensors constitute an important class of sensors capable of sensing biochemicals including glucose, estrogen, and urea. A very important type of biosensors is the electrochemical sensor, which can produce an electrical signal from a concentration of chemical species by use of enzymes. Nanocellulose has found use in nanocomposites for biosensor applications [11, 43]. Glucose biosensors attract extensive effort due to interest in diabetes treatment research. Here, glucose oxidase enzyme is used [40]. An active layer of this enzyme is used to react with glucose and transmits current into the electrode. Here, the strength of the electric signal produced at the electrode is proportional to the concentration of glucose. Another biosensor application of nanocellulose is in wound care via detecting elastase, a biomarker for inflammatory diseases [11].

Nag *et al.* [10] developed a flexible wearable sensor for monitoring respiration and other body motions. The multilayered sensor comprises a polydimethylsiloxane layer and a membrane of nanocomposite including functionalized multiwalled carbon nanotubes. The polydimethylsiloxane and the carbon nanotubes acted as electrodes. The sensor functioned based on strain sensitivity. The thickness was selected to give the best strain and conductivity behavior. Polydimethylsiloxane was selected based on its cost efficiency, hydrophobicity, inert structure, and nontoxic nature. Carbon nanotube was preferred based on thermal stability and tensile strength.

In the mentioned study, multiwalled carbon nanotubes were functionalized by introduction of carboxylic (–COOH) groups. Incorporation of these groups resulted in better dispersion in the polymer matrix. This, in turn, causes stronger interfacial bonding and, subsequently, better conductivity [10]. The electrodes were produced as patterns on the substrate. The selected method was laser ablation. Other techniques that can be used include inkjet printing, 3-D printing, and photolithography. The laser method is advantageous in terms of its ease of use as no template or extra material consumption is necessary [10].

The functionalized multiwalled carbon nanotubes were used as a conductor. They were embedded in the flexible substrate based on polydimethylsiloxane. The resultant structure acted as a sensor monitoring multiple physiological variables. The advantages of multiwalled carbon nanotubes include high electrical conductivity and aspect ratio. Incorporation of functional sites such as carboxyl (–COOH) groups leads to enhanced conducting and dispersing performance. Polydimethylsiloxane was selected for its low stiffness, which is required for wearable use. Polydimethylsiloxane is also more cost efficient in comparison to polyethylene naphthalate and polyethylene terephthalate, which have also been used for flexible sensor manufacturing. The hydrophobic nature of polydimethylsiloxane prevents interfering effect of sweat on sensor functionalization [10].

The mechanism behind the function of the sensor is the change in the capacitance induced by deformation of electrodes due to body motions. The capacitance of parallel-plate capacitor can be given as

$$C = (\varepsilon_o * \varepsilon_r * A)/d \tag{7.1}$$

where C is the capacitance in Farads, ε_o is the dielectric constant of free space, ε_r is the relative dielectric coefficient of the material, A is the area of one plate, and d is the distance between the plates. Difference in the area or the distance leads to change in the capacitance [10].

7.7.2 Antibacterial Activity

Another function that is expected from textiles is antibacterial activity. In a related study, nanocomposites of cellulose–*Cassia alata* leaf extract–copper nanoparticles were developed. Presence of copper nanoparticles improved thermal stability and tensile strength of cellulose. The nanocomposite exhibited good antibacterial activity [13].

7.7.3 Defense Applications

Nanocomposites are investigated for use in defense applications. One related study concerns body armor improvement. Shear thickening fluids are used in body armors. Shear thickening fluids exhibit rapid stiffening response when under shear. Recent work has been reported to improve shear stiffening effect using silica particles in polyethylene glycol fluid matrix. This fluid was applied on Kevlar (para-aramid) fabric and led to higher energy absorption under impact [87].

Nanocomposites also have found use in microwave protection. In a related study, a two-ply nanocomposite was prepared to have a total thickness of as low as 1 mm. One layer is polyaniline (absorbing layer) and the other is polyaniline–magnetite (Fe_3O_4; matching layer). The nanocomposite was reported to have better microwave absorption in comparison to single-layer absorbers [72]. In another work, composite nanofibers were produced from polyaniline–magnetite–polyvinyl pyrrolidone. In this nanocomposite structure, polyvinyl pyrrolidone acted as the matrix. The nanocomposite exhibits double functionality such as electrical conductivity and magnetic effect. The developed system may find use in electromagnetic interference shielding and microwave absorption [71].

7.7.4 Fire Protection

Polymers generally exhibit poor resistance to fire. Upon ignition, polymers mostly burn quickly emitting heat and toxic fumes. Incorporation of nanoclay leads to substantial increase in flame retardancy. Nanoclay addition results in substantial reduction in burning rate and hinders diffusion of volatiles and air [3].

7.7.5 Actuators

Nanocomposites enabled development of actuators that minimize material and energy consumption. Lower particle dimensions, higher surface areas, and lower nanofiller loading rates allow production of cost-efficient actuators [3]. Actuators have the capability to function in hostile environments that humans cannot withstand [9].

Enhanced piezoelectric performance can be achieved from composites of cellulose with nanotubes, chitosan, ionic liquid, polypyrrole, and polyaniline. An actuator of nanocomposites including cellulose, polypyrrole, and ionic liquid was developed. Here, wet cellulose films were produced from cotton pulp via spin coating. Then, pyrrole was polymerized and adsorbed on cellulose; further, activation was carried out in ionic liquid of 1-butyl-3-methylimidazolium chloride solution. The nanocomposite resulted in higher conductivity and mobility compared to pristine cellulose, conventional electroactive paper, and conducting polymer [70].

7.7.6 Self-Cleaning

Another function that nanocomposites can serve is self-cleaning. Self-cleaning mechanism can be induced by both UV and visible light via use of nanocomposites. TiO_2 presents promising effectivity for photocatalytic stain removal. However, long exposure to UV irradiation is necessary, whereas success under visible light is very limited. Thus, titania is imparted with nanoparticles of noble metals, dyes, and compounds including SiO_2 [12, 37, 81].

Nanocomposite systems of porphyrin and TiO_2 were developed to achieve photocatalytic self-cleaning. Self-cleaning effect is induced by visible light due to presence of porphyrin. Porphyrin dye molecules are excited under visible light and electrons are injected to the conduction band of titania. This leads to formation of superoxide anions, which take part in stain decomposing. The durability of porphyrin against light was improved via use of metals in the coating layer [33].

Superhydrophobicity also contributes to self-cleaning. As the water contact angle on a surface exceeds 150°, the surface is considered as superhydrophobic. Superhydrophobicity is attained via nano- and micro-roughness on the surface mimicking lotus leaves as well as using components of hydrophobic chemistry such as fluorosilane [27, 81].

7.7.7 Energy Harvesting

Energy harvesting is another advanced function that nanocomposites can serve. A group of smart textile components that can scavenge and store energy is referred to as nanogenerators [21]. Nanogenerators may be produced by printing nanocomposite inks on textile substrates. In such an example, a piezoelectric textile nanogenerator was developed by utilization of a nanocomposite film incorporating silver nanoparticles and polymers. The nanocomposite ink can be screen-printed on textile or plastic surfaces. Addition of silver nanoparticles improved piezoelectrical property. Through bending and compression motions, tens of microwatts were scavenged by this piezoelectrical nanogenerator [88].

Nanogenerators can be produced by integrating two electrodes in a single yarn even in a single fiber. In an example, a fiber-based nanogenerator was prepared by Zhong *et al.* [69]. In this example, two cotton threads were used, one of which was coated with carbon nanotubes, whereas the other was coated first with carbon nanotubes then with polytetrafluoroethylene. The treated cotton threads were twisted in a double helix. The twisted threads were woven into fabrics. These fiber-based nanogenerators can convert body motions and vibrations into electricity.

7.8 Future Outlook

Nanocomposites allow special potential for smart applications. They can serve smart textiles with different areas including electronics, protection, defense, electronics, and so on. Nanocomposites can replace and outperform rigid electronic devices in smart applications and wearables. Nanocomposites do not only come with advantages but also entail some challenges in terms of difficulties in production, characterization, and useful service stages. On the other hand, these difficulties offer new areas for the scientific community to plan further research areas. Further research may be devoted to overcome difficulties arising from the limited durability of nanocomposites as components of smart textiles as well as difficulties in dispersion, interfacial bonding, durability, characterization, and optimization.

An interdisciplinary approach (including engineering, materials science, chemistry, and physics) has to be conducted in the field of nanocomposite research in order to understand structure–property relations. Research at the nanoscale calls for advanced techniques to study the mechanics, interactions, as well as other properties of nanocomposites. Basic research is a must to understand structure–property relations, which will guide us in the development of novel nanocomposites with new characteristics. Characterization is another aspect that is indispensable for nanocomposite research [2].

The observed performance characteristics of nanocomposites are generally much lower than their theoretical values. The reasons can be given as lack of uniformity, weak interfacial adhesion, and insufficient orientation [5]. Determination of the optimum content of nanofillers as well as other nanocomponents in the nanocomposite is a critical aspect in nanocomposite research to achieve the best properties and to prevent agglomeration. Achieving effective dispersion of nanocomponents in the matrix as well as ensuring homogeneity in the nanocomposite are major issues in nanocomposite production. The challenges in terms of controlling elemental composition as well as stoichiometry in nanophases should also be solved. Mechanisms of failure modes constitute another aspect that should be closely investigated in order to design novel nanocomposites with the desired properties [2, 5].

Other future studies may be related to improvements in compatibility between the nanoparticle and the matrix material to achieve good interfacial bonding strength as well as prevention of agglomeration to obtain uniform nanocomposite structure. Further studies may be related to providing desired orientation of nanoparticles in the matrix. Modelling and simulating characteristics of nanocomposites are expected to be among the future trends in the nanocomposite research area [2].

7.9 Conclusion

Advancement in technology following increased requirements necessitates development of materials exhibiting novel characteristics and enhanced properties in comparison to conventional materials. Within this concept, nanocomposites offer properties surpassing those of monolithic materials as well as conventional microcomposites. Nanocomposites offer advanced multifunctions without interfering the comfort and aesthetics of textiles. As components of smart textiles, nanocomposites offer service for use in sensors, actuators, fire protection, defense applications, biosensing, self-cleaning, moisture management, thermoregulation, and energy storing and harvesting, among other sophisticated niche applications. While nanocomposites have already found use in various applications, there are numerous potential fields where nanocomposites can provide better service compared to conventional counterparts in the future.

References

1. Bratovčić, A., Odobašić, A., Ćatić, S., and Šestan, I., Application of polymer nanocomposite materials in food packaging, *Croat. J. Food Sci. Technol.*, 7, 2, 86–94, 2015.
2. Camargo, P. H. C., Satyanarayana, K. G., and Wypych, F., Nanocomposites: Synthesis, structure, properties and new application opportunities, *Mater. Res.*, 12, 1, 1–39, 2009.

3. Kurahatti, R. V., Surendranathan, A. O., Kori, S., Singh, N., Kumar, A. V. R., and Srivastava, S., Defence applications of polymer nanocomposites, *Def. Sci. J.*, 60, 5, 551–563, 2010.
4. Passador, F. R., Ruvolo-Filho, A., and Pessan, L. A., Nanocomposites of polymer matrices and lamellar lays, in *Nanostructures*, A. L. Da Róz, M. Ferreira, F. de L. Leite, and O. N. Oliveira Jr., (Eds.) William Andrew, pp. 187–201, 2017.
5. Hári, J. and Pukánszky, B., Nanocomposites: Preparation, structure, and properties, in *Applied Plastics Engineering Handbook*, M. Kutz, (Ed.) William Andrew, pp. 109–142, 2011.
6. Gashti, M. P., Alimohammadi, F., Song, G., and Kiumarsi, A., Characterization of nanocomposite coatings on textiles: A brief review on Microscopic technology, in *Current Microscopy Contributions to Advances in Science and Technology*, A. Méndez-Vilas, (Ed.) Formatex, pp. 1424–1437, 2012.
7. Majumer, D. D., Majumder, D. D., and Karan, S., Magnetic properties of ceramic nanocomposites, in *Ceramic Nanocomposites*, R. Banerjee and I. Manna, (Eds.) Woodhead Publishing, pp. 51–91, 2013.
8. Syduzzaman, M., Patwary, S. U., Farhana, K., and Ahmed, S., Textile science & engineering smart textiles and nano-technology: A general overview, *Text. Sci. Eng.*, 5, 1, 1–7, 2015.
9. Yilmaz, N. D., Multi-component, semi-interpenetrating-polymer-network and interpenetrating-polymer-network hydrogels: Smart materials for biomedical applications abstract, in *Functional Biopolymers*, V. K. Thakur and M. K. Thakur, (Eds.) Springer, 2017.
10. Nag, A., Mukhopadhyay, S. C., and Kosel, J., Flexible carbon nanotube nanocomposite sensor for multiple physiological parameter monitoring, *Sensors Actuators A. Phys.*, 251, November, 148–155, 2016.
11. Edwards, J. V., Prevost, N., French, A., Concha, M., DeLucca, A., and Wu, Q., Nanocellulose-based biosensors: Design, preparation, and activity of peptide-linked cotton cellulose nanocrystals having fluorimetric and colorimetric elastase detection sensitivity, *Engineering*, 5, 20–28, 2013.
12. Pakdel, E., Fang, J., Sun, L., and Wang, X., Nanocoatings for smart textiles, in *Smart Textiles: Wearable Nanotechnology*, N. D. Yılmaz, (Ed.) Wiley-Scrivener, 2018.
13. Sivaranjana, P., Nagarajan, E. R., Rajini, N., Jawaid, M., and Rajulu, A. V., Formulation and characterization of *in situ* generated copper nanoparticles reinforced cellulose composite films for potential antimicrobial applications, *J. Macromol. Sci. Part A*, 55, 1, 58–65, 2018.
14. Babar, A. A., Wang, X., Iqbal, N., Yu, J., and Ding, B., Tailoring differential moisture transfer performance of nonwoven/polyacrylonitrile–SiO_2 nanofiber composite membranes, *Adv. Mater. Interfaces*, 4, 15, 2017.
15. Huang, G., Yang, J., and Wang, X., Nanoclay, intumescent flame retardants, and their combination with chemical modification for the improvement of the flame retardant properties of polymer nanocomposites, *Macromol. Res.*, 21, 1, 27–34, 2013.

16. Kim, J. and Seo, Y. B., Electro-active paper actuators, *Smart Mater. Struct.*, 11, 355–360, 2002.
17. Khan, A., Abas, Z., Kim, H. S., and Kim, J., Recent progress on cellulose-based electro-active paper, its hybrid nanocomposites and applications, *Sensors*, 16, 8, 1172, 2016.
18. Yang, C., Kim, J. H., Jim, J., and Kim, H. S., Piezoelectricity of wet drawn cellulose electro-active paper, *Sensors Actuators A. Phys.*, 154, 117–122, 2009.
19. Qiao, Z., Yang, X., Yang, S., Zhang, L., and Cao, B., 3D hierarchical MnO_2 nanorod/welded Ag-nanowire-network composites for high-performance supercapacitor electrodes, *Chem. Commun.*, 52, 51, 7998–8001, 2016.
20. Lateef, A. and Nazir, R., Metal nanocomposites: Synthesis, characterization and their applications, in *Science and Applications of Tailored Nanostructures*, P. Di Sia, (Ed.) One Central Press, pp. 239–256, 2018.
21. Pu, X., Hu, W., and Wang, Z. L., Nanogenerators for smart textiles, in *Smart Textiles: Wearable Nanotechnology*, N. D. Yılmaz, (Ed.) Wiley-Scrivener, 2018.
22. Wan, L. Y., Nanofibers for smart textiles, in *Smart Textiles: Wearable Nanotechnology*, N. D. Yılmaz, (Ed.) Wiley-Scrivener, 2018.
23. Ibrahim, I. D., Jamiru, T., Sadiku, R. E., Kupolati, W. K., Agwuncha, S. C., and Ekundayo, G., The use of polypropylene in bamboo fibre composite and their mechanical properties—A review, *J. Reinf. Plast. Compos.*, 34, 16, 1347–1356, 2015.
24. Altenbach, H., Altenbach, J., and Kissing, W., *Mechanics of Composite Structural Elements. Springer*, Berlin, Heidelberg: Springer, 2004.
25. Yilmaz, N. D., Khan, G. M. A., and Yilmaz, K., Biofiber reinforced acrylated epoxidized soybean oil (AESO) composites, in *Handbook of Composites from Renewable Materials, Physico-Chemical and Mechanical Characterization*, V. K. Thakur and M. K. Thakur, (Eds.) Wiley Scrivener, pp. 211–251, 2017.
26. Bristy, S. S., Rahman, M. A., Tauer, K., Minami, H., and Ahmad, H., Preparation and characterization of magnetic γ-Al_2O_3 ceramic nanocomposite particles with variable Fe_3O_4 content and modification with epoxide functional polymer, *Ceram. Int.*, 44, 4, 3951–3959, 2018.
27. Nguyen-Tri, P., Nguyen, T. A., Carriere, P., and Xuan, C. N., Nanocomposite coatings: Preparation, characterization, properties, and applications, *Int. J. Corros.*, 2018, 1–19, 2018.
28. Khan, G. M. A., Yilmaz, N. D., and Yilmaz, K., Okra bast fiber as potential reinforcement element of biocomposites: Can it be the flax of the future, in *Handbook of Composites from Renewable Materials Volume 4: "Functionalization"*, Wiley-Scrivener, pp. 379–406, 2017.
29. Yilmaz, N. D., Introduction to smart nanotextiles, in *Smart Textiles: Wearable Nanotechnology*, N. D. Yilmaz, (Ed.) Wiley-Scrivener, 2018.
30. Huang, C.-T., Tang, C.-F., and Shen, C.-L., A wearable textile for monitoring respiration, using a yarn-based sensor, in *Tenth IEEE International Symposium on Wearable Computers*, 2006.

31. Stoppa, M. and Chiolerio, A., Wearable electronics and smart textiles: A critical review, *Sensors*, 14, 11957–11992, 2014.
32. Saba, N., Tahir, P. M., and Jawaid, M., A review on potentiality of nano filler/natural fiber filled polymer hybrid composites, *Polymers (Basel)*, 6, 2247–2273, 2014.
33. Gashti, M. P., Pakdel, E., and Alimohammadi, F., Nanotechnology-based coating techniques for smart textiles, in *Active Coatings for Smart Textiles*, Elsevier Ltd, pp. 243–268, 2016.
34. Tripathi, K. M., Vincent, F., Castro, M., and Feller, J. F., Flax fibers—Epoxy with embedded nanocomposite sensors to design lightweight smart biocomposites, *Nanocomposites*, 2, 3, 125–134, 2016.
35. Gorss, J., *High Performance Carbon Fibers*, American Chemical Society, 2003.
36. Zannotti, M., Giovannetti, R., D'Amato, C. A., and Rommozzi, E., Spectroscopic studies of porphyrin functionalized multiwalled carbon nanotubes and their interaction with TiO_2 nanoparticles surface, *Spectrochim. Acta Part A Mol. Biomol. Spectrosc.*, 153, 22–29, 2016.
37. Lee, H. J., Kim, J., and Park, C. H., Fabrication of self-cleaning textiles by TiO_2–carbon nanotube treatment, *Text. Res. J.*, 84, 3, 267–278, 2013.
38. Zhang, X., *et al.*, Optically- and thermally-responsive programmable materials based on carbon nanotube-hydrogel polymer composites, *Nano Lett.*, 11, 3239–3244, 2011.
39. Chowdhury, S., Olima, M., Liu, Y., Saha, M., Bergman, J., and Robison, T., Poly dimethylsiloxane/carbon nanofiber nanocomposites: Fabrication and characterization of electrical and thermal properties, *Int. J. Smart Nanomater.*, 7, 4, 236–247, 2016.
40. Yilmaz, N. D., Koyundereli Cilgi, G., and Yilmaz, K., Natural polysaccharides as pharmaceutical excipients, in *Handbook of Polymers for Pharmaceutical Technologies, Volume 3, Biodegradable Polymers*, V. K. Thakur and M. K. Thakur, (Eds.) Wiley Scrivener, pp. 483–516, 2015.
41. Yilmaz, N. D., Konak, S., Yilmaz, K., Kartal, A. A., and Kayahan, E., Characterization, modification and use of biomass: Okra fibers, *Bioinspired, Biomim. Nanobiomaterials*, 5, 3, 85–95, 2016.
42. Lv, P., *et al.*, A highly flexible self-powered biosensor for glucose detection by epitaxial deposition of gold nanoparticles on conductive bacterial cellulose, *Chem. Eng. J.*, 351, 177–188, 2018.
43. Abdi, M. M., Abdullah, L. C., Tahir, P. M., and Zaini, L. H., Cellulosic nanomaterials for sensing applications, in *Handbook of Green Materials 3 Self- and Direct Assembling of Bionanomaterials*, K. Oksman, A. P. Mathew, A. Bismarck, O. Rojas, and M. Sain, (Eds.) World Scientific Publishing, pp. 197–212, 2014.
44. Yilmaz, N. D., Biomedical applications of microbial cellulose nanocomposites, in *Biodegradable Polymeric Nanocomposites: Advances in Biomedical Applications*, Springer, pp. 231–249, 2015.

45. Yilmaz, N. D., Agro-residual fibers as potential reinforcement elements for biocomposites, in *Lignocellulosic Polymer Composites: Processing, Characterization and Properties*, V. K. Thakur, (Ed.) Wiley-Scrivener, pp. 233–270, 2015.
46. Thielemans, W., Warbeya, C. R., and Walsh, D. A., Permselective nanostructured membranes based on cellulose nanowhiskers, *Green Chem.*, 4, 11, 531–537, 2009.
47. Hubbe, M. A., Sensing the electrokinetic potential of cellulosic fiber surfaces, *BioResources*, 1, 1, 116–149, 2006.
48. Nyström, G., Mihranyan, A., Razaq, A., Lindström, T., Nyholm, L., and Strømme, M., A nanocellulose polypyrrole composite based on microfibrillated cellulose from wood, *J. Phys. Chem. B*, 114, 12, 4178–82, 2010.
49. Toomadj, F., *et al.*, Strain sensitivity of carbon nanotubes modified cellulose, *Procedia Eng.*, 25, 1353–1356, 2011.
50. Kim, J.-H., *et al.*, Review of nanocellulose for sustainable future materials, *Int. J. Precis. Eng. Manuf. Technol.*, 2, 2, 197–213, 2015.
51. Razaq, A., Mihranyan, A., Welch, K., Nyholm, L., and Strømme, M., Influence of the type of oxidant on anion exchange properties of fibrous *Cladophora* cellulose/polypyrrole composites, *J. Phys. Chem. B*, 113, 2, 426–433, 2009.
52. Esmaeili, C., Abdi, M. M., Mathew, A. P., Jonoobi, M., Oksman, K., and Rezayi, M., Synergy effect of nanocrystalline cellulose for the biosensing detection of glucose, *Sensors*, 15, 10, 24681–24697, 2015.
53. Wang, Y., *et al.*, Highly stable Al-doped ZnO by ligand-free synthesis as general thickness-insensitive interlayers for organic solar cells, *Sci. China Chem.*, 61, 1, 127–134, 2018.
54. Song, J., Nanowires for smart textiles, in *Smart Textiles: Wearable Nanotechnology*, N. D. Yılmaz, (Ed.) Wiley-Scrivener, 2018.
55. Catenacci, M. J., Reyes, C., Cruz, M. A., and Wiley, B. J., Stretchable conductive composites from Cu–Ag nanowire felt, *ACS Nano*, 12, 4, 3689–3698, 2018.
56. Barako, M. T., *et al.*, Dense vertically aligned copper nanowire composites as high performance thermal interface materials, *ACS Appl. Mater. Interfaces*, 9, 48, 42067–42074, 2017.
57. Devarayan, K., Kim, D. L.-Y., and Kim, B.-S., Flexible transparent electrode based on PANi nanowire/nylon nanofiber reinforced cellulose acetate thin film as supercapacitor, *Chem. Eng. J.*, 273, 603–609, 2015.
58. Agarwal, M., Lvov, Y. M., and Varahramyan, K., Conductive wood microfibres for smart paper through layer-by-layer nanocoating, *Nanotechnology*, 17, 21, 5319–5325, 2006.
59. Chalco-Sandoval, W., Fabra, M. J., López-Rubio, A., and Lagaron, J. M., Development of an encapsulated phase change material via emulsion and coaxial electrospinning, *J. Appl. Polym. Sci.*, 133, 36, 43903, 2016.
60. Das, S., *et al.*, *In vivo* studies of silk based gold nano-composite conduits for functional peripheral nerve regeneration, *Biomaterials*, 62, 66–75, 2015.

61. Khan, G. M. A., Yilmaz, N. D., and Yilmaz, K., Recent developments in design and manufacturing of biocomposites of *Bombyx mori* silk fibroin, in *Handbook of Composites from Renewable Materials vol 2 Design and Manufacturing*, V. K. Thakur and M. K. Thakur, (Eds.) Wiley Scrivener, p. 377, 2017.
62. Kang, M., Jung, R., Kim, H., Ji, H. Y., and Jin, H. J., Silver nanoparticles incorporated electrospun silk fibers, *J. Nanosci. Nanotechnol.*, 7, 11, 3888–3891, 2007.
63. Luchnikov, V., Sydorenko, O., and Stamm, M., Self-rolled polymer and composite polymer/metal micro- and nanotubes with patterned inner walls, *Adv. Mater.*, 17, 1177–1182, 2005.
64. Thielemans, W., McAninch, I. M., Barron, W., Blau, W. J., and Wool, R. P., Impure carbon nanotubes as reinforcements for acrylated epoxidized soy oil composites, *J. Appl. Polym. Sci.*, 98, 3, 1325–1338, 2005.
65. van Deventer, N. and Mallon, P. E., Electrospun nanocomposite nanofibres with magnetic nanoparticle decorated carbon nanotubes, *Macromol. Symp.*, 378, 1, 1600140, 2018.
66. Soliman, E., Kandil, U., and Taha, M. R., Improved strength and toughness of carbon woven fabric composites with functionalized MWCNTs, *Materials*, 7, 6, 4640–4657, 2014.
67. Jeong, B., Choi, Y. K., Bae, Y. H., Zentner, G., and Kim, S. W., New biodegradable polymers for injectable drug delivery systems, *J. Control. Release*, 62, 109–114, 1999.
68. Zhang, K. and Choi, H. J., Smart polymer/carbon nanotube nanocomposites and their electrorheological response, *Materials (Basel)*, 7, 5, 3399–3414, 2014.
69. Zhong, J., *et al.*, Fiber-based generator for wearable electronics and mobile medication, *ACS Nano*, 8, 6273, 2014.
70. Mahadeva, S. K., Yang, S. Y., and Kim, J., Electrical and electromechanical properties of cellulose-polypyrrole-ionic liquid nanocomposite: Effect of polymerization time, *IEEE Trans. Nanotechnol.*, 10, 3, 445–450, 2011.
71. Yang, M., *et al.*, Single flexible nanofiber to simultaneously realize electricity–magnetism bifunctionality, *Mater. Res.*, 19, 2, 2016.
72. Xu, F., Ma, L., Huo, Q., Gan, M., and Tang, J., Microwave absorbing properties and structural design of microwave absorbers based on polyaniline and polyaniline/magnetite nanocomposite, *J. Magn. Magn. Mater.*, 374, 311–316, 2015.
73. Xing, X., Wang, Y., and Li, B., Nanofiber drawing and nanodevice assembly in poly(trimethylene terephthalate), *Opt. Express*, 16, 14, 2018.
74. Gulrajani, M. L., Gupta, D., Periyasamy, S., and Muthu, S. G., Preparation and application of silver nanoparticles on silk for imparting antimicrobial properties, *J. Appl. Polym. Sci.*, 108, 614–623, 2008.
75. Soin, N., *et al.*, Novel '3-D spacer' all fibre piezoelectric textiles for energy harvesting applications, *Energy Environ. Sci.*, 7, 1670, 2014.

76. Lian, X., Wang, S., Xu, G., Lin, N., Li, Q., and Zhu, H., The application with tetramethyl pyrazine for antithrombogenicity improvement on silk fibroin surface, *Appl. Surf. Sci.*, 255, 2, 480–482, 2008.
77. Qin, Y., Wang, X., and Wang, Z. L., Microfibre-nanowire hybrid structure for energy scavenging, *Nature*, 451, 809, 2008.
78. Pathirane, M. K., Khaligh, H. H., Goldthorpe, I. A., and Wong, W. S., Al-doped ZnO/Ag-nanowire composite electrodes for flexible 3-dimensional nanowire solar cells, *Sci. Rep.*, 7, 2017.
79. Dastjerdi, R., Mojtahedi, M. R. M., Shoshtari, A. M., and Khosroshani, A., Investigating the production and properties of Ag/TiO_2/PP antibacterial nanocomposite filament yarns, *J. Text. Inst.*, 101, 3, 204–213, 2010.
80. Morawski, A. W., Kusiak-Nejman, E., Przepiorski, J., Kordala, R., and Pernak, J., Cellulose–TiO_2 nanocomposite with enhanced UV-Vis light absorption, *Cellulose*, 20, 3, 1293–1300, 2013.
81. Xu, Q. F., Liu, Y., Lin, F.-J., Mondal, B., and Lyons, A. M., Superhydrophobic TiO_2–polymer nanocomposite surface with UV induced reversible wettability and self-cleaning properties, *ACS Appl. Mater. Interfaces*, 5, 18, 8915–8924, 2013.
82. Nateghi, M. and Shateri-Khalilabad, M., Silver nanowire-functionalized cotton fabric, *Carbohydr. Polym.*, 117, 160–168, 2015.
83. Wang, Z. L. and Song, J., Piezoelectric nanogenerators based on zinc oxide nanowire arrays, *Science*, 312, 242, 2006.
84. Lim, H. S., *et al.*, Superamphiphilic janus fabric, *Langmuir*, 26, 24, 19159–19162, 2010.
85. Nayak, R., Production methods of nanofibers for smart textiles, in *Smart Textiles: Wearable Nanotechnology*, N. D. Yılmaz, (Ed.) Wiley-Scrivener, 2018.
86. Niiyama, E., Fulati, A., and Ebara, M., Responsive polymers for smart textiles, in *Smart Textiles: Wearable Nanotechnology*, N. D. Yılmaz, (Ed.) Wiley-Scrivener, 2018.
87. Talreja, K., Chauhan, I., Ghosh, A., Majumdar, A., and Butola, B. S., Functionalization of silica particles to tune the impact resistance of shear thickening fluid treated aramid fabrics, *RSC Adv.*, 7, 78, 49787–49794, 2017.
88. Almusallam, A., *et al.*, Flexible piezoelectric nano-composite films for kinetic energy harvesting from textiles, *Nano Energy*, 33, 146, 2017.

8

Nanocoatings for Smart Textiles

Esfandiar Pakdel[1], Jian Fang[1,2], Lu Sun[1] and Xungai Wang*[1,2]

[1]*Deakin University, Geelong, Australia, Institute for Frontier Materials*
[2]*ARC Centre of Excellence for Electromaterials Science (ACES)*

Abstract
This chapter presents a general overview of recent advances in developing smart textiles through application of nanostructured materials. It focuses on different synthesis approaches of nanoparticles, immobilization methods, and coating techniques reported in the literature. Different techniques of treating fabric surfaces, including sol–gel, cross-linking, plasma, and nanocomposite coating methods, are explored. In addition, the application of different types of nanoparticles in textile modification processes along with the resultant functionalities such as self-cleaning, UV protection, thermal regulation, antimicrobial activity, fire retardancy, and conductivity as well as associated mechanisms are discussed.

Keywords: Nanoparticles, coating, smart textiles, sol–gel

8.1 Introduction

The concept of textiles coating dates back to antiquity era. The simplest coating process was born when our primary ancestors realized that they could get waterproofing features on their outfits by simply smearing the animal fats on clothes. Since then, humans have always been trying to introduce novel materials to textiles by coating [1]. The main objective of the finishing process in the textile industry is adding or improving the functions of current textile-based products or introducing new properties to the substrate. The application of numerous materials, chemicals, and methods to coat the fabrics has led to introduction of many new products in the market with novel features. Finishing, coating, and laminating processes cover

**Corresponding author*: xungai.wang@deakin.edu.au

Nazire D. Yilmaz (ed.) Smart Textiles, (247–300) © 2019 Scrivener Publishing LLC

various fields, aspects, and contexts in the textile industry, each of which warrants its own technical prerequisites, knowledge, expertise, and equipment. The coated textiles have several layers, and the overall properties of each layer determine the final characteristics of the product. In general, the coating process is defined as a process in which a thick polymer solution, paste, or other substances are applied to a textile substrate to form a continuous, durable, and uniform layer of coating formulation on the substrate. In the laminating process, a thin pre-prepared polymer film is affixed to the surface of a fabric using heat, mechanical bonding, pressure, binders, and adhesives [2]. There are two common types of coating in the textile industry including fluid coating and dry compound coating [3]. The final application, nature of substrate, polymer type, and viscosity of coating paste and solution define the methods and equipment to be employed in coating [3]. For each of these coating methods, different types of polymers, solvents, and instruments should be used. The main steps of a textile coating process in industry are applying the coating formulation to the fabric surface, stabilizing the coating layer in the curing step, cooling, and winding up the coated products to rolls [3]. The first step of the coating process is spreading a viscous paste or solution on textiles and allowing the solvent to evaporate, leaving a polymer network on textiles [4]. In a coated product, the textile substrate provides strength and mechanical properties and the coating layer introduces novel functions to the surface [4]. Some main parameters such as the type of polymer used in the formulation, the substrate type, and the method of coating all have a direct impact on the ultimate features of the coated products. The coating is not limited to woven or knitted fabrics, and it can be applied to any form of fibrous materials such as fibers, yarns, and nonwovens [1]. The most commonly used polymers in the textile coating industry are polyvinyl chloride (PVC), polyvinyl acetate (PVA), acrylics, polyurethanes (PUs), and polyvinylidene chloride (PVDC), among others [4].

Nowadays, textiles with high performance are required in different aspects of life, and therefore, researchers continue to seek new approaches to integrate nanoscience into textile productions. The functionalized textiles should provide some advantages such as easy care, comfort, health, and hygiene for the end users; however, this should not be at the expense of sacrificing the intrinsic features of common textiles [5]. Most of the research carried out in this field focuses on growing and introducing a thin layer of nanomaterials and nanocrystals on textiles as one of the most feasible methods of surface coating [6]. In general, a smart textile is categorized as a new generation of products that can actively detect and sense various external stimuli including environmental, chemical, mechanical, thermal, pH, and electrical changes and then react to them according to the defined functionalities. The main hurdle in producing such smart textiles is the lack of acceptable washing fastness of coating layers. Different categories of materials are used to fabricate smart textiles such as phase change materials (PCMs), shape memory materials (SMMs), conductive materials, hollow glass microspheres, photocatalytic structures, chromic materials, mechanical responsive materials, polymers, nanowires, and noble metals, to name but a few. Depending on the type of nanomaterials used in the coating process, different functionalities can be introduced to textiles such as self-cleaning, antimicrobial property, UV protection, wrinkle resistance, oil and water repellence, and flame retardancy. Also, different coating methods such as sol–gel, spraying, immersion, plasma, chemical and physical vapor evaporation, and sono-processing have widely been used by researchers to coat the textiles [2].

This chapter provides a general overview on some of the novel aspects of nanocoatings that are currently used to produce smart textiles. To this end, some fundamental approaches to prepare nanocoatings will be assessed, and the obtained novel functionalities on textiles along with their associated mechanisms will be discussed.

8.2 Fabrication Methods of Nanocoatings

8.2.1 Sol–Gel

The main idea behind using the sol–gel method is creating an inorganic or organic network from a colloidal solution synthesized from precursors [7]. The sol in general is defined as a stable suspension of metal oxide particles in a liquid medium [7]. In a colloidal sol, the sizes of dispersed particles are too small in a way that the gravity force on them is negligible and the van der Waals forces and surface chemistry are influential parameters [8]. The gel is an interconnected porous 3D rigid network of suspended matters in the sol that are developed throughout the liquid medium. Depending on the type of initial sol, the formed gel is named as polymeric or colloidal gel. In the gel structure, there is a thermodynamic equilibrium between the solid gel networks and the liquid that exists in their structure. Depending on the nature of liquid media in the system, the gel is categorized as either aquagel (hydrogel) or alcogel. In an aquagel, water is the main liquid medium, while in an alcogel, the dominant liquid medium is alcohol. The existence of liquid in the structure of gels leads to a soft structure of gels. After drying the gels by removing the liquid from the network, a xerogel will be obtained [7]. The most commonly used precursors for synthesizing the metal oxide colloidal sols are metal alkoxides.

The alkoxides are classified as a group of materials in which organic ligands have linked to metal or metalloid atoms [8]. The synthesized process of sols starts with a hydrolysis of precursors in the presence of water molecules. The amount of water and catalyst in the system determines the progress rate of hydrolysis. The next step is polycondesation process, which occurs by the reaction of hydrolyzed precursors with each other generating either water or alcohol by-products [8]. The influential factors in these reactions are pH, time, temperature, and concentrations of precursors and catalysts. The metal alkoxides react instantly with water due to their reactive alkoxide groups (–OR), which exist in their structure. During the hydrolysis process, the alkoxide groups are replaced with hydroxyl groups, which, in turn, results in formation of metaloxane links (M–O–M) [9]. Figure 8.1 illustrates the hydrolysis and condensation reactions of metal alkoxides in a sol–gel method.

The sol–gel process has been used in coating different substrates including textiles [10]. Application of sols to a substrate and then drying it results in formation of a thin and transparent layer of nanomaterials on fibrous matters. The common deposition methods of sols to substrates are dip coating, spray coating, and spin coating [11]. The hydrolysis and drying conditions are effective on some parameters such as density, porosity, cracking, and the mechanical properties of the coating layers [11].

Sol–gel method has been considered as one of the most feasible methods for surface functionalization of textiles. There are some advantages in this method including (i) obtaining a thin and transparent film; (ii) protecting the substrate against dirt, heat, and microorganisms; (iii) bolstering the mechanical characteristics of substrates; (iv) developing multifunctional coatings; and (v) feasibility [12]. After applying the sols to the textile surface, a padding step should be introduced to remove the excess uptake of sols and create a thin layer of three-dimensional network of nanoparticles on the surface. After deposition of nanoparticles on the textile surface, a curing process is recommended to improve the washing fastness and

Hydrolysis

$$\equiv M-OR + H_2O \Rightarrow \equiv M-OH + ROH$$

Condensation

$$\equiv M-OH + \equiv M-OH \Rightarrow \equiv M-O-M\equiv + H_2O$$

or

$$\equiv M-OH + \equiv M-OR \Rightarrow \equiv M-O-M\equiv + ROH$$

Figure 8.1 Hydrolysis and condensation reactions of metal alkoxide precursors. (The figure has been prepared by the authors.)

mechanical stability of the coating layer (Figure 8.2) [12, 13]. During the curing step, the M–OH groups remained from the precursor can establish hydrogen bonds with the hydroxyl groups of substrate as well as covalent bonds with the functional groups of fiber [9]. Figures 8.3 and 8.4 show

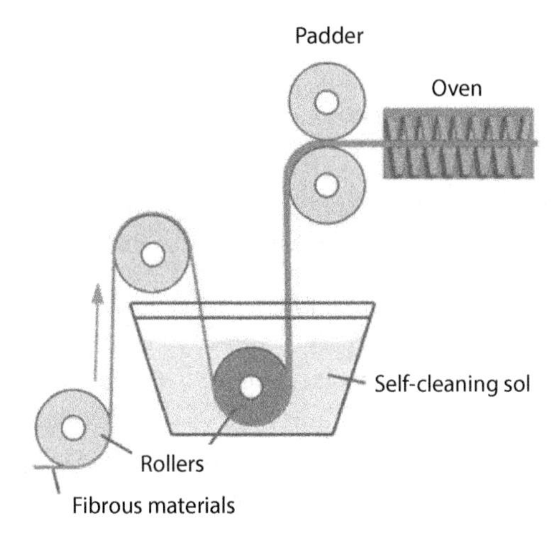

Figure 8.2 Coating process of textiles with the prepared sol. (Reprinted from reference [13], with permission of The Royal Society of Chemistry.)

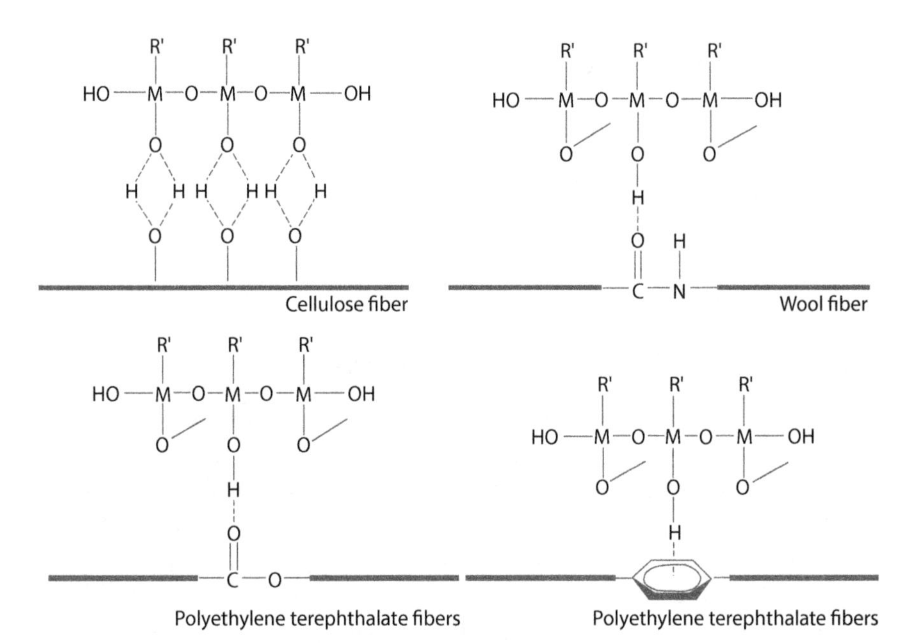

Figure 8.3 Hydrogen bonds between M—OH groups of precursor and fiber surface. (Reprinted from reference [9] with permission of Springer, © Springer 2016.)

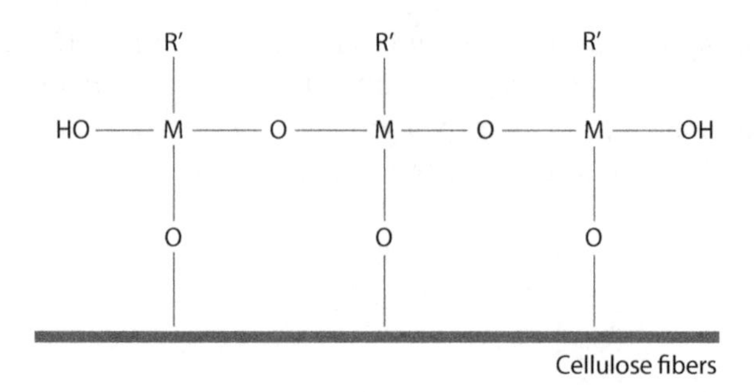

Figure 8.4 Establishing covalent bonds between sol and fiber surface. (Reprinted from reference [9] with permission of Springer, © Springer 2016.)

the mechanisms of cross-linking processes, which take place during the curing step.

The final features of synthesized sol–gel coating on textiles mainly depend on some parameters such as rate of hydrolysis and condensation reactions of precursors, pH, temperature, time of reaction, precursor concentration, catalyst type, ageing temperature, aging duration, and drying conditions [9]. Textiles can be coated with sols simply through a dip-coating process. This technique is based on soaking the substrate into a sol and then removing it based on a predefined speed under certain pressure and temperature. The applied layer of nanoparticles is transformed to gel after drying and stabilized during the curing process. The thickness of the coating on substrates can be controlled by the number of coating layers and also the viscosity of sols and the concentration of nanoparticles [14].

8.3 Sol–Gel Coatings on Textiles

8.3.1 Self-Cleaning Coatings

8.3.1.1 Photocatalytic Self-Cleaning Nanocoatings

One of the new functionalities that can be produced through the sol–gel method on textiles is the self-cleaning property. There are two types of self-cleaning surfaces including photocatalytic and superhydrophobic, each of which has its own mechanisms to clean the substrate surface. For the former one, the photocatalytic activity of TiO_2 nanoparticles plays the main role in eliminating the pollutions from the textile surface. Titanium dioxide is a semiconductor that is widely used in different fields and products such as paints, cosmetic materials, and waste water treatment systems [15]. TiO_2 nanoparticles show photocatalytic activity under the illumination of UV ray with wavelengths greater than TiO_2's band gap energy, which is 388 nm. Under UV light, the electrons of the valence band of TiO_2 absorb the energy of UV light and promote to the conduction band producing negative electrons and positive holes. These negative and positive species, in turn, react with water and oxygen molecules producing superoxide anions and hydroxyl radicals. These two products are considered as active species that can decompose dirt and pollution adsorbed on the surface of textiles. Figure 8.5 illustrates the photocatalytic mechanism of TiO_2 nanoparticles under UV light.

Daoud and Xin [16, 17] in a series of publications investigated different aspects of cellulosic substrates coated with TiO_2 nanosols. They synthesized TiO_2 sol from titanium tetra isopropoxide (TTIP) in the media of ethanol and water mixture in the presence of acetic acid and nitric acid. The sol was synthesized after refluxing the ingredients together at 60°C for 16 h. The prepared sol was applied to cotton fabric surfaces through a conventional dip–pad–dry–cure process. Some features of fabrics such as

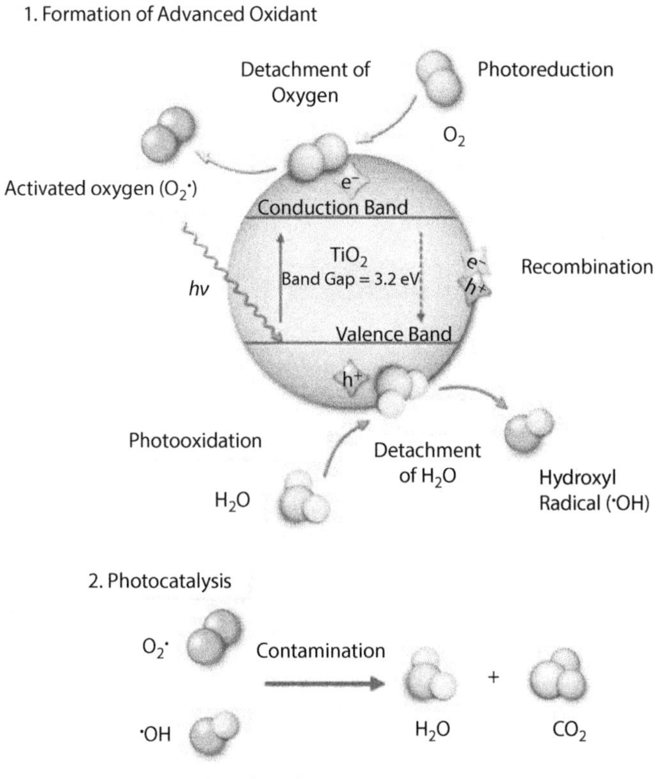

Figure 8.5 Photocatalytic activity of TiO_2 nanoparticles. (Reproduced from reference [13] with permission of The Royal Society of Chemistry.)

ultraviolet protection factor (UPF) and antimicrobial activity against different pathogen microorganisms such as *Klebsiella pneumoniae* (a gram-negative bacterium) and *Staphylococcus aureus* (a gram-positive bacteria) were investigated [16, 18]. Qi *et al.* [19] synthesized TiO_2 sol at different temperatures and applied it to the surface of cotton fabrics. It was demonstrated that coated cotton fabrics possessed photocatalytic self-cleaning property due to the presence of anatase TiO_2 nanoparticles. The coated fabrics successfully decomposed the coffee and red wine stains under simulated sunlight (Figures 8.6 and 8.7).

Daoud *et al.* [20] applied TiO_2 sol on keratin fibers and investigated its self-cleaning property. The main challenge in applying nanosols to the proteinous fibers was the lack of adequate functional groups in the wool fiber structure to bond with TiO_2 nanoparticles. Therefore, they modified the surface of wool fibers through an acylation process using succinic anhydride and DMF to introduce new functional groups to wool [20]. Tung *et al.* [21–24] investigated the self-cleaning property of wool fabrics coated with TiO_2 nanoparticles, optimizing the TiO_2 sol preparation procedure. They assessed and optimized the influential parameters in TiO_2 sol preparation such as temperature, stirring time, TTIP concentration, and acid catalyst. They synthesized two types of sols using hydrochloric acid (H-sol) and nitric acid (N-sol) and compared the crystallinity, particle size, and photocatalytic property of nanoparticles coated on textiles (Figures 8.8 and 8.9).

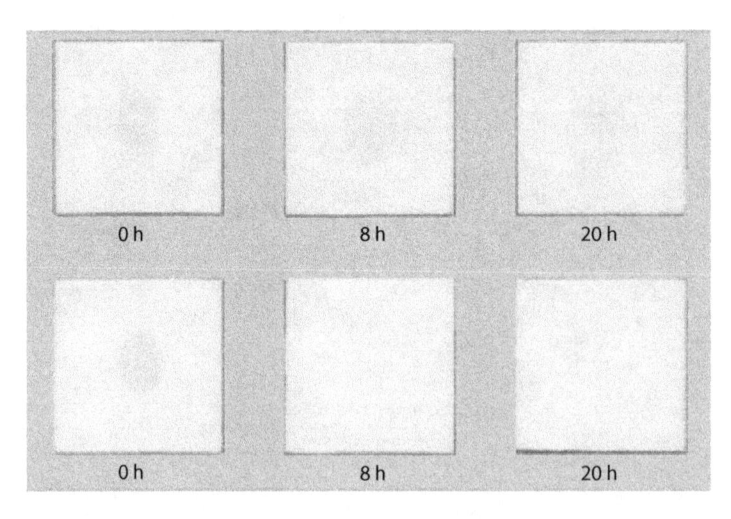

Figure 8.6 Red wine stain removal on raw (upper row) and TiO_2 sol-coated (lower row) cotton fabrics. (Reproduced from reference [19] with permission of The Royal Society of Chemistry.)

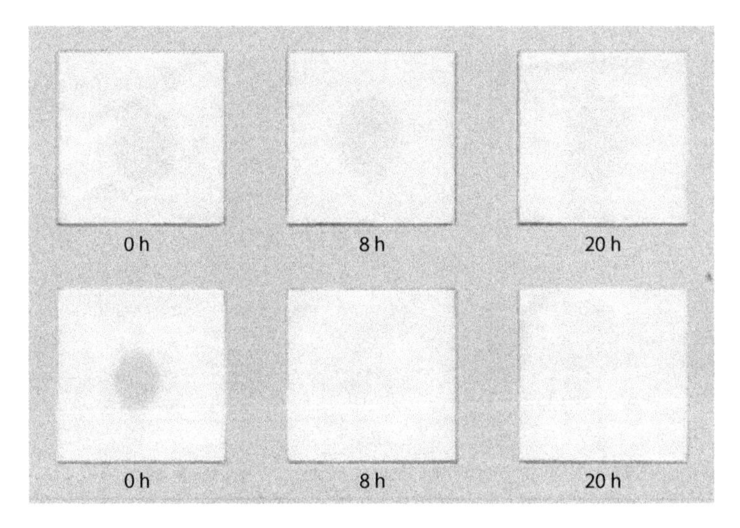

Figure 8.7 Coffee stain removal on raw (upper row) and TiO_2 sol-coated cotton fabrics (lower row). (Reproduced from reference [19] with permission of The Royal Society of Chemistry.)

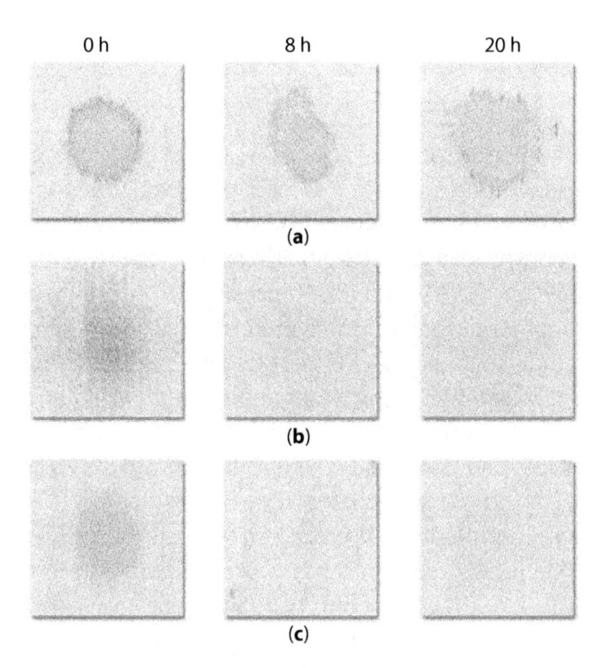

Figure 8.8 Coffee stain removal on wool fabrics: (a) pristine wool fibers, (b) fabrics coated with N-sol, and (c) fabrics coated with H-sol. (Reprinted from reference [21] with permission from Elsevier.)

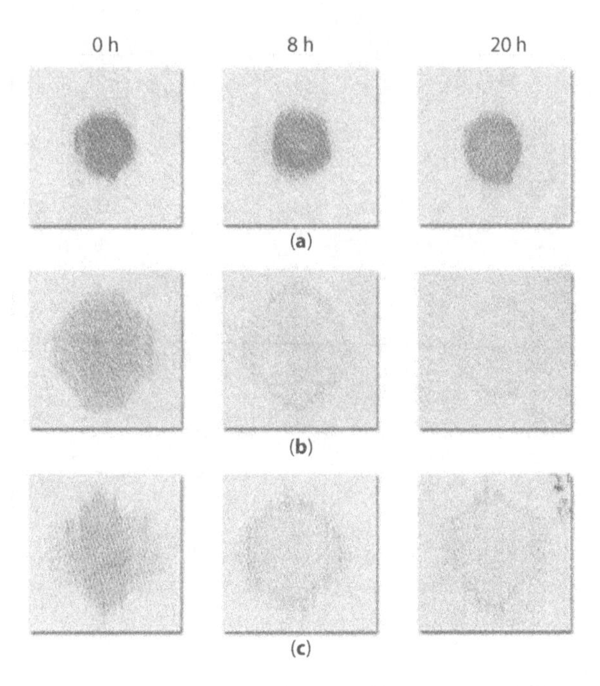

Figure 8.9 Red wine stain removal on wool fabrics: (a) pristine wool fibers, (b) fabrics coated with N-sol, and (c) fabrics coated with H-sol. (Reprinted from reference [21] with permission from Elsevier.)

Although the applications of TiO_2 sols to the textiles were promising in terms of decomposing stains, it was observed that photocatalytical removal of stains needed a long period of UV irradiation. Also, it had a very low efficiency under visible light. Therefore, researchers tried to boost the photocatalytic activity of TiO_2 nanoparticles through modifying sol synthesis through doping with noble metals and dyes and combining with other compounds such as SiO_2. In a research conducted by Pakdel *et al.* [25, 26], the self-cleaning property of nanocoatings was improved through integrating different concentrations of silica in the synthesis of sols. They prepared the TiO_2/SiO_2 sols with the compositions of 30/70 (1:2.33), 50/50 (1:1), and 70/30 (1:0.43) and coated on wool and cotton fabrics through a dip–pad–dry–cure method. It was reported that the presence of SiO_2 in the coating system rendered the surface of wool and cotton fabrics super-hydrophilic [25, 26]. The prepared TiO_2/SiO_2 coatings particularly TiO_2/SiO_2 30/70 showed higher efficiency compared with pure TiO_2 on both wool and cotton fabrics (Figures 8.10 and 8.11). The synergistic role of silica was linked to its role in increasing the surface acidity of nanocoatings,

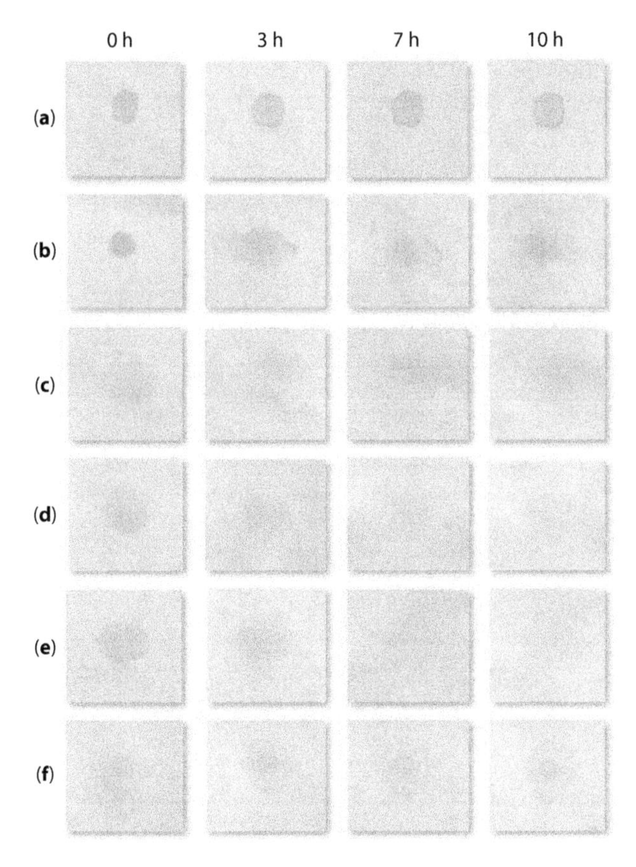

Figure 8.10 Degradation of coffee stain on wool samples: (a) pristine wool; fabric treated with (b) TiO_2, (c) TiO_2/SiO_2 70:30, (d) TiO_2/SiO_2 50:50, (e) TiO_2/SiO_2 30:70, and (f) SiO_2. (Reprinted from reference [25] with permission from Elsevier.)

increasing the surface area in the surrounding of the photocatalyst, and lowering the overall refractive index of nanocomposites [26].

Further research in this field was carried out to increase photoefficiency of the coating on textiles. One of the main research objectives was expanding the activation territory of nanoparticles applied to textiles towards the visible light region. Incorporating noble metals was considered as an effective remedy to improve the photoactivity of TiO_2 nanoparticles under UV and shift the photocatalytic activity threshold of TiO_2 nanoparticles to the visible region [27]. There are some associated mechanisms regarding the role of noble metals in increasing the photocatalytic activity of TiO_2 nanoparticles. It was confirmed that the presence of metals on TiO_2 surface

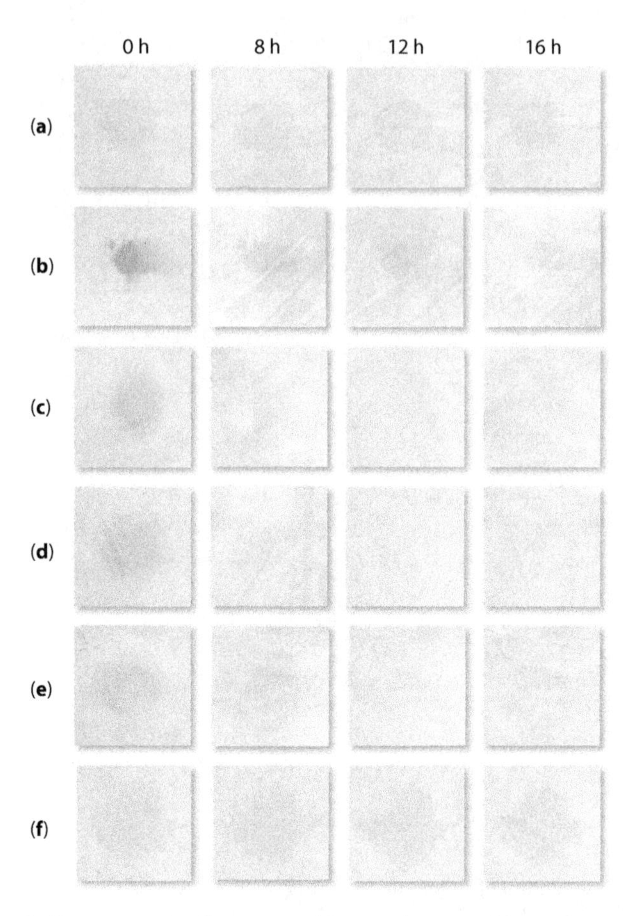

Figure 8.11 Coffee stain removal on cotton fabric under UVA: (a) pristine cotton, cotton functionalized with (b) TiO_2, (c) TiO_2/SiO_2 1:0.43 (70:30), (d) TiO_2/SiO_2 1:1 (50:50), (e) TiO_2/SiO_2 1:2.33 (30:70), and (f) SiO_2. (Reprinted from reference [26] with permission from Elsevier.)

can reduce the recombination rate of generated electrons and holes. This, in turn, would increase the production of active species of superoxide anions and hydroxyl radicals, hence a higher photocatalytic activity. At the same time, TiO_2/metal systems can show a higher photocatalytic activity under visible light due to the synergistic role of metals in producing a surface plasmon resonance and narrowing the semiconductor's band gap [27, 28]. Wang *et al.* [29] synthesized the ternary nanocomposite system of $Au/TiO_2/SiO_2$ through the sol–gel method and applied it to cotton fabrics. It was observed that the $Au/TiO_2/SiO_2$ nanocomposites showed higher photoefficiency compared with bare TiO_2 on cotton surface. The efficiency of

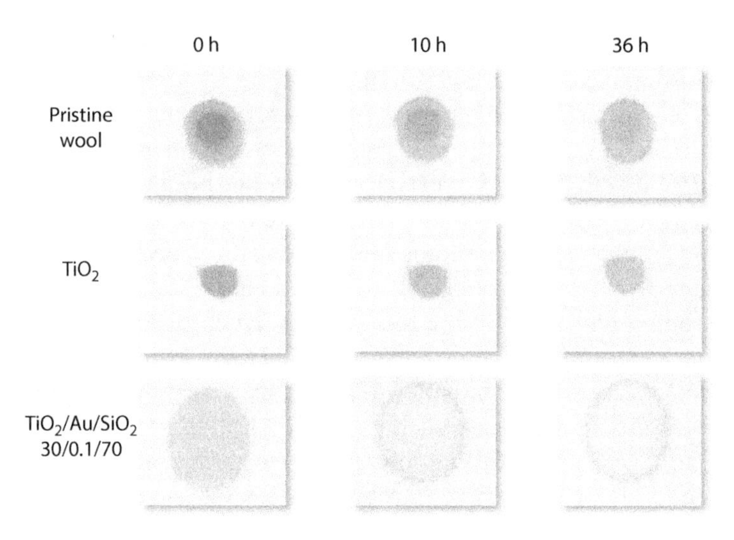

Figure 8.12 Self-cleaning property of wool fabrics coated with TiO_2 and ternary Au-modified sols under filtered simulated sunlight. (Reprinted from reference [30] by permission of Taylor & Francis Ltd.)

coated samples was compared based on the degradation rate of red wine stains under visible light [29]. Pakdel *et al.* [30, 31] synthesized TiO_2/metal/SiO_2 through the sol–gel method and investigated the role of metal type and concentration on the photocatalytic activity of coated wool and cotton fabrics. The effect of metal type (Pt, Au, and Ag) and their concentration on photocatalytic and antibacterial activities were investigated. They assessed the photoinduced efficiency of nanocoatings based on the degradation of red wine stains on wool and decomposition of methylene blue solution under visible light sources [30]. Figure 8.12 shows the self-cleaning property of wool samples coated with $TiO_2/Au/SiO_2$ nanocomposites under filtered simulated sunlight.

8.3.1.2 *Self-Cleaning Surface Based on Superhydrophobic Coatings*

Another type of self-cleaning surface relies on superhydrophobic properties of coated surfaces. A surface is defined with superhydrophobic property when contact angle is $90° < \theta \leq 180°$ and contact angle hysteresis is very low (typically $<10°$) [32]. The superhydrophobic property is also recognized as *lotus effect* because this behavior was first observed in nature on the leaves of a lotus plant, a symbol of purity [33]. It has been proved that the geometric structure and chemical composition of surface are two main factors that contribute to superhydrophobicity. In the main, there are two major approaches to confer a superhydrophobic property to a surface including introducing micro/nanostructure roughness to the surface or using chemicals with low surface free energy to modify the nanostructured surfaces [34, 35]. The electron microscopy investigation of lotus leaves demonstrated that the surface of leaves has been covered with a lot of nubs sticking out of the surface, which were covered with smaller rough scale epicuticular wax crystalloids [36]. Due to the waxy and rough structure, water droplets will not be able to spread on the surface and therefore will be rolling off on top of the hierarchical structures on coated surface and remove any contaminating particles [37]. Understanding the reason of lotus effect has spawned many new research investigations to expand this property to different surfaces such as textiles, building materials, and car paints. Utilizing chemicals that lower the free surface energy along with making microlevel and nanolevel roughness structures on the surface can even result in higher contact angles on coated samples [38]. Some methods such as sol–gel, electrochemical deposition, chemical vapor deposition, template synthesis, and hydrothermal approach, among others, can be used to construct the rough coating surface [35]. In order to reduce the free surface energy, some compounds such as alkyl- and fluoro-alkyl-substituted silanes and fluorine in the form of fluorocarbon polymers can be added to the coating system [34]. In the sol–gel method, the process of preparing sols can be modified to incorporate the hydrophobic compounds in the coating systems. To this end, some compounds of silane monomers with long alkyl chains such as hexadecyltriethoxysilane and polymer additives such polysiloxanes can be used [39]. Also, fluorinated additives such as fluoroalkylsilane monomers and polymers with perfluorinated alkyl side chains can be used to produce oleophobic properties on the textiles [39]. However, the interest in using fluorinated compounds in coating systems is diminishing due to its potential risks to human health and environment.

Mahltig and Böttcher [40] used the sol–gel method to produce pure and modified silica sols and assessed their water repellency after modifying with different additives on polyamide and polyester/cotton fabrics. Silica sols were modified with alkyltrialkoxysilanes, polysiloxane derivatives, and a fluorine-containing silane. It was observed that fabrics coated with sols containing hexadecyltrimethoxysilane and triethoxytridecafluorooctylsilane showed suitable water repellency. Shorter alkyl-chain length did not result in acceptable hydrophobicity [40]. Wang *et al.* [41] demonstrated that through applying the fluoro-containing silica nanoparticles to the surface of textiles such as wool, cotton, and polyester, a superhydrophobic property can be obtained. A co-hydrolysis of tetra-ethylorthosilicate together with fluorinated alkyl silane in $NH_3{\cdot}H_2O$–ethanol media was carried out to produce the solution. The contact angle of water droplets on the fabrics treated by this method was larger than 170°. The size of nanoparticles on the surface of polyester ranged within the region of 50–150 nm [41]. Daoud *et al.* [42] coated cotton fabrics with a superhydrophobic layer of silica synthesized by low-temperature sol–gel method. A mixture of hexadecyltrimethoxysilane (HDTMS), tetraethylorthosilicate (TEOS), and 3-glycidyloxypropyltrimethoxysilane (GPTMS) was used to prepare the finishing sol. The coated sample showed a contact angle of around 141°. A slight reduction of water contact angle from 141° to 105° after 30 cycles of washing was observed highlighting the suitable stability of the coating layer on the cotton surface. It was claimed that the suitable stability of the hydrophobic layer is due in large measure to the linkages created between GPTMS and cotton fabric. Moreover, it was observed that the bursting strength and air permeability were improved; conversely, the tensile strength declined by 5% [42]. Gao *et al.* [43] produced hydrophobic cotton and polyester fabrics through incorporating hexadecyltrimethoxysilane (HDTMS) in the coating process. The samples were dip-coated in silica sol and then treated with hydrolyzed HDTMS for 1 h. It was observed that applying silica layer to the surface of fabrics before coating with hydrolyzed HDTMS led to higher water contact angle due to the increased surface roughness. They achieved water contact angles of 155.1° and 143.8° on cotton and polyester fabrics, respectively [43]. Xue *et al.*, [44] produced UV-protective-superhydrophobic cotton fabrics through coating TiO_2 sol to cotton. Then, hydrophobization process was carried out through immersing the coated samples into a 1% solution of 1H, 1H, 2H, 2H-perfluorodecyltrichlorosilane (PFTDS). Afzal *et al.* [45] reported a superhydrophobic cotton fabric with photocatalytic activity that was fabricated through the low-temperature sol–gel method. In the first step, they synthesized the TiO_2 sol through the sol–gel method, applied to the fabrics surface, and then modified the coating with TTCP (*meso*-tetra(4-carboxyphenyl)porphyrin) to shift the photocatalytic activity threshold to the visible region. The coated samples were further functionalized with trimethoxy(octadecyl)silane (OTMS) solution to get the superhydrophobic property. The samples coated with OTMS/TTCP/TiO_2 showed a water contact angle of 156° (Figure 8.13) [45].

Some of the main drawbacks of liquid-repellent coatings on textiles are poor washing durability, low resistance to physical abrasion, and in some cases high stiffness of the coated samples. These issues have made these types of functionalized products inappropriate options for daily use [46]. Therefore, some methods such as covalent bonding with the substrate, crosslinking the coating layers, and using an elastomeric nanocomposite as the coating material, or introducing a self-healing function, have been

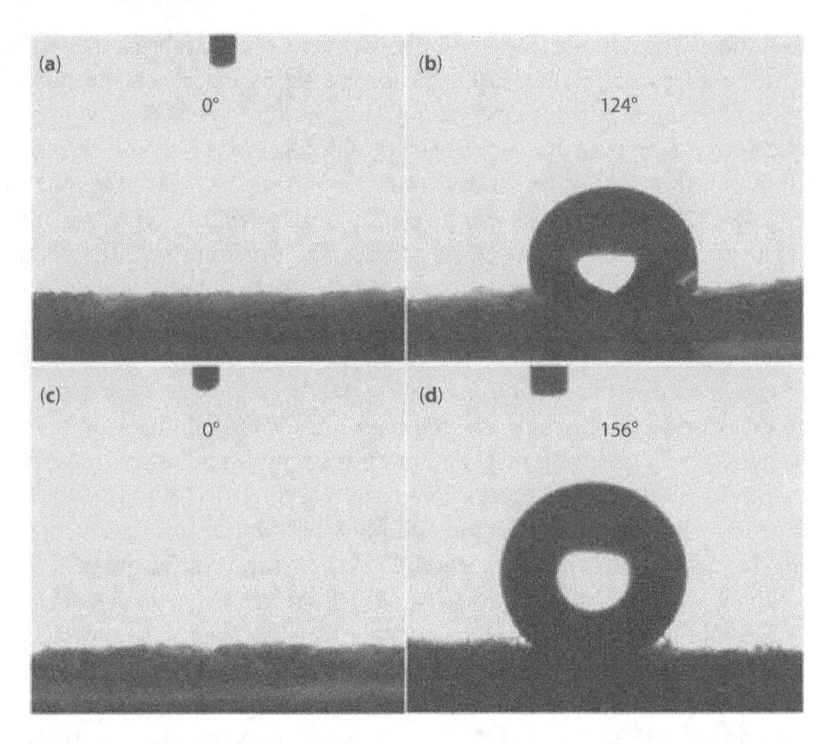

Figure 8.13 Water droplets on (a) pristine cotton, (b) TiO_2-coated cotton, (c) TCPP/TiO_2-coated cotton, and (d) OTMS/TCPP/TiO_2-coated cotton. (Reprinted from reference [45] with permission of The Royal Society of Chemistry.)

recommended to tackle these drawbacks [46]. In order to improve the durability of coatings, Zhou *et al.* [47] filled polydimethylsiloxane (PDMS) with fluorinated alkyl silane (FAS) functionalized silica nanoparticles to fabricate a superhydrophobic coating on polyester fabric. The hydrophobic silica particles were synthesized through co-hydrolysis and co-condensation of tetra ethyl orthosilicate under an alkaline condition in the presence of a fluorinated alkyl silane [47]. They reported a water contact angle of 171° and a sliding angle of 2° for coated samples, which can be considered as a very high superhydrophobicity on samples. The coated samples underwent 500 cycles of wash, and it was observed that the water contact angle and sliding angle variations were less than 5° indicating the excellent washing fastness. They also tested the resistance of coated samples against boiling water condition as well as abrasion, and it was found out that the superhydrophobicity was still stable. In another research, Zeng *et al.* [32] produced hydrophobic silica particles and then applied to cotton fabrics along with SU-8, which is an epoxy-based photoresist compound (Figure 8.14).

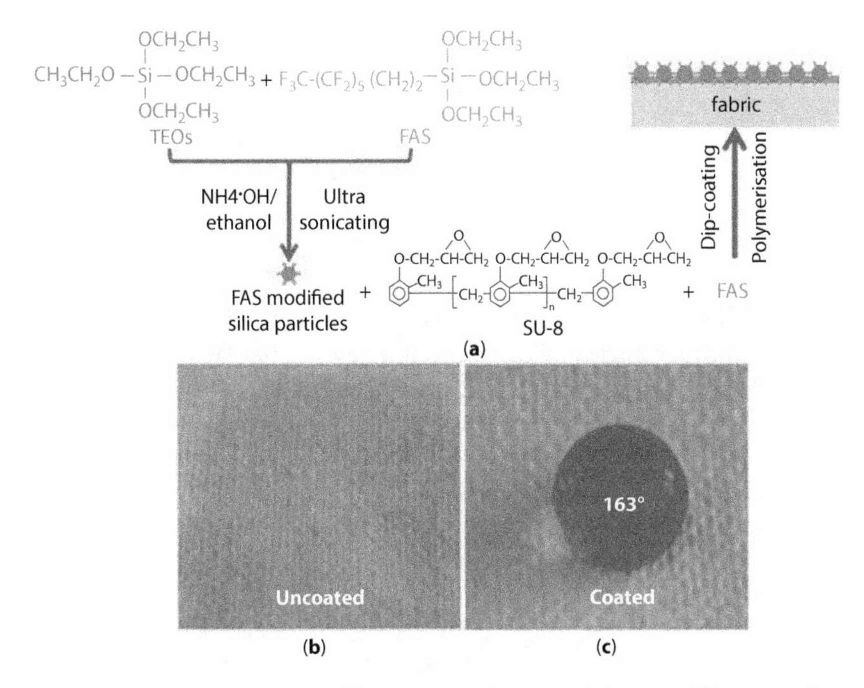

Figure 8.14 (a) Coating process of fabrics; water absorption behavior of (b) uncoated fabric and (c) coated fabric. (Reprinted from reference [32] with permission of The Royal Society of Chemistry.)

The coating was stabilized on the fabric by UV irradiation, followed by heat treatment. They obtained a coating layer that had a contact angle of 163° and was durable against organic solvents, acid and base solutions, as well as multiple washing cycles.

8.3.2 Antimicrobial Sol–Gel Nanocoatings

In recent years, the population growth and the diversity of microorganisms in living environments have motivated researchers to introduce advanced antimicrobial features to fibrous materials. Textiles can provide suitable breeding grounds for microorganisms such as warmth, moisture, and required nutrients, facilitating their proliferation. The growth of bacteria on textiles can negatively impact fabric's physical and aesthetic features such as strength, color, and smell, among others [31, 48]. There is a great demand in the market for some antimicrobial textile products such as sportswear, shoe linings, socks, and underwear. Moreover, these functional textiles can be introduced to different types of textiles such as upholstery covers, outdoor textiles, air filters, automotive textiles, carpets, and medical textiles [49]. The antimicrobial property on textiles can be achieved by different types of organic and inorganic antimicrobial agents. The application of inorganic nanoparticles such as TiO_2, Ag, ZnO, Au, and CNT and their nanocomposites to coat textiles have been widely used by researchers [48]. The coated antimicrobial textiles should be efficient enough to eliminate different types of microorganisms while posing no toxicity to the consumers. At the same time, applied coatings need to be durable enough to withstand numerous cycles of launderings and hot pressing processes [49]. Also, the antimicrobial coating should not hamper the intrinsic features of the textile substrate. Due to the continuous contact between the textiles and human skin, it is important that the finished products do not impose any change to the bacterial ecology of human skin. This can interfere with the normal functionality of bacteria species, which usually inhabit on human skin, and their existence is necessary for balanced human health [49].

Among different coating methods to fabricate antimicrobial textiles, the sol–gel technique is considered as a versatile and feasible approach. Through the sol–gel method, the photoactive nanomaterials such as TiO_2 can be synthesized and applied to textiles [12]. Also, it is possible to incorporate some bioactive compounds into the inorganic matrices through this method [12]. Nanocoatings containing metal oxides or metals show antimicrobial properties based on different mechanisms. The antimicrobial efficiency of metal oxides such as TiO_2 nanoparticles results from their photocatalytic activity. The active species of super oxide anions and hydroxyl radicals generated from TiO_2 particles can react with the cell wall of bacteria; as a result, they would be able to destroy the cell wall and then the cell membrane, endotoxin, of *E. coli* [50]. The destruction of cell structure causes leakage of macromolecular compounds, such as proteins, minerals, and genetic materials causing the death of cells [50–52]. Further AFM analysis of antimicrobial activity of TiO_2 nanoparticles demonstrated that following the decomposition of endotoxin, a lipopolysaccharide, the intracellular material of cells leaks out, causing the death of bacteria [53]. However, heavy metals such as silver follow different strategy to deactivate the bacteria. The antimicrobial property of silver results from the silver ions generated from the silver particles in the presence of moisture. The antimicrobial efficiency of silver depends on its concentration, surface area, and the interaction rate of ions with fungus and bacteria [54]. The silver ions (Ag^+) are generated in the presence of water and body fluid and can instantly react with proteins, amino acid residues, free anions, and receptor groups on cell membranes and deactivate them by denaturizing [55]. The efficiency of silver in eradicating the bacteria and fungus is also pertinent to its role in disrupting the enzymatic systems of microorganisms [54]. The Ag^+ cations attach to electron donor groups containing sulfur (—SH), oxygen, phosphates, carboxylates, and nitrogen in the cell structure [54, 56] The possible alterations caused by silver in cell permeability, respiration, and DNA replication of microorganisms such as *E. coli* all are contributing factors that can deplete the growth of bacteria [57–59].

Mahltig *et al.* [60] thoroughly investigated the antimicrobial finishing of textiles using the sol–gel method. In one of their research studies, they produced antimicrobial viscose fabric through embedding the organic and inorganic biocidal compounds such as silver and copper compounds and hexadecyltrimethyl-ammonium-p-toluolsulfonat (HTAT) in silica sol [60]. They assessed the efficiency of coating in hindering the growth of *Aspergillus niger* fungus and two types of bacteria including *Bacillus subtilis* (a gram positive) and *Pseudomonas putida* (a gram negative) [60]. The silver modified coatings were successful in eliminating both bacteria and fungus; however, the coating containing copper did not show a good performance against fungus. They reported that 100% efficiency against bacteria and fungus was observed for samples coated with sols containing silver/copper/HTAT [60]. Busila *et al.* [61] coated cotton/polyester fabric by Ag:ZnO/chitosan colloids and assessed the antibacterial activity against *S. aureus*, *E. coli*, and *M. luteus* bacteria. It was observed that chitosan–metal ion chelation enhanced the positive charge density of coatings leading to a higher interaction with the negatively charged cell surface [62]. They employed two functionalizing agents of GPTMS and TEOS to have a better dispersion of nanoparticles in chitosan and better adhesion of hybrid coatings to the substrates. The results showed that Ag-doped ZnO/chitosan composite systems had better efficiency compared with Ag/chitosan and ZnO/chitosan. All samples showed very good antimicrobial activity with the ability to reduce up to 50–95% of the viability of bacteria [61]. Daoud and Xin [17] synthesized the TiO_2 colloid and applied it to cotton fabric and analyzed the antimicrobial property of samples against *K. pneumoniae*, a gram-negative bacterium, and *S. aureus*, a gram-positive bacterium [19, 63]. They incubated the plates for 24 h at 37°C under ambient cool white fluorescent light having a slight portion of UV. It was reported that TiO_2 coating layer efficiently inhibited the growth of bacteria. Pakdel *et al.* [30, 31] investigated the impact of different types of noble metals in introducing permanent antimicrobial activity to wool and cotton fabrics. Their research compared the efficiency of three types of noble metals including Ag, Au, and Pt in improving the functionality of TiO_2/SiO_2 coatings against *E. coli* bacteria [30, 31]. It was observed that silver was more efficient in improving the antimicrobial activity of coatings. Au and

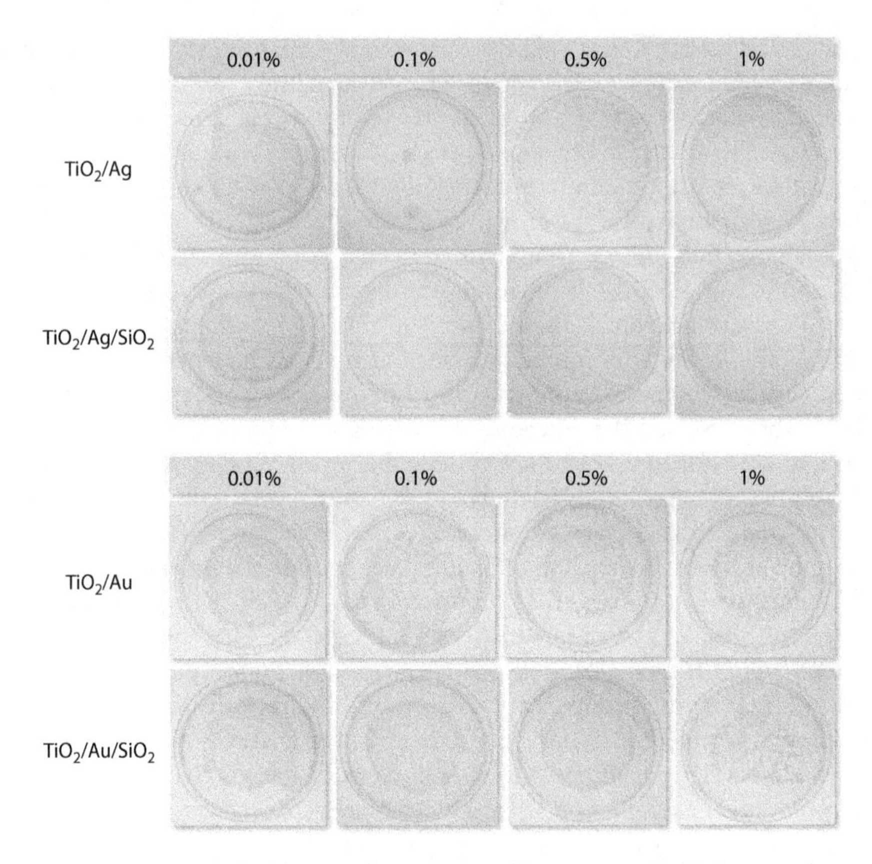

Figure 8.15 Antimicrobial activity of coated cotton fabrics against *E. coli*. (Reprinted from reference [30] with permission of Springer, © Springer Science 2017.)

Pt improved the antimicrobial property of coatings only in their highest concentrations. Figure 8.15 shows the antimicrobial activity of cotton samples coated with Ag- and Au-modified colloids [30].

8.3.3 UV-Protective Nanocoatings

Improving the UV protection features of textiles has attracted great attention in recent years. It is proven that the UV section of solar spectrum can trigger some unfavorable degradation reactions in polymeric structures such as fibers [64]. Solar spectrum can be classified into three main UV wavelength sections including UVC (220–280 nm), UVB (280–320 nm), and UVA (320–400 nm) [65]. The ozone layer of our planet prevents the UVC rays to reach the earth surface protecting creatures against the harmful impacts. Therefore, only UVA and UVB can pass through the stratosphere and reach the earth. But some parts of UVB with lower wavelengths than 310 nm can still be harmful for human skin causing some problems such as skin reddening, skin burning, and even skin cancer [66]. More and more recently, due to the depletion of the earth's ozone layer, an increasing rate of harmful solar radiation hits the earth. Therefore, numerous research has been devoted to enhancing the UV protection property of textiles [67]. Some factors such as the presence of dyes, textile structure, and finishing history, among others, can impact on the ultimate UV protection characteristic of fabrics [68–70]. The UV protection capability of fabrics is evaluated by ultraviolet protection factor (UPF) according to Australia and New Zealand standards [27]. This scale defines the amount of UV radiation that can pass through the fabrics and reach the skin. This simply means only 1/UPF of UV radiation can pass through the fabric. Also, it gives an idea of how long a person can stay under sunlight while wearing protective clothes without skin reddening. UPF is classified in three main groups including good (15<UPF<24), very good (25<UPF<39), and excellent (40<UPF<50, +50) UV protection levels [27].

There are numerous publications regarding the UV protective clothes particularly cotton [71]. Daoud *et al.* [16, 18] demonstrated that the anatase crystallite structures of TiO_2 nanoparticles grown through the low-temperature sol–gel method on cotton fabric surface can give birth to an excellent (+50) UV protection. Likewise, Abidi *et al.* [72] applied titania nanosols to the cotton fabrics and produced self-cleaning and UV-protective cotton fabrics. Apart from cotton, the UV absorption property of TiO_2 (P-25) nanoparticles was used to increase the photostability of other types of fabrics such as wool, polyester, and polypropylene against harmful impacts of UV [73–75]. Pakdel *et al.* [76] studied the impact of increasing photocatalytic activity of TiO_2 nanoparticles applied to wool on fabric's photoyellowing. Based on the obtained results, an inverse relation was established between the photocatalytic activity of TiO_2-based nanoparticles and the photoyellowing rate of wool [76]. They employed the photoinduced chemiluminescence (PICL) method to compare the efficiency of synthesized coating in blocking UV irradiation [76]. PICL is a delicate and feasible approach to study the photodecomposition of polymeric structures such as textiles and fibers. It measures the population of free radicals generated in polymeric structures exposed to light illumination at a controllable atmosphere. The samples were exposed to the light source under the N_2 gas atmosphere. The initial sharp peak was related to phosphorescence emission and charge recombination of free radicals [76]. After decaying the first peak and switching the test environment to oxygen, the second sharp peak appeared, which was related to PICL. The intensity of PICL

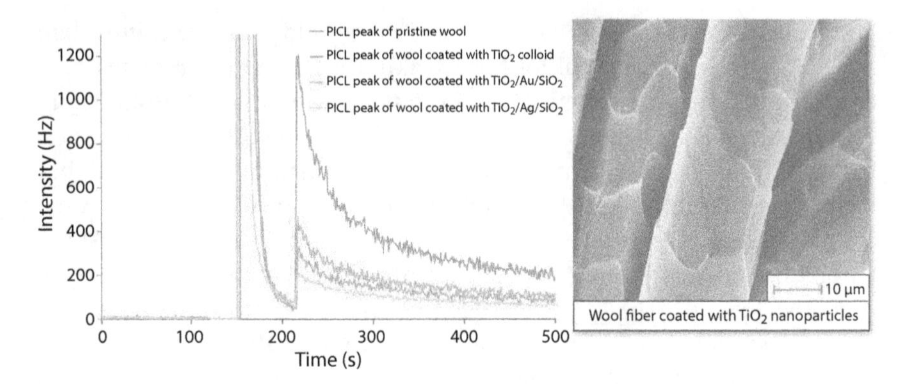

Figure 8.16 PICL spectra of wool fabrics before and after coating with TiO_2 nanoparticles, and SEM image of wool fibers coated with TiO_2 nanoparticles. (Reprinted from reference [76], with permission from Elsevier.)

peak was used to study the photodegradation pace of polymeric substrates [76]. Figure 8.16 shows that TiO_2-based coatings have suppressed the PICL peak intensity implying a very good UV blocking of nanocoating applied to wool fabrics through the sol–gel method.

It was reported that reducing the photocatalytic activity of photoactive nanoparticles such as ZnO can make them suitable choices for UV protection applications [77, 78]. Zhang *et al.* [77] reduced the photocatalytic activity of ZnO nanoparticles by providing a silica shield and applied them to the wool fabric surface. It was realized that the modified nanoparticles could further retard the photoyellowing rate of wool fabrics through reducing the free radicals in the system [77]. This method prevented any potential damage caused by the presence of photoactive nanoparticles on wool [77].

8.4 Impregnation and Cross-Linking Method

One of the most important aspects of nanocoatings on textiles is their endurance against washing processes and mechanical tensions. Using cross-linking agents and chemical spacers in the impregnating bath has been reported effective to improve the nanoparticle stability through facilitating the establishment of linkages with functional groups on textile surface [79]. Using some polycarboxylic compounds such as 1,2,3,4-butane tetracarboxylic acid (BTCA), maleic acid, succinic acid, citric acid, and 1,2,3-propanetricarboxylic acid as cross-linking agents was effective in improving the stability of nanoparticles on textile surfaces [2, 80]. The most commonly used substrates for this method are cellulosic and proteinous fibers thanks to their intrinsic functional groups that exist in their structure. Natural cotton possesses plenty of hydroxyl groups in its structure, all of which can be considered as potential sites for reaction with cross-linking agents. However, proteinous fibers have fewer functional groups on their surfaces compared to cellulosic ones. Therefore, a pretreatment step should be introduced prior to surface coating. Some pretreatment methods such as surface oxidation with $KMnO_4$ in mild acidic conditions or treatment with succinic acid solution in DMF have been suggested to increase the population of negatively charged functional groups on the surface and in turn increase the durability of nanoparticles [20, 71]. The reaction between the cross-linking agents and substrate was mostly performed in the presence of sodium hypophosphite (SHP), which played a role as a catalyst. Also, it was observed that the photoinduced radicals generated from TiO_2 can trigger the cross-linking reaction by activating the carboxylic groups and the vinyl double bond of the cross-linking agents and facilitate the reaction with cellulose chains [81].

The cross-linking method has been used to apply different types of nanoparticles such as TiO_2, CNT, ZrO_2 and silica nanoparticles to the fabric surface [71, 82–84]. Meilert *et al.* [85] applied TiO_2 nanoparticles to cotton fabrics and affixed them to cellulose fibers using chemical spacers. They used three types of chemical spacers including succinic acid, 1,2,3-propanetricarboxylic acid, and BTCA in their research. The chemical spacers had at least two free carboxylic acids, one of which could be involved in the esterification process with hydroxyl groups that exist in the cellulose fiber surface. TiO_2 nanoparticles were attached to the second free carboxylic acid through the electrostatic interaction. Figure 8.17 shows the mechanisms of esterification reactions between cellulose and chemical spacers. The commercial TiO_2 nanoparticles were employed for coating the cotton samples. The self-cleaning property of fabrics was compared through monitoring the red wine stain degradation under simulated sunlight and the amount of released CO_2 gas from each sample.

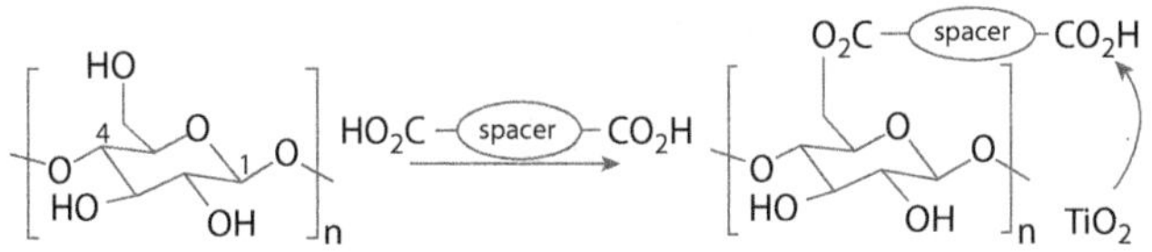

Figure 8.17 Formation of covalent ester bonds between chemical spacer and cellulose. (Reprinted from reference [85] with permission from Elsevier.)

In a research study conducted by Montazer *et al.* [71, 86], the TiO_2 Degussa P-25 nanoparticles were stabilized on the surface of wool samples using cross-linking agents of citric acid and BTCA under sonication. The former one had three carboxylic acid groups and the latter one had four carboxylic acid groups in their structures. Different aspects of coated samples were assessed including photoyellowing, self-cleaning, wettability, antimicrobial property, and antifelting [86–89]. Prior to coating, the wool samples underwent a surface oxidation process in a mild acidic solution of $KMnO_4$. Figures 8.18 and 8.19 show the possible route of stabilizing TiO_2 nanoparticles on wool using BTCA and SHP. It was observed that the presence of TiO_2 (P-25) coating brought about self-cleaning, antifelting, and reduced photoyellowing for wool samples [71]. A photoinduced hydrophilicity was also observed on coated wool samples [88]. A similar study was carried out to stabilize the TiO_2/Ag nanoparticles synthesized through the photoreduction method on wool samples [90]. A significant improvement in the antimicrobial activity of samples against *E. coli* and *S. aureus* was observed after incorporating silver in the impregnating bath [90]. The application of this approach was tried for other types of nanoparticles such carbon nanotubes (CNTs) [91]. Conductive wool samples were produced through applying CNTs to fabrics in an ultrasonic bath in the presence of cross-linking agents [91].

Gashti *et al.* [84] used the cross-linking method to stabilize dimethyldichlorosilane modified silica nanoparticles on the surface of cotton to

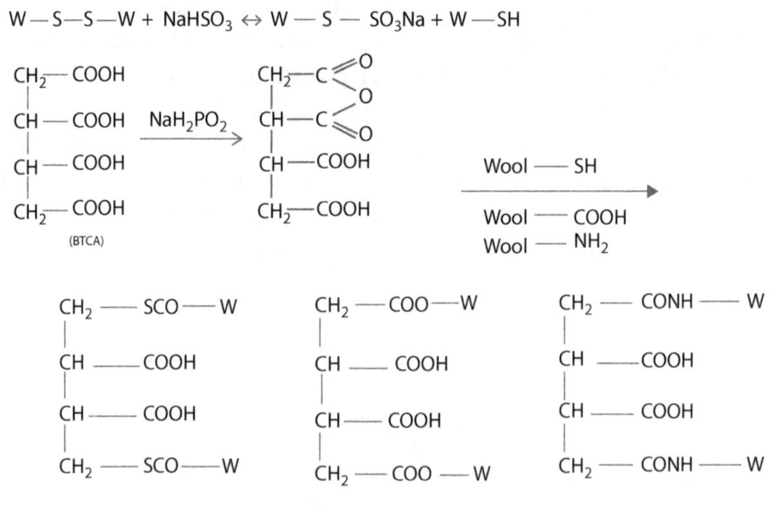

$$W{-}S{-}S{-}W + NaHSO_3 \leftrightarrow W{-}S{-}SO_3Na + W{-}SH$$

$CH_2{-}COOH$ / $CH{-}COOH$ / $CH{-}COOH$ / $CH_2{-}COOH$ (BTCA) $\xrightarrow{NaH_2PO_2}$ cyclic anhydride ($CH_2{-}C(=O){-}O{-}C(=O){-}CH$) / $CH{-}COOH$ / $CH_2{-}COOH$ $\xrightarrow{\text{Wool—SH, Wool—COOH, Wool—NH}_2}$

$CH_2{-}SCO{-}W$ / $CH{-}COOH$ / $CH{-}COOH$ / $CH_2{-}SCO{-}W$

$CH_2{-}COO{-}W$ / $CH{-}COOH$ / $CH{-}COOH$ / $CH_2{-}COO{-}W$

$CH_2{-}CONH{-}W$ / $CH{-}COOH$ / $CH{-}COOH$ / $CH_2{-}CONH{-}W$

Figure 8.18 The cross-linking mechanism between wool and BTCA. (Reprinted from reference [88] with permission from Elsevier.)

$$
\begin{array}{l}
CH_2-SCO-W \\
| \\
CH-COO^- \quad {}^+Ti \\
| \\
CH-COO^- \quad {}^+Ti \\
| \\
CH_2-SCO-W
\end{array}
\longrightarrow
\begin{array}{l}
CH_2-SCO-W \\
| \\
CH-COO^- + TiO_2 \\
| \\
CH-COO^- + TiO_2 \\
| \\
CH_2-SCO-W
\end{array}
$$

$$
\begin{array}{l}
CH_2-COO-W \\
| \\
CH-COO^- \quad {}^+Ti \\
| \\
CH-COO^- \quad {}^+Ti \\
| \\
CH_2-COO-W
\end{array}
\longrightarrow
\begin{array}{l}
CH_2-COO-W \\
| \\
CH-COO^- + TiO_2 \\
| \\
CH-COO^- + TiO_2 \\
| \\
CH_2-COO-W
\end{array}
$$

$$
\begin{array}{l}
CH_2-CONH-W \\
| \\
CH-COO^- \quad {}^+Ti \\
| \\
CH-COO^- \quad {}^+Ti \\
| \\
CH_2-CONH-W
\end{array}
\longrightarrow
\begin{array}{l}
CH_2-CONH-W \\
| \\
CH-COO^- + TiO_2 \\
| \\
CH-COO^- + TiO_2 \\
| \\
CH_2-CONH-W
\end{array}
$$

Figure 8.19 Interactions between TiO_2 and carboxylic acid groups. (Reprinted from reference [88] with permission from Elsevier.)

produce hydrophobic fabric. They used BTCA and SHP in their research as a cross-linking agent and catalyst, respectively. It was reported that the water contact angle on treated cotton samples was 132.4°. In another study, ZrO_2 nanoparticles were applied to wool samples by the cross-linking method, and some features such as self-cleaning, electromagnetic reflection, and fire retardant of samples were studied [82, 92]. The presence of nanoparticles on the surface of wool enhanced the fire-retarding property and electromagnetic reflection of wool samples [92]. The positive impact of CNTs, stabilized on cotton by BTCA, on thermal properties, flammability, and antimicrobial activity of cotton samples were demonstrated [93].

8.5 Plasma Surface Activation

In order to achieve an efficient durable coating of nanoparticles, some methods such as laser irradiation, electron beam, UV irradiation, ion beam, microwave irradiation, and plasma have been used to physically modify the structures of textile substrates. Among these techniques, plasma surface modification of textiles is mostly considered as an effective pretreatment approach in the nanocoating field. In general, plasma is considered as the fourth state of matter and is defined as an excited, ionized, and equally charged state of gas that consists of photons, electrons, atoms, molecules, and ions [94, 95]. Plasma by itself has some applications in the textile finishing field [96]. Some features such as enhanced mechanical characteristics, antistatic finish, hydrophilic treatment, hydrophobic finishing, improving the dye uptake, bleaching, and flame retardant have been reported [97]. There are several advantages for plasma processing such as being clean, dry technology, and no need to use solvent in comparison with common conventional treatment methods. Depending on the gas pressure, two methods of low-pressure plasma and atmosphere-pressure plasma can be categorized. The latter, in turn, is divided into three categories of corona discharge, dielectric barrier discharge, and atmospheric-pressure glow discharge [98]. Low-pressure plasma equipment usually uses more energy and less gas compared with atmosphere-pressure plasma. It is much easier to obtain a uniform treatment on textiles using low-pressure plasma equipment [98]. Basically, the impact of plasma processing on textiles depends on some factors such as the type of textiles, the type of used gas, the power and frequency of electrical supply, the temperature of the process, and the duration. The generated plasma has the capability to alter the composition of chemical groups on polymeric substrates by introducing free radicals that resulted from dissociating the chemical bonds [99]. One of the advantages of plasma method is modifying and optimizing the features of superficial layers of substrates while sustaining the intrinsic features of the bulk material [100]. Usually, plasma process impacts the depth of <100 nm of substrate surface, and it can modify the surface morphology as well as chemical composition of the material surface. The generated free radicals react with the gaseous compounds in the atmosphere leading to introducing new functional groups on the substrate. New functional groups on the surface will increase the adhesive strength of nanoparticles promoting the durability of nanocoating layer. Cold plasma is employed for surface modification of textiles. The plasma is named cold or low temperature because the overall temperature of plasma is at the ambient level. The generated electrons have very high temperature, but due to their low heat capacity, the plasma is not hot [95, 100]. Therefore, it can be used for surface modification of textiles. Depending on the employed gas, there are four major plasma processes on textiles: cleaning (ablation), activation, grafting, and deposition [95, 100]. In a cleaning process, an inert gas such as He, Ar, and N_2 plasmas will be used. In this case, the generated plasma can break apart the polymeric structure of contaminants such as oil adsorbed on the substrates and then completely get them removed in the vacuum condition [100, 101]. The ablation process can be done based on either physical sputtering or chemical etching depending on the type of employed gas [95, 101]. For surface activation purposes, some gases without any carbon such as oxygen and ammonia can be employed. The generated species of these gases can react with the outer layer of substrates and produce functional groups such as hydroxyl, carbonyl, peroxyl, carboxylic, amino, and amines on the surface [100, 101]. The activated surfaces have better interactions with the coating layers applied to their surface, hence more stable composites and coating structures. In grafting, an inert gas such as Ar reacts with the outer surface of polymeric substrate to introduce some free radicals on the surface of the material. Then by introducing some allyl alcohol to the system, the grafting of monomers to the active sites of substrate will be completed [98, 100, 101]. By using other types of gases such as methane or carbon tetrafluoride, the plasma process can be used for material deposition. This method is also called as plasma polymerization or plasma-enhanced chemical vapor deposition (PECVD) method [98]. Through this method, a very thin film of polymers deposits on the textile surface [98, 101].

Plasma processing has widely been used by researchers as a pretreatment to introduce active sites on different types of textiles such as polyester, cotton, wool, cotton/polyester, and wool–polyamide before coating with nanomaterials. This process increases interactions between the applied nanoparticles and substrate surface. It has been reported that through some modification processes such as radio frequency plasma and vacuum–UV light irradiation, some negatively charged groups such as —COO and —O—O can be introduced to the fabric surfaces increasing the tendency of positively charged nanoparticles for deposition on fabric surfaces (Figure 8.20) [102]. The TiO_2 nanoparticles can be anchored to modified fabrics through ionic interaction between the negatively charged

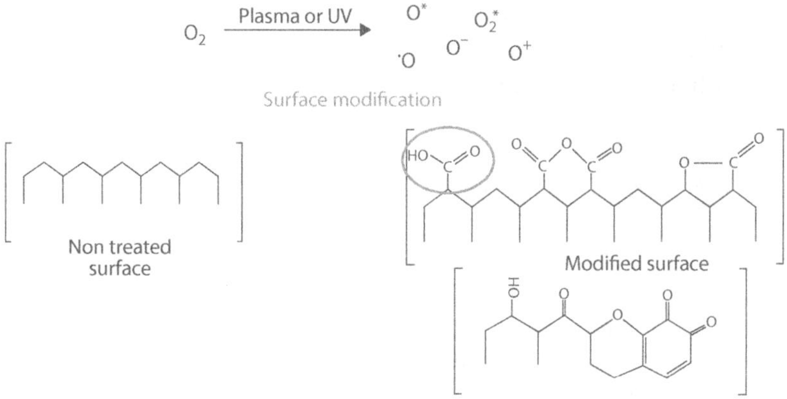

Figure 8.20 Surface modification of cotton textiles by plasma or vacuum–UV pretreatment. (Reprinted from reference [102] with permission from Elsevier.)

groups and positively charged Ti^{4+} [103]. Despite the plasma processing, the UV light does not introduce any cationic and anionic groups on the textile surfaces. However, the exposure of textiles to UV with wavelength below 241 nm results in breakage of O=O bonds leading to producing new reactive sites on the fiber surface.

The application of different types of plasma has been reported in literature [104, 105]. Qi *et al.* [103] used the low-temperature plasma of oxygen gas to modify the surface properties of polyester fabrics prior to coating with TiO_2 colloid. The application of oxygen gas resulted in introducing negatively charged groups of COO^- and —O—O^- on the polyester surface. Bozzi *et al.* [102, 106] employed radio frequency plasma (RF-plasma), microwave plasma (MW-plasma), and vacuum–UV light irradiation for pretreatment of polyester, cotton, and wool–polyamide fabrics. Interactions between RF-plasma and the carbon of the substrate resulted in producing new functional groups on the surface such as C—O, C=O, —O—C=O, —COH, and —COOH. The pretreated samples showed higher hydrophilicity and were able to degrade red wine and coffee stains [106]. Yuranova *et al.* [107] used RF-plasma and vacuum–UV to activate the surface of polyester–polyamide fabric. They applied silver nanoparticles on the modified substrate and assessed the antimicrobial activity against *E. coli* bacteria. Tung *et al.* [108] employed a microwave-generated plasma afterglow (MWGPA) treatment to modify wool fiber surface prior to coat with TiO_2 nanoparticles. They investigated the impact of some parameters such as gas mixtures, gas flow, treatment distance, treatment duration, and power flow of the plasma treatment on wool by evaluating the photocatalytic activity of coated fabrics [108]. They used different mixtures of gases such as argon/oxygen, argon/hydrogen, and argon/oxygen/hydrogen to produce plasma. The plasma pretreatment significantly increased the uptake of TiO_2 nanoparticles on wool surfaces leading to 70% improvement in photocatalytic activity compared with untreated wool sample.

8.6 Polymer Nanocomposite Coatings

The application of nanoparticles to textiles mostly does not result in a robust coating layer to withstand multiple washing cycles. All treatment methods mentioned in previous sections aimed at increasing the stability of nanocoatings on fabrics to provide safe and eco-friendly products for end users. One of the most efficient methods of fabricating a durable nanocoating layer on textiles is applying the functional coatings in the form of polymer nanocomposites. In this method, nanoparticles will be embedded into the dispersion of polymers, which play a role as carriers in surface coating process. This method will significantly enhance the stability of coatings on substrates. Through selecting appropriate functional polymer matrices as medium and nanoparticles as fillers, some features such as desired wettability, ultraviolet (UV) resistance, antimicrobial, conductivity, and flame retarding can be imparted to textiles [109]. In a sense, the polymer nanocomposites containing nanoparticles can possess more functionalities compared with each individual component of composite systems. The ultimate functionalities of polymer nanocomposites depend largely on the type of polymer, type of nanoparticles, and their shape and size [110]. The nanoparticles intensely tend to agglomerate during the production of nanocomposites [110]. However, polymers can be effective in reducing the aggregations among the fillers [109]. The surface modification of inorganic nanoparticles with some polymer surfactants or other types of modifiers such as silane coupling agents has been suggested to induce a repulsion among the dispersed nanoparticles. Figure 8.21 demonstrates surface modification of an inorganic nanoparticle with 3-methacryloxypropyl trimethoxysilane molecules [110]. This will result in a better compatibility between nanoparticles and polymer dispersion.

The nanoparticles that exist in the composite systems can react with functional groups of textile substrates covalently resulting in good wash fastness of applied coatings. Various techniques have been reported to prepare the polymer nanocomposites such as sol–gel, *in situ* polymerization, and blending the polymer and fillers together. There are two ways to use polymer nanocomposites: incorporating them into the fiber polymer in

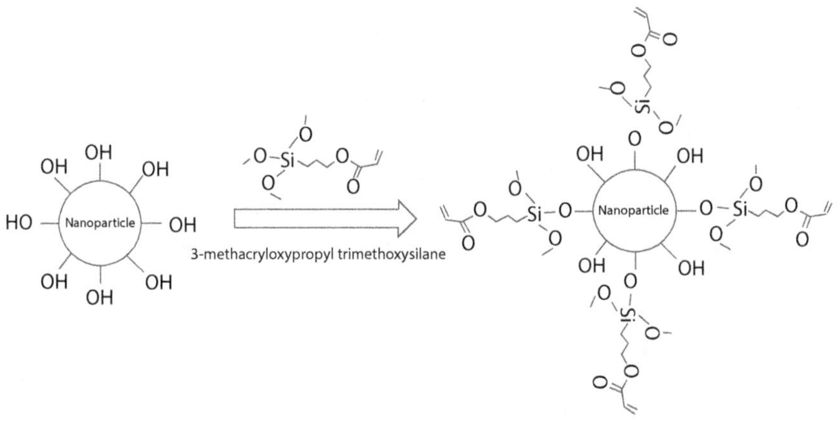

Figure 8.21 Modification of a nanoparticle with 3-methacryloxypropyl trimethoxysilane. (The figure has been prepared by the authors.)

melt spinning or developing functional coatings. Among coating methods, the most common methods are dip coating and blade coating followed by drying and curing. Using polymer nanocomposites as a coating can produce stable, thin, transparent, and multifunctional layers on textiles. Different types of micro- and nanosize materials such as metals, metal oxides (TiO_2, ZnO), graphene, carbon nanotubes (CNTs), and phase change materials (PCMs) can be used in preparing the polymer nanocomposites. In sections below, some of the reported applications of nanocomposite coatings on textiles will be highlighted.

8.6.1 Flame-Retardant Coatings

Introducing flame-retardant functionality to textiles using polymer nanocomposite systems has been a promising research area in recent years. Fire-retardant property is one of the pivotal requirements of textiles applicable in flooring, carpets, drapes, and upholstery. The main purpose of a flame-retardant treatment is reducing the flammability of textiles through hindering their ignition. There are two main categories of fire-retardant materials, which can be used in textiles, including intumescent and non-intumescent compounds [111]. There are three main methods through which the flame-retardant materials can be incorporated in textile products: (i) mixing with the fiber polymers during the melt spinning process, (ii) using copolymerization technique to graft fire retardants to the structure of polymers, and (iii) surface coating and treatment of textiles with fire-retardant materials. Three methods of cone calorimetry, limiting oxygen index (LOI), and thermogravimetric analysis (TGA) are currently used to evaluate the flame retardancy of a textile material [111]. Each of intumescent and non-intumescent compounds has its unique mechanisms to retard the ignition of textiles. The intumescent systems usually are composed of a char former or carbon source, an acid source, and a blowing agent [112, 113]. The acid source is decomposed to produce a mineral acid that plays a role as a dehydrating catalyst of char former or carbonizing agent. The blowing agent releases gas during the combustion process to form a swelled barrier layer on burning surface against oxygen and heat transmission [113]. Some examples of non-intumescent materials are halogenated, phosphorus, nitrogen, silicone, or inorganic metal compounds, the acting mechanisms of which are based on free radical scavenging or char layer forming. The application of halogenated and phosphorus compounds has recently been restricted mostly due to associated concerns in producing toxic gases during the burning process [111].

Using organic and inorganic nanoparticles as fillers in nanocomposite coatings has been found effective in reducing the flammability of coatings. Some nanoparticles such as TiO_2, silica, graphene, and nanoclays have extensively been used in the coating systems of textiles. The nanoparticles can be added to the fiber polymer solution prior to the melt-spinning process to generate the nanocomposite filaments. The produced fibers can subsequently be woven or knitted into products with fire-retardant features [114]. However, achieving uniform homogeneity and distribution of nanoparticles in the polymer solution is a great challenge for researchers and producers. Also, it is possible that the nanofillers and polymer matrix cannot establish the required strong bonds with each other due to chemical incompatibility issues. This problem can be resolved through surface modification of nanoparticles before mixing with polymer [115]. There are some important parameters such as sample ignition trend, the released heat during combustion, the total heat released, the diffusion rate of flame, and the production of smoke and its toxicity, which should be considered to evaluate the fire hazard of a substance [114]. However, some parameters that should be considered in incorporating the nanofillers to polymers are compatibility of nanoparticles and polymer, impacts of nanoparticle addition on polymer rheology, dispersibility of nanoadditives into polymer systems, and flame-retardant efficiency [116].

It has been reported that for achieving an efficient flame-retardant coating layer on textiles, the concentration of ingredients should be very high [116]. This results in a thick coating that can hamper the intrinsic characteristics of fabrics. Therefore, adding 20–100% of nanoadditives is required plus using an appropriate polymer type such as polyvinyl alcohol, which can contribute to the fire-retardant efficiency of coating [116].

Devaux *et al.* [114] incorporated nanoparticles of synthetic clay montmorillonite and two polyhedral oligomeric silsequioxanes (POSS) into polyurethane matrix and coated woven polyester and cotton fabrics. They used TGA and Stanton Redcroft Cone Calorimeter to measure the rate of heat release, time to ignition, total heat evolved, fire index of growth rate, CO and CO_2 production, and volume of smoke production. They concluded that the type of nanoparticles and the order of their addition in the preparation step of coating formulation should be controlled to yield the optimum fire-retardant coating [114]. Li *et al.* [117] coated the surface of cotton fabrics through a layer-by-layer assembly method by branched polyethylenimine (BPEI) polymer and sodium montmorillonite (MMT) clay. The flame retardancy of cotton is important because of its very low limiting oxygen index (LOI) and combustion temperature (360–425°C).

Different concentrations of MMT clay (0.2 and 1% wt) were used at different pH levels (pH = 7 and 10) of coating mixture. The obtained results demonstrated that all of the coated cotton fabrics had a lower afterglow time and heat release capacity. The samples coated with 1 wt% MMT treated at pH 7 showed the most effective performance. It was observed that coating process in higher pH and concentrations of clay resulted in a thicker film on cotton samples. The results of flame retardancy test revealed that the fabrics coated with 5 and 20 bilayers (BL) of BPEI and MMT left 7% and 13% ash residues, respectively, after conducting combustion test at 500°C. Moreover, the coatings were effective in reducing the afterglow time and heat release and also were helpful in maintaining the weave structure of fabrics [117]. Lessan *et al.* [118] investigated the flame retardancy of cotton fabrics coated with sodium hypophosphite (SHP), maleic acid (MA), triethanol amine (TEA), and TiO_2 nanoparticles through a conventional pad–dry–cure method. They characterized the coated fabrics based on thermal gravimetric analysis (TGA) and differential thermal analysis (DTA). Also, the char length, char yield before and after five washing cycles, limited oxygen index (LOI), and whiteness index of the treated cotton fabrics were assessed. They reported that the SHP, which was a phosphorous compound, improved the flame-retardant functions of coated cotton fabrics. TiO_2 nanoparticles increased the char formation implying the self-extinguishing property of coated cotton samples [118]. Gashti *et al.* [119], used polypyrrole (PPy) with $AgNO_3$ to establish a UV-induced polymerization process on wool fabrics. Their TGA results revealed that the thermal stability of coated fabrics improved by the nanocomposite coating.

Graphene oxide (GO), as a carbon-based nanomaterial, is one of the candidates as a nanofiller in the coating systems. Huang *et al.* [120] produced an intumescent flame-retardant cotton fabric through a layer-by-layer assembly method using GO and intumescent flame-retardant polyacrylamide. They modified the synthesis process of polyacrylamide by introducing N_1-(5,5-dimethyl-1,3,2-dioxaphosphinyl-2-yl)-acrylamide (DPAA) as a phosphorus–nitrogen containing compound and an intumescent flame retardant [120]. During burning, DPAA produced a swollen char layer on polymer, which played a role to prevent heat transmission, and pyrolysis of polymers to volatile gas products [120]. The thermal functionality of coated fabrics was dependent on the number of coatings applied to them. Based on their TGA results, it was revealed that the coating layer containing GO and polymer bolstered the thermal stability of fabrics. The cotton fabrics coated with 20 bilayers of GO and intumescent flame-retardant polymer were less flammable compared to control fabric, where 50% less tendency to ignition was observed. Also, their peak heat release level reduced by 23 s according to cone calorimeter testing.

8.6.2 Thermal Regulating Coatings

Heating the indoor living environment takes approximately around 42% of the overall consumed residential energy bills [121]. This underlines the necessity of introducing new inventory measures and products to mitigate the load of energy consumption in the residential heating systems [121]. Therefore, new insulation techniques for building materials should be introduced such as new products with high thermal resistance (R-values) and low emissivity [122]. One of the alternative methods of reducing the household heating can be using textiles that are capable of preventing the loss of body heat or in some cases regulating the body temperature. The main aim of producing textiles with personal thermal management (PTM) capability is preventing the loss of thermal radiative energy in the winter and facilitating its release in the summer [123]. These types of textiles should be wearable, stretchable, and mechanically robust and at the same time efficient in preventing the energy loss. Sections below focus on some of the methods that have been reported in the literature to fabricate textiles with PTM function.

8.6.2.1 Phase Change Materials (PCMs)

One of the materials that have been used in surface coating of textiles to produce thermoregulating smart textiles are phase change materials (PCMs). PCMs, which are also called latent heat storage materials, are capable of storing and releasing the heat by changing their physical state [124]. Different types of PCMs such as solid–solid, liquid–gas, and solid–gas have been reported, but the solid–liquid type is the most common version of PCMs. The use of solid–liquid PCMs has been hailed mostly due to their lower volume change during phase transition and at the same time higher heat storage capacity [125]. When the temperature of environment is higher than the melting point of PCMs, the solid material encapsulated in the thin polymeric shell starts absorbing the heat energy and therefore transforms to the liquid form. The liquid form of PCMs again changes to the solid form when the surrounding environment temperature falls to lower than its melting point [125]. Through the phase transformations, a substantial amount of heat can be stored and released [125]. In a sense, these materials can absorb heat in the warm condition, store it through transition of their physical state from solid to liquid, and then release the stored energy to the environment during the cooling process by changing their phase from liquid to solid. When the wearer's body temperature increases, the PCMs absorb the extra heat, and when the wearer's body temperature drops, they can release the absorbed heat (Figure 8.22). The insulation effect that PCMs can provide relies largely on temperature change and its pace, which takes place over the narrow temperature range [126].

There are two main types of PCMs including organic (paraffin and nonparaffin-based materials) and inorganic [127]. Paraffin waxes can be used as PCMs and have a changing phase temperature of 18–36°C [128]. Different types of paraffin with various numbers of carbon in their structure and characteristics such as melting temperature and crystallinity can be utilized [124]. Of these alkyl hydrocarbon paraffin PCMs, heptadecane, hexadecane, octadecane, nonadecane, and eicosane are noteworthy [128]. Fatty acids, alcohols, and glycolic acids are among the common non-paraffin materials, and hydrated inorganic salts are classified as inorganic PCMs [127, 129]. These materials are confined in a thin layer of polymer shell through microencapsulation (for particles between 1 and 1000 μm) and nanoencapsulation (for sizes <1 μm) processes [128]. It is worth mentioning that a higher efficiency can be obtained from nanocapsules compared to microcapsules due mostly to their faster heat transfer rate and smaller particle size [127]. There are several methods for incorporating the PCMs in textiles such as adding PCMs in coating formulation, spinning PCMs–fiber polymer mixture solution, cross-linking, and laminating pre-prepared PCM–polymer film to the substrate [130]. PCMs can be embedded into different smart textile coatings including various types of binders such as acrylic and polyurethane solution/foam before applying to textiles. There are some influential parameters that determine the final

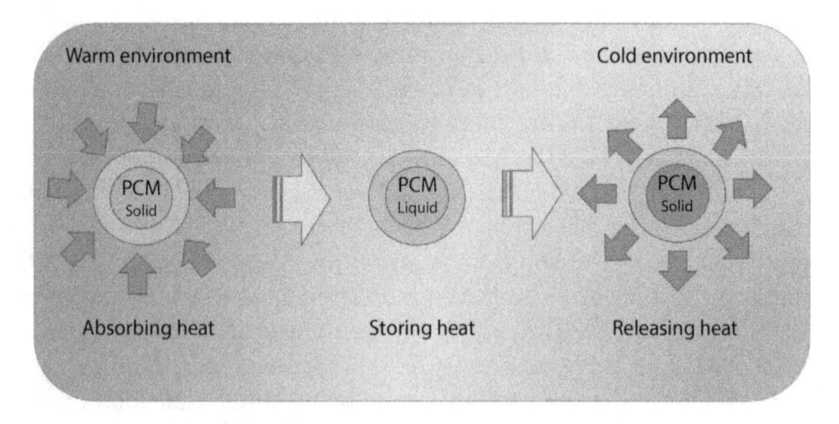

Figure 8.22 Thermal regulating mechanism of PCM on coated textiles. (The figure has been prepared by the authors.)

quality of coating and its efficiency. These factors are polymer binder type, the mass ratio between the binder and PCM, the type and mechanical stability of the PCM shell, affinity between the fabric and the binder, and curing conditions [131]. Efficiency of PCMs in producing thermoregulating sport products such as ski wear, hunting clothing, boots, gloves, and ear warmers has already been reported [128].

In a research carried out by Shin *et al.* [128], the thermoregulating polyester fabric was prepared. They synthesized melamine–formaldehyde microcapsules containing eicosane through an *in situ* polymerization. The polyurethane binder was used to apply the prepared PCMs to polyester fabrics through a conventional pad–dry–cure process [128]. They added 5–23% wt. of prepared microcapsules into the coating formulation and reported that the coated fabrics had the heat storage capacity of 0.91–4.44 J/g. However, the washing test results revealed that the synthesized microcapsules did not have good durability on fabrics, and only 40% of heat storage capacity could be retained after five cycles of laundering. With increasing amount of add-ons on fabrics, it was observed that some characteristics such as air and vapor permeability, flexibility, and shear properties of fabrics were adversely affected. However, the coating layer increased the tensile linearity, roughness, and moisture regain of polyester fabrics. Therefore, there should be a balance between the formulation of coating that is applied to fabrics and the expected properties of products considering their ultimate performance and application [132]. Sánchez *et al.* [133] synthesized PCM microcapsules of paraffin wax with polystyrene shell and applied to cotton fabrics using different types of commercial binders. Their results demonstrated that using 35 wt% microcapsules to binder ratio in the coating formulation provided a thermal storage capacity of 7.6 J/g. The performance of coated fabrics dropped to 3.6 J/g after undergoing a washing durability test [133].

Specific types of fabrics containing PCMs can also be employed as tensile structures in architecture and construction applications [134]. The thermal insulation capacity of architecture fabrics plays a very important role in preventing the structure's interior space from being overheated by solar radiation. Some practical applications of these types of fabrics are the covers over sport halls, greenhouses, tennis arenas, stadiums, military shelters and tents, and airport ceilings and roofs, [134]. Therefore, the PCM embedded textiles are important due to their insulating properties. When the outside temperature rises and the construction is exposed to the solar radiation, PCMs that exist in the structure of fabrics absorb the latent heat energy by changing their phase; as a result, the surrounding temperature variation will be negligible [134]. For these types of fabrics, the silicon rubber binders are appropriate carriers for applying PCMs to the surface of fabrics [134]. The salt hydrate PCM can be mixed with silicone rubber binder and be applied to the fabric surface by a simple knife-over-roll coating approach. The smart coated fabrics with PCM-silicon rubber are able to reduce the heat flux into the interior space of buildings and alleviate the overheating of structures significantly. Regulating the heat flux in and out of constructions, the PCM-silicone rubber coatings on fiberglass fabrics can contribute to saving consumed energy for cooling and heating systems [134].

8.6.2.2 Nanowire Composite Coatings

Coating textiles with metal nanowires has been suggested as an effective approach to prevent the waste of body heat. The nanowires are right candidates for this purpose because of their high aspect ratio leading to high electrical conductivity and mechanical characteristics. The nanowire coatings on textiles can reflect around 40% of heat radiation generated by body inwards leading to keeping the wearer warm. This product is suitable to be used in winters when keeping the body heat is a crucial factor. Conversely, the coatings that are transparent to infrared radiation and opaque to the visible light can be a suitable option for summer thermal clothes [123]. The spaces between the metallic nanowires applied to textiles can be controlled to less than the wavelength of the infrared emitted from human body to prevent the loss of thermal energy. These textiles with personal thermal management capabilities will retain their flexibility, breathability, and wearability like a normal cloth [121].

In a paper published by Hsu *et al.* [121], different aspects of personal thermal management textiles have been thoroughly discussed. Among various types of nanowires, silver nanowires (AgNWs) have been found promising for personal thermal management applications in textiles. The advantages of AgNWs are their high conductivity and yield strength. What is more, applying an electricity source to the AgNWs can provide the function of Joule heating capability for textiles, and the wearer can even feel warmer by this method. CNT nanowires have also been used for this purpose, and it was reported that they had lower rate of IR reflection and higher emissivity compared to the products coated with AgNWs. The assessment of coatings demonstrated that the AgNW-coated textiles can provide a better insulation compared with CNT-coated samples [121]. Figure 8.23 shows the thermal images of fabrics coated with AgNWs and CNT as well as the Joule heating functionality of coated fabrics. It can be seen that the AgNW-coated samples are shown in dark blue implying their

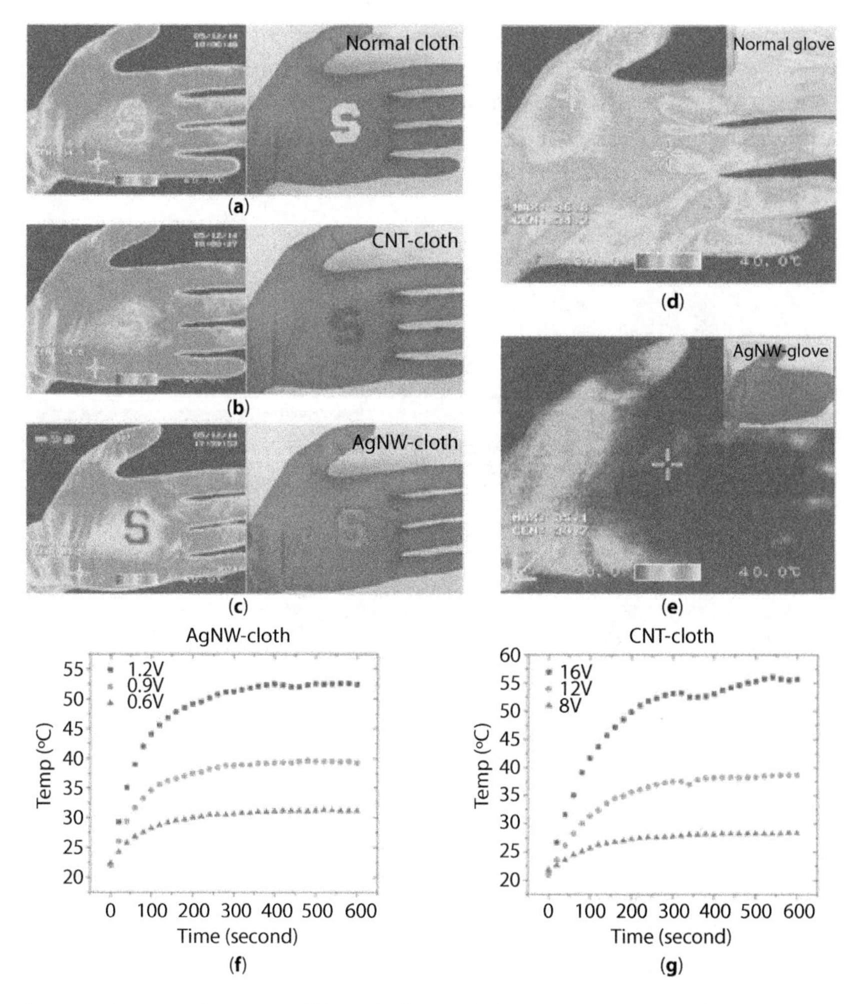

Figure 8.23 Thermal images of fabrics before and after coating process with nanowires. (a) Normal cloth, (b) CNT-coated cloth, (c) AgNW-coated cloth. Thermal images of human hand with (d) normal glove and (e) AgNW-coated glove. Temperature variation of 1 in. × 1 in. sample after applying different voltage to (a) AgNW-coated cloth and (b) CNT-coated cloth. (Reprinted with permission from reference [121]. Copyright © 2015 American Chemical Society.)

lower temperature. This indicates that the AgNWs can prevent the emittance of body IR radiation to the ambient environment, hence providing a lower emissivity and better personal thermal management functionality. At the same time, fabrics coated with CNT needed more voltage (12 V) to generate heat to reach to 38°C compared with AgNW-coated fabrics that needed just 0.9 V [121].

Stabilizing the nanowires on textiles is a challenging issue mostly due to their weak bonds with the substrate [121]. Among different reported methods for improving the adhesion of nanowires on substrates, the application of coatings to textiles in the form of nanocomposites has been found promising and resulted in a durable layer of nanomaterials on textiles. In

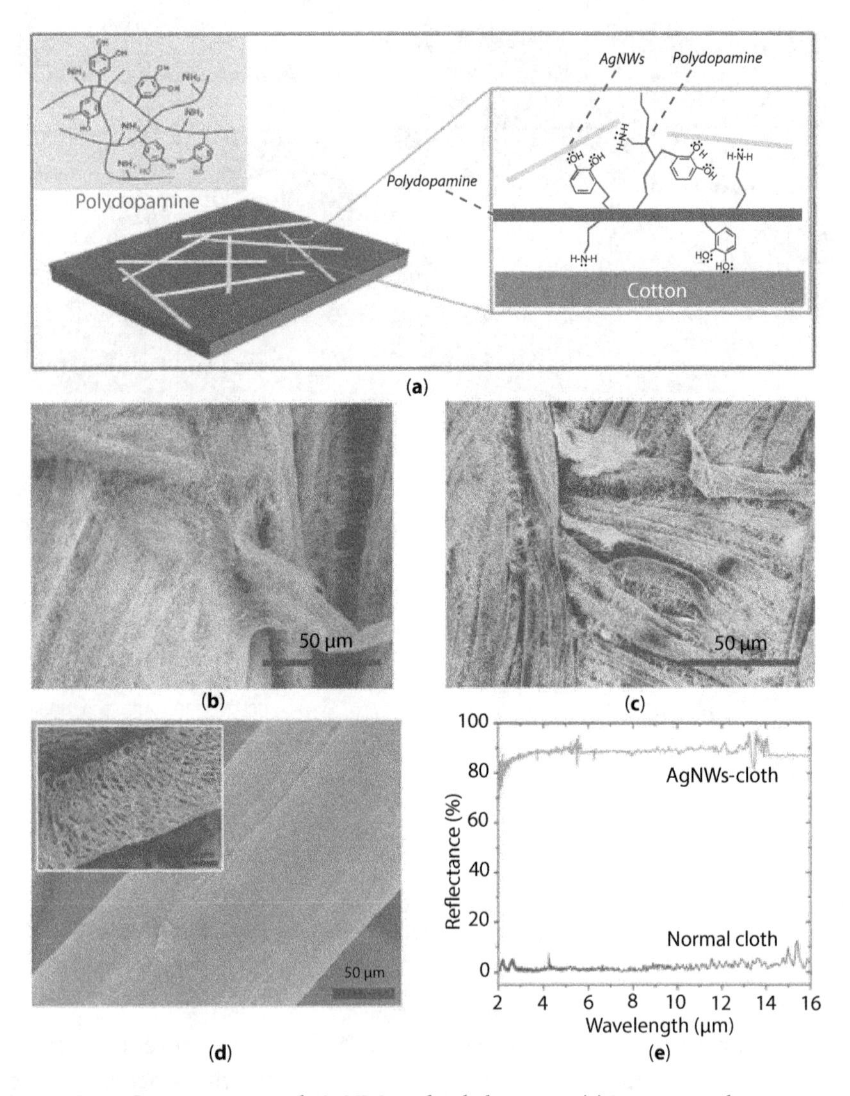

Figure 8.24 Coating cotton with AgNWs and polydopamine (a) interactions between polydopamine and cotton surface. SEM images of (b) cotton fiber coated with AgNW/polydopamine, (c) AgNW-cloth, (d) untreated cotton fiber and ANDC fiber (inset), and (e) reflectance measurement of normal cloth and ADNC. (Reprinted from reference [135] with permission from The Royal Society of Chemistry.)

a research study conducted by Yu *et al.* [135], the AgNWs were applied to cotton surface in the form of AgNW/polydopamine nanocomposite through dip-coating method. They reported fabrication of a flexible and washable cotton cloth, which was capable of reflecting the middle-to-far IR radiation emitted from the wearer's body by 86%. Also, the coatings provided a rapid Joules heating function to the products. They first soaked the cotton fabric into the polydopamine solution and then immersed into AgNW dispersion for three times. Figure 8.24 shows the surface morphology of coated samples and IR reflectance spectra of samples coated with AgNW/polydopamine (ANDC) [135]. In another research, Guo *et al.* [136] prepared a textile-based product with two functionalities of energy harvesting and energy saving. They coated the nylon fabrics (FPAN) by multilayer coating composed of silver nanowires (AgNWs), polydimethylsiloxane (PDMS), and fluoroalkylsilanes (FAS) (Figure 8.25) [136]. The gap between the AgNWs was adjusted around 300 nm, which was much narrower than the wavelength of the IR radiation (9 μm) emitted from

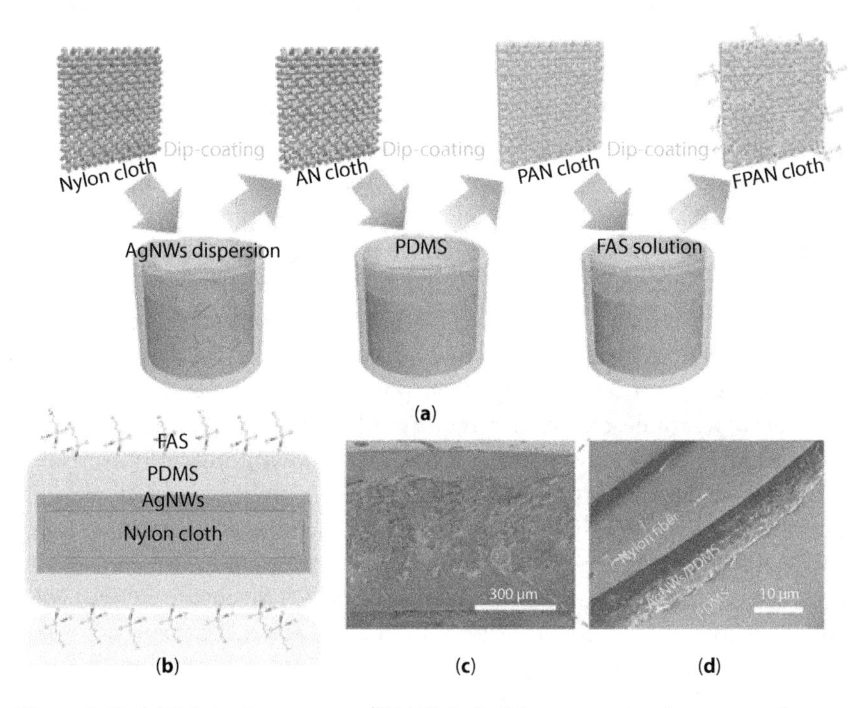

Figure 8.25 (a) Fabrication process of FPAN cloth, (b) cross-sectional structure of coating on the FPAN, and SEM images of the (c) cross-section and (d) FPAN cloth. (Reprinted with permission from reference [136], Copyright © 2016, American Chemical Society.)

human body. PDMS played a role as a triboelectric material and protective layer on AgNWs preventing them to peel off. FAS, which had a low surface energy, was applied simply through a dip-coating method and bounded to the PDMS via silane groups and increased the output performance by enhancing the surface charge. The devised multilayer fabric could behave like a triboelectric material of a triboelectric nanogenerator (TENG) by saving the movement energy and using it for powering the LEDs.

8.6.3 Conductive Coatings

With the emerging application of a vast number of personal electronics, sensing, and health care devices, wearable electronics have attracted wide attention from both academia and industry [137]. In recent years, many new wearable products (such as smart watches, soft displays, wearable sensors, etc.) have been introduced to the consumer market. It has been forecasted that the wearable technology market will be worth more than US$50 billion by 2022. Among these wearable electronic devices, conductive textiles will be playing more and more important roles along with current and future developments [138].

Traditionally, conductive textiles are produced mainly for antistatic and electromagnetic shielding applications. Common methods of preparing conductive textiles include adding conductive fillers into the polymer solution before synthetic fiber spinning, spinning metal fibers into normal yarns, metal plating on yarns or fabrics, converting polymer filaments to carbon fibers, and dip-coating conductive materials onto fabrics. However, the conductive textiles resulted from these conventional fabrication processes and microsized conductive materials normally suffer the disadvantages of low electrical conductivity, poor structure stability, or severe performance deterioration under wearing and washing conditions.

Modern wearable electronics require all functional components with high electrical conductivity to reduce power consumption. For integrated textile-based devices, structural and electrical stabilities should be high enough to maintain a high performance level during practical wearing conditions and survive a certain number of laundry cycles. In some of the cases, stretchability is also important to corporate stretchable devices with human body movements. Traditional conductive textiles have not been able to meet the requirements for most of the wearable applications; therefore, the development of new materials and structures will be critical for the future of wearable electronics. In this regard, various nanomaterials that have been developed for other uses can find their application in conductive textiles.

8.6.3.1 Carbon-Based Conductive Coating

Carbon is an earth-rich and highly conductive material that can be fabricated into many formats and dimensions, such as buckyballs, carbon quantum dots, carbon nanotubes, carbon fibers/nanofibers, graphene, and various three-dimensional porous carbon structures [139]. Compared with other conductive nanomaterials, carbon nanomaterials have the advantages of lower cost and relatively simpler material preparation.

Hu *et al.* [140] used a simple dip-coating process to prepare highly conductive cotton fabrics using single-walled carbon nanotubes (SWNTs), as shown in Figure 8.26a. By immersing cotton fabrics into a well-dispersed

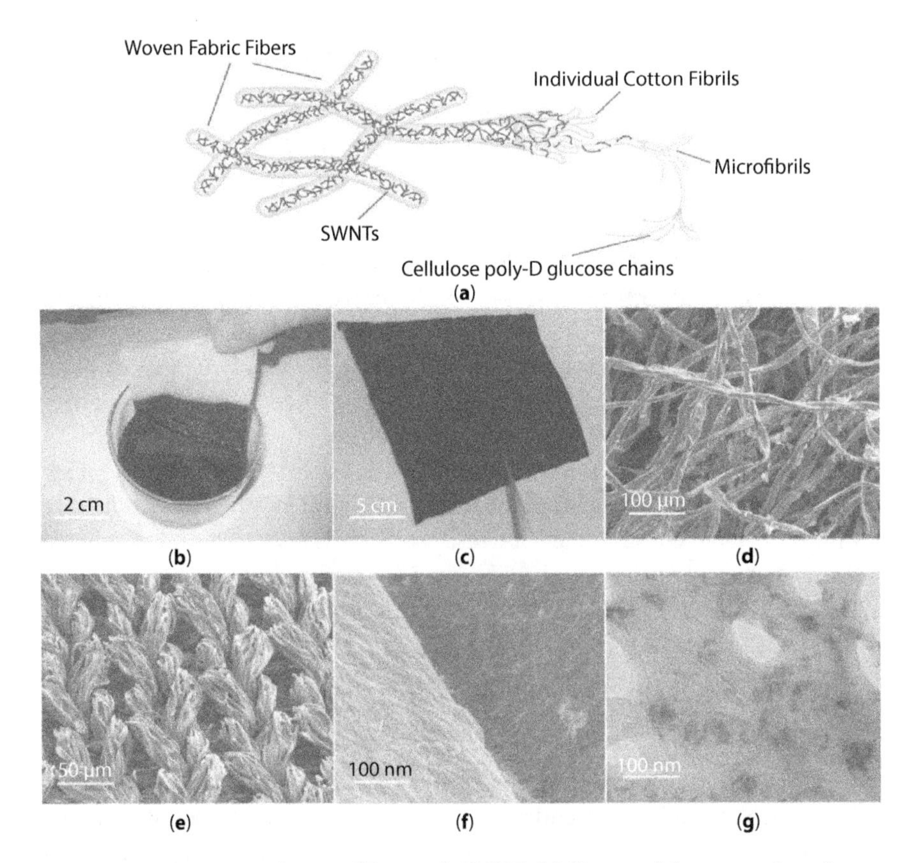

Figure 8.26 Treatment of cotton fabric with SWNTs (a) illustrated demonstration of SWNTs coating on cotton fibers for preparing a conductive fibrous structure; (b) coating process; (c) coated fabric; (d–f) SEM images of the SWNTs on cotton fiber surface; and (g) TEM image of SWNTs on cotton fibers. (Reprinted with permission from reference [140]Copyright © 2010, American Chemical Society.)

aqueous SWNT solution (Figure 8.26b), a black-colored conductive fabric can be formed after drying (Figure 8.26c). Due to the large van der Waals forces between carbon nanotubes and textile fibers, strong hydrogen bonding formed between carboxyl groups on nanotube surface and hydroxyl groups of cotton. Also an ideal adhesion of flexible nanotubes to fiber surface (Figure 8.26d) was achieved, and the coated conductive fabric exhibited great mechanical properties. The carbon nanotubes withstood both tape and washing tests, without obvious degradation of electrical conductivity. Through multiple coating cycles, acid treatment, and mechanical pressing, electrical conductivity of the coated fabric reached as high as 125 S/cm, with a sheet resistance lower than 1 Ω/sq. Unusually, it has also been found that the fabric conductivity increased by stretching the coated fabric to 2.4 times the original length. This phenomenon is probably contributed by the enhanced physical contact between interconnected carbon nanotubes among stretching. Supercapacitor devices were fabricated using CNT coated cotton fabrics as both electrodes and current collectors. With a 0.24 mg/cm^2 CNT loading on the fabric, the device achieved a specific capacitance of 140 F/g at 20 $\mu A/cm^2$.

In wearable electronics, conductive textiles can work not only as electrodes but also as stretchable electrical conductors in some cases. For these applications, a minimum possible conductivity variation is desirable to maintain stable function of electronic devices. Using a superelastic fiber made of styrene–ethylene–butylene)–styrene block copolymer as the core, carbon nanotube sheets were wrapped on the stretched fiber to form a core-sheath conducting fiber [141]. A multilevel buckling effect of the carbon nanotube sheet after releasing the tensile strain was observed on the fiber surface. This stretchable conductive fiber maintained efficient conductive pathway to enable a less than 5% resistance change with a 1000% tensile deformation, which makes it an ideal candidate for elastic electronics and artificial muscles.

8.6.3.2 Metal-Based Conductive Coating

Compared with carbon material, metals have much higher electrical conductivity; therefore, they are more effective to form conductive layers. Many metals, such as gold, silver, and platinum, have been made to nanowires for the application of transparent conductive coating, and mainly benefited from mesh like nanowire network and partial coverage on the substrates. Using silver nanowires through a dip-coating process, conductive nylon, cotton, and polyester threads were prepared (Figure 8.27) [142]. Due to the hydrophilic feature of cotton surface, silver nanowires from the aqueous coating solution

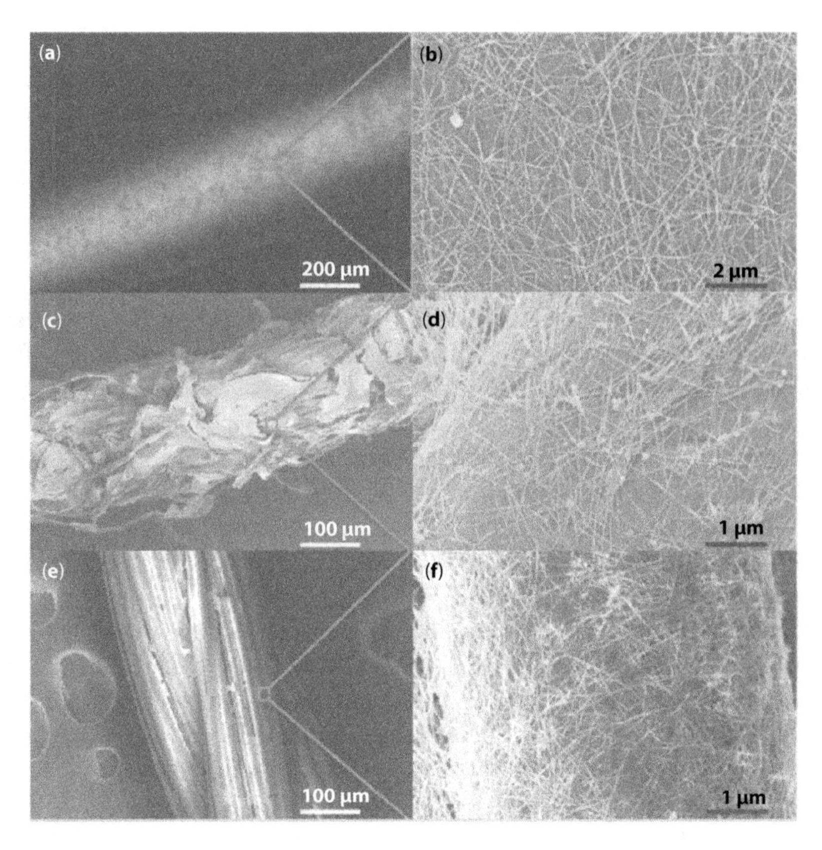

Figure 8.27 SEM images of silver-nanowire-coated (a, b) nylon thread, (c, d) cotton thread, and (e, f) polyester thread. (Reprinted from open access reference [142] published by The Royal Society of Chemistry.)

had good interaction with cotton thread, while the other two hydrophobic fibers required chemical treatment to achieve ideal nanowire coverage. By increasing the coating cycle number to adjust silver nanowire loading, the unit resistance of 0.8 Ω/cm was achieved on nylon thread.

It has been found that a thermal annealing post-treatment at 150°C would effectively reduce electrical conductivity of the coated threads, through removing organic surfactant existing in the coating solution and creating interwire junctions with thermal fusing of the nanowires. The excellent mechanical stability of the silver nanowire coated threads was demonstrated by a bending test. In this test, both nanowire-coated nylon threads and commercial silver-coated conductive yarns were repeatedly bent on a metal rod with a radius of 6 mm. The results revealed that the resistance of the commercial conductive yarn increased from 2.8 to 12.2 Ω/cm after 200 bending cycles; at the same time, the nanowire-coated thread had only a slight resistance increase to 3.2 Ω/cm, from the same starting resistance of 2.8 Ω/cm. It should be pointed out that due to significantly lower coverage on thread surface than conventional particle or dense film-based metal coating, metal nanowires can be more efficient to achieve flexible and low-cost conductive coating on textiles.

Through using a novel template method, conductive coating has been applied on single fibers to form parallel electrode pairs for fabricating dual functional electronic devices [143]. As shown in Figure 8.28a, four parallel stripes of sticky tape were used to closely wrap a single fiber. After removing two stripes from the fiber surface, gold sputter coating was applied to introduce a conductive layer onto the fiber. By unwrapping the remaining two stripes, a parallel electrode pair was formed on the fiber. The single fiber device was completed after depositing a layer of poly(3,4-ethylenedioxythiophene) (PEDOT) (Figure 8.28b) on the gold electrodes through electrochemical polymerization and finally covering the fiber with a layer of gel electrolyte. The as-prepared single fiber device showed electrochromic effect by applying a low voltage (±0.6 V) across the electrodes. Clear color change of the PEDOT-covered gold layer to dark blue can be found in Figures 8.28c and d after switching the applied voltage from 0 to 0.6 V, while the color change happened on the neighboring electrode once the voltage changed to –0.6 V, with a fast response time shorter than 5 s. In

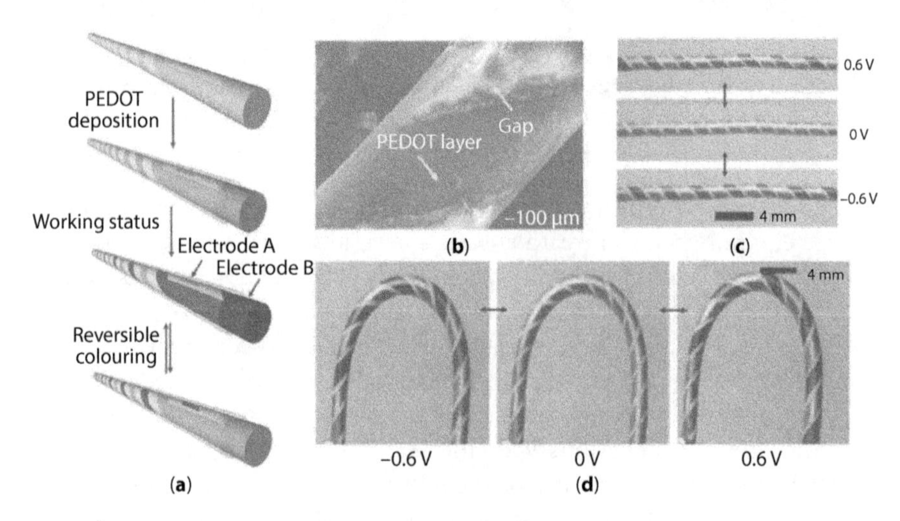

Figure 8.28 (a) Illustrated demonstration of fiber-shaped device fabrication; (b) SEM image of PEDOT-coated gold electrode and uncoated fiber surface; and digital images of (c) straight and (d) curved fiber device at decolorized and colorized states. (Reproduced in part from reference [143] with permission of The Royal Society of Chemistry.)

addition to stable electrochromic behavior, the single fiber device can also function as a supercapacitor with a specific capacitance of 20.3 F/g. It has also been reported that both electrochromic and supercapacitor functions can work simultaneously without any interruption.

8.7 Conclusion and Future Prospect

This chapter tried to shed light on some aspects of nanocoatings that are commonly employed to fabricate smart textiles. Different methods of coating and associated mechanisms in introducing novel functionalities to textiles were discussed. The focused methods include sol–gel, cross-linking, plasma, and nanocomposite coatings, which are among the most used methods reported in literature. The mechanisms of different functionalities of smart fabrics including self-cleaning, UV protection, hydrophobicity, antimicrobial activity, flame retardancy, personal thermal management systems, and conductive coatings were thoroughly discussed. There are numerous publications in each of these fields, all of which could not be covered in one book chapter. There are significant achievements on the concept of smart textiles, but yet again researchers are exploring new methods to promote the efficiency and durability of novel functionalities and nanocoatings. Also, investigating the safety aspects of nanomaterials on human health and their potential impacts on environment is among the important fields that scientists should explore in their future work.

Acknowledgements

We wish to acknowledge support of an Alfred Deakin Postdoctoral Fellowship for the first author. This work was also supported (partially) by the Australian Government through the Australian Research Council's Industrial Transformation Research Hub scheme (IH140100018) and Centre of Excellence scheme (CE140100012).

References

1. Smith, W.C. (Ed.), *Smart textile coatings and laminates*. Woodhead Publishing, Elsevier, Cambridge, 2010.
2. Gashti, M.P., Pakdel, E., Alimohammadi, F., Nanotechnology-based coating techniques for smart textiles. In: *Active coatings for smart textiles*, J. Hu (Ed.), pp. 243–268, Woodhead Publishing, Duxford, 2016.

3. Sen, A.K., *Coated textiles: Principles and applications*. pp. 75–103, Technomic Publishing Company, Lancaster, 2007.
4. Akovali, G., Thermoplastic polymers used in textile coatings. In: *Advances in polymer coated textiles*, G. Akovali (Ed.), pp. 1–6, Smithers Rapra Technology Ltd., Shawbury, 2012.
5. Yetisen, A.K., Qu, H., Manbachi, A., Butt, H., Dokmeci, M.R., Hinestroza, J.P. *et al.*, Nanotechnology in textiles, *ACS Nano*, 10, 3042, 2016.
6. Joshi, M., Bhattacharyya, A., Nanotechnology—A new route to high-performance functional textiles, *Text. Prog.*, 43, 155, 2011.
7. Pierre, A.C., *Introduction to sol–gel processing*. pp. 1–9, Springer Science & Business Media, LLC, New York, 2013.
8. Brinker, C.J., Scherer, G.W., *Sol–gel science: The physics and chemistry of sol–gel processing*. pp. 235–303, Academic Press, Inc., Boston, USA, 1990.
9. Ismail, W.N.W., Sol–gel technology for innovative fabric finishing—a review, *J. Sol–Gel Sci. Technol.*, 78, 698, 2016.
10. Daoud, W.A. (Ed.), *Self-cleaning materials and surfaces: A nanotechnology approach*. John Wiley & Sons, Chichester, 2013.
11. Böttcher, H., Bioactive sol–gel coatings, *J. Prakt. Chem.*, 342, 427, 2000.
12. Mahltig, B., Haufe, H., Böttcher, H., Functionalisation of textiles by inorganic sol–gel coatings, *J. Mater. Chem.*, 15, 4385, 2005.
13. Tung, W.S., Daoud, W.A., Self-cleaning fibers via nanotechnology: A virtual reality, *J. Mater. Chem.*, 21, 7858, 2011.
14. Fateh, R., Dillert, R., Bahnemann, D., Self-cleaning coatings on polymeric substrates. In: *Self-cleaning coatings: Structure, fabrication and application*, J. He (Ed.), pp. 142–165, Royal Society of Chemistry, Cambridge, UK, 2016.
15. Pakdel, E., Wang, J., Allardyce, B.J., Rajkhowa, R., Wang, X., Functionality of nano and 3D-microhierarchical TiO_2 particles as coagulants for sericin extraction from the silk degumming wastewater, *Sep. Purif. Technol.*, 170, 92, 2016.
16. Daoud, W.A., Xin, J.H., Nucleation and growth of anatase crystallites on cotton fabrics at low temperatures, *J. Am. Ceram. Soc.*, 87, 953, 2004.
17. Daoud, W.A., Xin, J.H., Low temperature sol–gel processed photocatalytic titania coating, *J. Sol–Gel Sci. Technol.*, 29, 25, 2004.
18. Xin, J.H., Daoud, W.A., Kong, Y.Y., A new approach to UV-blocking treatment for cotton fabrics, *Text. Res. J.*, 74, 97, 2004.
19. Qi, K., Daoud, W.A., Xin, J.H., Mak, C.L., Tang, W., Cheung, W.P., Self-cleaning cotton, *J. Mater. Chem.*, 16, 4567, 2006.
20. Daoud, W.A., Leung, S.K., Tung, W.S., Xin, J.H., Cheuk, K., Qi, K., Self-cleaning keratins, *Chem. Mater.*, 20, 1242, 2008.
21. Tung, W.S., Daoud, W.A., Photocatalytic self-cleaning keratins: A feasibility study, *Acta Biomater.*, 5, 50, 2009.
22. Tung, W.S., Daoud, W.A., Self-cleaning surface functionalisation of keratins: Effect of heat treatment and formulation preparation time on photocatalysis and fibres mechanical properties, *Surf. Eng.*, 26, 525, 2010.

23. Tung, W.S., Daoud, W.A., Photocatalytic formulations for protein fibers: Experimental analysis of the effect of preparation on compatibility and photocatalytic activities, *J. Colloid Interface Sci.*, 326, 283, 2008.
24. Tung, W.S., Daoud, W.A., Leung, S.K., Understanding photocatalytic behavior on biomaterials: Insights from TiO_2 concentration, *J. Colloid Interface Sci.*, 339, 424, 2009.
25. Pakdel, E., Daoud, W.A., Wang, X., Self-cleaning and superhydrophilic wool by TiO_2/SiO_2 nanocomposite, *Appl. Surf. Sci.*, 275, 397, 2013.
26. Pakdel, E., Daoud, W., Self-cleaning cotton functionalized with TiO_2/SiO_2: Focus on the role of silica, *J. Colloid Interface Sci.*, 401, 1, 2013.
27. Pakdel, E., Daoud, W.A., Wang, X., Assimilating the photo-induced functions of TiO_2-based compounds in textiles: Emphasis on the sol–gel process, *Text. Res. J.*, 85, 1404, 2014.
28. Sung-Suh, H.M., Choi, J.R., Hah, H.J., Koo, S.M., Bae, Y.C., Comparison of Ag deposition effects on the photocatalytic activity of nanoparticulate TiO_2 under visible and UV light irradiation, *J. Photochem. Photobiol., A*, 163, 37, 2004.
29. Wang, R., Wang, X., Xin, J.H., Advanced visible-light-driven self-cleaning cotton by $Au/TiO_2/SiO_2$ photocatalysts, *ACS Appl. Mater. Interfaces*, 2, 82, 2010.
30. Pakdel, E., Daoud, W.A., Afrin, T., Sun, L., Wang, X., Self-cleaning wool: Effect of noble metals and silica on visible-light-induced functionalities of nano TiO_2 colloid, *J. Text. Inst.*, 106, 1348, 2015.
31. Pakdel, E., Daoud, W.A., Afrin, T., Sun, L., Wang, X., Enhanced antimicrobial coating on cotton and its impact on UV protection and physical characteristics, *Cellulose*, 24, 4003, 2017.
32. Zeng, C., Wang, H., Zhou, H., Lin, T., Self-cleaning, superhydrophobic cotton fabrics with excellent washing durability, solvent resistance and chemical stability prepared from an SU-8 derived surface coating, *RSC Adv.*, 5, 61044, 2015.
33. Guo, Z., Liu, W., Su, B.-L., Superhydrophobic surfaces: From natural to biomimetic to functional, *J. Colloid Interface Sci.*, 353, 335, 2011.
34. Taurino, R., Fabbri, E., Messori, M., Pilati, F., Pospiech, D., Synytska, A., Facile preparation of superhydrophobic coatings by sol–gel processes, *J. Colloid Interface Sci.*, 325, 149, 2008.
35. Shi, Y., Wang, Y., Feng, X., Yue, G., Yang, W., Fabrication of superhydrophobicity on cotton fabric by sol–gel, *Appl. Surf. Sci.*, 258, 8134, 2012.
36. Barthlott, W., Neinhuis, C., Purity of the sacred lotus, or escape from contamination in biological surfaces, *Planta*, 202, 1, 1997.
37. Yamamoto, M., Nishikawa, N., Mayama, H., Nonomura, Y., Yokojima, S., Nakamura, S. *et al.*, Theoretical explanation of the lotus effect: Superhydrophobic property changes by removal of nanostructures from the surface of a lotus leaf, *Langmuir*, 31, 7355, 2015.
38. Zhang, X., Shi, F., Niu, J., Jiang, Y., Wang, Z., Superhydrophobic surfaces: From structural control to functional application, *J. Mater. Chem.*, 18, 621, 2008.

39. Mahltig, B., Smart hydrophobic and soil-repellent protective composite coatings for textiles and leather. In: *Smart composite coatings and membranes*, M.F. Montemor (Ed.), pp. 261–292, Woodhead Publishing, Tokyo, 2016.
40. Mahltig, B., Böttcher, H., Modified silica sol coatings for water-repellent textiles, *J. Sol–Gel Sci. Technol.*, 27, 43, 2003.
41. Wang, H., Fang, J., Cheng, T., Ding, J., Qu, L., Dai, L. *et al.*, One-step coating of fluoro-containing silica nanoparticles for universal generation of surface superhydrophobicity, *Chem. Commun.*, 877, 2008.
42. Daoud, W.A., Xin, J.H., Tao, X., Superhydrophobic silica nanocomposite coating by a low-temperature process, *J. Am. Ceram. Soc.*, 87, 1782, 2004.
43. Gao, Q., Zhu, Q., Guo, Y., Yang, C.Q., Formation of highly hydrophobic surfaces on cotton and polyester fabrics using silica sol nanoparticles and non-fluorinated alkylsilane, *Ind. Eng. Chem. Res.*, 48, 9797, 2009.
44. Xue, C.-H., Jia, S.-T., Chen, H.-Z., Wang, M., Superhydrophobic cotton fabrics prepared by sol–gel coating of TiO_2 and surface hydrophobization, *Sci. Technol. Adv. Mater.*, 9, 1, 2008.
45. Afzal, S., Daoud, W.A., Langford, S.J., Superhydrophobic and photocatalytic self-cleaning cotton, *J. Mater. Chem. A.*, 2, 18005, 2014.
46. Zhou, H., Zhao, Y., Wang, H., Lin, T., Recent development in durable super-liquid-repellent fabrics, *Adv. Mater. Interfaces*, 3, 1, 2016.
47. Zhou, H., Wang, H., Niu, H., Gestos, A., Wang, X., Lin, T., Fluoroalkyl silane modified silicone rubber/nanoparticle composite: A super durable, robust superhydrophobic fabric coating, *Adv. Mater.*, 24, 2409, 2012.
48. Dastjerdi, R., Montazer, M., A review on the application of inorganic nano-structured materials in the modification of textiles: Focus on anti-microbial properties, *Colloids Surf., B*, 79, 5, 2010.
49. Gao, Y., Cranston, R., Recent advances in antimicrobial treatments of textiles, *Text. Res. J.*, 78, 60, 2008.
50. Fu, G., Vary, P.S., Lin, C.-T., Anatase TiO_2 nanocomposites for antimicrobial coatings, *J. Phys. Chem. B*, 109, 8889, 2005.
51. Blake, D.M., Maness, P.-C., Huang, Z., Wolfrum, E.J., Huang, J., Jacoby, W.A., Application of the photocatalytic chemistry of titanium dioxide to disinfection and the killing of cancer cells, *Sep. Purif. Rev.*, 28, 1, 1999.
52. Pagnout, C., Jomini, S., Dadhwal, M., Caillet, C., Thomas, F., Bauda, P., Role of electrostatic interactions in the toxicity of titanium dioxide nanoparticles toward *Escherichia coli*, *Colloids Surf., B*, 92, 315, 2012.
53. Lu, Z.-X., Zhou, L., Zhang, Z.-L., Shi, W.-L., Xie, Z.-X., Xie, H.-Y. *et al.*, Cell damage induced by photocatalysis of TiO_2 thin films, *Langmuir*, 19, 8765, 2003.
54. Hipler, U.-C., Elsner, P. (Eds.), *Biofunctional textiles and the skin*. Karger Medical and Scientific Publishers, Basel, 2006.
55. Sondi, I., Salopek-Sondi, B., Silver nanoparticles as antimicrobial agent: A case study on *E. coli* as a model for gram-negative bacteria, *J. Colloid Interface Sci.*, 275, 177, 2004.

56. Ahearn, D.G., May, L.L., Gabriel, M.M., Adherence of organisms to silver-coated surfaces, *J. Ind. Microbiol.*, 15, 372, 1995.
57. Li, W.-R., Xie, X.-B., Shi, Q.-S., Zeng, H.-Y., Ou-Yang, Y.-S., Chen, Y.-B., Antibacterial activity and mechanism of silver nanoparticles on *Escherichia coli*, *Appl. Microbiol. Biotechnol.*, 85, 1115, 2010.
58. Feng, Q., Wu, J., Chen, G., Cui, F., Kim, T., Kim, J., A mechanistic study of the antibacterial effect of silver ions on *Escherichia coli* and *Staphylococcus aureus*, *J. Biomed. Mater. Res.*, 52, 662, 2000.
59. Lakshmanan, A., Chakraborty, S., Coating of silver nanoparticles on jute fibre by *in situ* synthesis, *Cellulose*, 24, 1563, 2017.
60. Mahltig, B., Fiedler, D., Fischer, A., Simon, P., Antimicrobial coatings on textiles—modification of sol–gel layers with organic and inorganic biocides, *J. Sol–Gel Sci. Technol.*, 55, 269, 2010.
61. Busila, M., Musat, V., Textor, T., Mahltig, B., Synthesis and characterization of antimicrobial textile finishing based on Ag:ZnO nanoparticles/chitosan biocomposites, *RSC Adv.*, 5, 21562, 2015.
62. Kong, M., Chen, X.G., Liu, C.S., Liu, C.G., Meng, X.H., Yu, L.J., Antibacterial mechanism of chitosan microspheres in a solid dispersing system against *E. coli*, *Colloids Surf., B*, 65, 197, 2008.
63. Daoud, W.A., Xin, J.H., Zhang, Y.-H., Surface functionalization of cellulose fibers with titanium dioxide nanoparticles and their combined bactericidal activities, *Surf. Sci.*, 599, 69, 2005.
64. Millington, K.R., Deledicque, C., Jones, M.J., Maurdev, G., Photo-induced chemiluminescence from fibrous polymers and proteins, *Polym. Degrad. Stab.*, 93, 640, 2008.
65. Millington, K.R., Photoyellowing of wool. Part 1: Factors affecting photoyellowing and experimental techniques, *Color. Technol.*, 122, 169, 2006.
66. Abidi, N., Hequet, E., Tarimala, S., Dai, L.L., Cotton fabric surface modification for improved UV radiation protection using sol–gel process, *J. Appl. Polym. Sci.*, 104, 111, 2007.
67. Sawhney, A.P.S., Condon, B., Singh, K.V., Pang, S.S., Li, G., Hui, D., Modern applications of nanotechnology in textiles, *Text. Res. J.*, 78, 731, 2008.
68. Riva, A., Algaba, I., Pepió, M., Action of a finishing product in the improvement of the ultraviolet protection provided by cotton fabrics. Modelisation of the effect, *Cellulose*, 13, 697, 2006.
69. Urbas, R., Kostanjšek, K., Dimitrovski, K., Impact of structure and yarn colour on UV properties and air permeability of multilayer cotton woven fabrics, *Text. Res. J.*, 81, 1916, 2011.
70. Dubrovski, P.D., Golob, D., Effects of woven fabric construction and color on ultraviolet protection, *Text. Res. J.*, 79, 351, 2009.
71. Montazer, M., Pakdel, E., Functionality of nano titanium dioxide on textiles with future aspects: Focus on wool, *J. Photochem. Photobiol., C*, 12, 293, 2011.

72. Abidi, N., Cabrales, L., Hequet, E., Functionalization of a cotton fabric surface with titania nanosols: Applications for self-cleaning and UV-protection properties, *ACS Appl. Mater. Interfaces*, 1, 2141, 2009.
73. Erdem, N., Erdogan, U.H., Cireli, A.A., Onar, N., Structural and ultraviolet-protective properties of nano-TiO_2-doped polypropylene filaments, *J. Appl. Polym. Sci.*, 115, 152, 2010.
74. Zhang, H., Millington, K.R., Wang, X., The photostability of wool doped with photocatalytic titanium dioxide nanoparticles, *Polym. Degrad. Stab.*, 94, 278, 2009.
75. Mihailović, D., Šaponjić, Z., Radoičić, M., Radetić, T., Jovančić, P., Nedeljković, J. *et al.*, Functionalization of polyester fabrics with alginates and TiO_2 nanoparticles, *Carbohydr. Polym.*, 79, 526, 2010.
76. Pakdel, E., Daoud, W.A., Sun, L., Wang, X., Photostability of wool fabrics coated with pure and modified TiO_2 colloids, *J. Colloid Interface Sci.*, 440, 299, 2015.
77. Zhang, M., Tang, B., Sun, L., Wang, X., Reducing photoyellowing of wool fabrics with silica coated ZnO nanoparticles, *Text. Res. J.*, 84, 1840, 2014.
78. Wang, J., Tsuzuki, T., Sun, L., Wang, X., Reducing the photocatalytic activity of zinc oxide quantum dots by surface modification, *J. Am. Ceram. Soc.*, 92, 2083, 2009.
79. Gashti, M.P., Nanocomposite coatings: State of the art approach in textile finishing, *J. Textile Sci. Eng.*, 4, 1, 2014.
80. Harifi, T., Montazer, M., Past, present and future prospects of cotton cross-linking: New insight into nano particles, *Carbohydr. Polym.*, 88, 1125, 2012.
81. Nazari, A., Montazer, M., Rashidi, A., Yazdanshenas, M., Moghadam, M.B., Optimization of cotton crosslinking with polycarboxylic acids and nano TiO_2 using central composite design, *J. Appl. Polym. Sci.*, 117, 2740, 2010.
82. Parvinzadeh Gashti, M., Almasian, A., Citric acid/ZrO_2 nanocomposite inducing thermal barrier and self-cleaning properties on protein fibers, *Composites Part B*, 52, 340, 2013.
83. Alimohammadi, F., Parvinzadeh Gashti, M., Shamei, A., Functional cellulose fibers via polycarboxylic acid/carbon nanotube composite coating, *J. Coat. Technol. Res.*, 10, 123, 2013.
84. Gashti, M.P., Alimohammadi, F., Shamei, A., Preparation of water-repellent cellulose fibers using a polycarboxylic acid/hydrophobic silica nanocomposite coating, *Surf. Coat. Technol.*, 206, 3208, 2012.
85. Meilert, K.T., Laub, D., Kiwi, J., Photocatalytic self-cleaning of modified cotton textiles by TiO_2 clusters attached by chemical spacers, *J. Mol. Catal. A: Chem.*, 237, 101, 2005.
86. Montazer, M., Pakdel, E., Self-cleaning and color reduction in wool fabric by nano titanium dioxide, *J. Text. Inst.*, 102, 343, 2011.
87. Montazer, M., Pakdel, E., Reducing photoyellowing of wool using nano TiO_2, *Photochem. Photobiol.*, 86, 255, 2010.

88. Montazer, M., Pakdel, E., Moghadam, M.B., The role of nano colloid of TiO_2 and butane tetra carboxylic acid on the alkali solubility and hydrophilicity of proteinous fibers, *Colloids Surf., A*, 375, 1, 2011.
89. Montazer, M., Pakdel, E., Moghadam, M., Nano titanium dioxide on wool keratin as UV absorber stabilized by butane tetra carboxylic acid (BTCA): A statistical prospect, *Fibers Polym.*, 11, 967, 2010.
90. Montazer, M., Behzadnia, A., Pakdel, E., Rahimi, M.K., Moghadam, M.B., Photo induced silver on nano titanium dioxide as an enhanced antimicrobial agent for wool, *J. Photochem. Photobiol., B*, 103, 207, 2011.
91. Montazer, M., Ghayem Asghari, M.S., Pakdel, E., Electrical conductivity of single walled and multiwalled carbon nanotube containing wool fibers, *J. Appl. Polym. Sci.*, 121, 3353, 2011.
92. Parvinzadeh Gashti, M., Almasian, A., Parvinzadeh Gashti, M., Preparation of electromagnetic reflective wool using nano-ZrO_2/citric acid as inorganic/organic hybrid coating, *Sens. Actuators, A*, 187, 1, 2012.
93. Alimohammadi, F., Parvinzadeh Gashti, M., Shamei, A., A novel method for coating of carbon nanotube on cellulose fiber using 1,2,3,4-butanetetracarboxylic acid as a cross-linking agent, *Prog. Org. Coat.*, 74, 470, 2012.
94. Virk, R.K., Ramaswamy, G.N., Bourham, M., Bures, B.L., Plasma and antimicrobial treatment of nonwoven fabrics for surgical gowns, *Text. Res. J.*, 74, 1073, 2004.
95. Kan, C.W., Plasma surface treatments for smart textiles. In: *Active coatings for smart textiles*, J. Hu (Ed.), pp. 221–241, Woodhead Publishing, Duxford, 2016.
96. Naebe, M., Cookson, P.G., Rippon, J.A., Wang, X.G., Effects of leveling agent on the uptake of reactive dyes by untreated and plasma-treated wool, *Text. Res. J.*, 80, 611, 2010.
97. Gashti, M.P., Willoughby, J., Agrawal, P., Surface and bulk modification of synthetic textiles to improve dyeability. In: *Textile dyeing*, P. Hauser (Ed.), pp. 261–298, InTech, Rijeka, 2011.
98. Mather, R., Surface modification of textiles by plasma treatments. In: *Surface modification of textiles*, Q. Wei (Ed.), pp. 296–317, Woodhead Publishing, Cambridge, 2009.
99. Morent, R., De Geyter, N., Verschuren, J., De Clerck, K., Kiekens, P., Leys, C., Non-thermal plasma treatment of textiles, *Surf. Coat. Technol.*, 202, 3427, 2008.
100. Sparavigna, A., *Plasma treatment advantages for textiles*, Cornell University Library, https://arxiv.org/abs/0801.3727v1, 1, 2008.
101. Shishoo, R. (Ed.), *Plasma technologies for textiles*. Woodhead Publishing, Cambridge, 2007.
102. Bozzi, A., Yuranova, T., Guasaquillo, I., Laub, D., Kiwi, J., Self-cleaning of modified cotton textiles by TiO_2 at low temperatures under daylight irradiation, *J. Photochem. Photobiol., A*, 174, 156, 2005.
103. Qi, K., Xin, J.H., Daoud, W.A., Mak, C.L., Functionalizing polyester fiber with a self-cleaning property using anatase TiO_2 and low-temperature plasma treatment, *Int. J. Appl. Ceram. Technol.*, 4, 554, 2007.

104. Kiwi, J., Pulgarin, C., Innovative self-cleaning and bactericide textiles, *Catal. Today*, 151, 2, 2010.
105. Mihailović, D., Šaponjić, Z., Radoičić, M., Lazović, S., Baily, C., Jovančić, P. *et al.*, Functionalization of cotton fabrics with corona/air RF plasma and colloidal TiO_2 nanoparticles, *Cellulose*, 18, 811, 2011.
106. Bozzi, A., Yuranova, T., Kiwi, J., Self-cleaning of wool-polyamide and polyester textiles by TiO_2-rutile modification under daylight irradiation at ambient temperature, *J. Photochem. Photobiol. A*, 172, 27, 2005.
107. Yuranova, T., Rincon, A.G., Bozzi, A., Parra, S., Pulgarin, C., Albers, P. *et al.*, Antibacterial textiles prepared by RF-plasma and vacuum-UV mediated deposition of silver, *J. Photochem. Photobiol., A*, 161, 27, 2003.
108. Tung, W.S., Daoud, W.A., Henrion, G., Enhancement of anatase functionalization and photocatalytic self-cleaning properties of keratins by microwave-generated plasma afterglow, *Thin Solid Films*, 545, 310, 2013.
109. Gowri, S., Almeida, L., Amorim, T., Carneiro, N., Pedro Souto, A., Fátima Esteves, M., Polymer nanocomposites for multifunctional finishing of textiles—A review, *Text. Res. J.*, 80, 1290, 2010.
110. Kango, S., Kalia, S., Celli, A., Njuguna, J., Habibi, Y., Kumar, R., Surface modification of inorganic nanoparticles for development of organic–inorganic nanocomposites—A review, *Prog. Polym. Sci.*, 38, 1232, 2013.
111. Norouzi, M., Zare, Y., Kiany, P., Nanoparticles as effective flame retardants for natural and synthetic textile polymers: Application, mechanism, and optimization, *Polym. Rev.*, 55, 531, 2015.
112. Yew, M.C., Ramli Sulong, N.H., Yew, M.K., Amalina, M.A., Johan, M.R., Fire propagation performance of intumescent fire protective coatings using eggshells as a novel biofiller, *Sci. World. J.*, 2014, 1, 2014.
113. Giraud, S., Bourbigot, S., Rochery, M., Vroman, I., Tighzert, L., Delobel, R., Flame behavior of cotton coated with polyurethane containing microencapsulated flame retardant agent, *J. Ind. Text.*, 31, 11, 2001.
114. Devaux, E., Rochery, M., Bourbigot, S., Polyurethane/clay and polyurethane/POSS nanocomposites as flame retarded coating for polyester and cotton fabrics, *Fire Mater.*, 26, 149, 2002.
115. Erdem, N., Cireli, A.A., Erdogan, U.H., Flame retardancy behaviors and structural properties of polypropylene/nano-SiO_2 composite textile filaments, *J. Appl. Polym. Sci.*, 111, 2085, 2009.
116. Horrocks, A.R., Flame retardant challenges for textiles and fibres: New chemistry versus innovatory solutions, *Polym. Degrad. Stab.*, 96, 377, 2011.
117. Li, Y.-C., Schulz, J., Mannen, S., Delhom, C., Condon, B., Chang, S. *et al.*, Flame retardant behavior of polyelectrolyte–clay thin film assemblies on cotton fabric, *ACS Nano*, 4, 3325, 2010.
118. Lessan, F., Montazer, M., Moghadam, M.B., A novel durable flame-retardant cotton fabric using sodium hypophosphite, nano TiO_2 and maleic acid, *Thermochim. Acta*, 520, 48, 2011.

119. Gashti, M.P., Ghehi, S.T., Arekhloo, S.V., Mirsmaeeli, A., Kiumarsi, A., Electromagnetic shielding response of UV-induced polypyrrole/silver coated wool, *Fibers Polym.*, 16, 585, 2015.
120. Huang, G., Yang, J., Gao, J., Wang, X., Thin films of intumescent flame retardant-polyacrylamide and exfoliated graphene oxide fabricated via layer-by-layer assembly for improving flame retardant properties of cotton Fabric, *Ind. Eng. Chem. Res.*, 51, 12355, 2012.
121. Hsu, P.-C., Liu, X., Liu, C., Xie, X., Lee, H.R., Welch, A.J. *et al.*, Personal thermal management by metallic nanowire-coated textile, *Nano Lett.*, 15, 365, 2015.
122. Sadineni, S.B., Madala, S., Boehm, R.F., Passive building energy savings: A review of building envelope components, *Renew. Sust. Energ. Rev.*, 15, 3617, 2011.
123. Hsu, P.-C., Song, A.Y., Catrysse, P.B., Liu, C., Peng, Y., Xie, J. *et al.*, Radiative human body cooling by nanoporous polyethylene textile, *Science*, 353, 1019, 2016.
124. Celcar, D., Influence of phase-change materials on thermo-physiological comfort in warm environment, *J. Text.*, 2013, 1, 2013.
125. Salunkhe, P.B., Shembekar, P.S., A review on effect of phase change material encapsulation on the thermal performance of a system, *Renew. Sust. Energ. Rev.*, 16, 5603, 2012.
126. Mondal, S., Phase change materials for smart textiles—An overview, *Appl. Therm. Eng.*, 28, 1536, 2008.
127. Harifi, T., Montazer, M., Application of nanotechnology in sports clothing and flooring for enhanced sport activities, performance, efficiency and comfort: A review, *J. Ind. Text.*, 46, 1147, 2017.
128. Shin, Y., Yoo, D.-I., Son, K., Development of thermoregulating textile materials with microencapsulated phase change materials (PCM). II. Preparation and application of PCM microcapsules, *J. Appl. Polym. Sci.*, 96, 2005, 2005.
129. Karthikeyan, M., Ramachandran, T., Sundaram, O.S., Nanoencapsulated phase change materials based on polyethylene glycol for creating thermoregulating cotton, *J. Ind. Text.*, 44, 130, 2014.
130. Erkan, G., Enhancing the thermal properties of textiles with phase change materials, *Res. J. Text. Apparel*, 8, 57, 2004.
131. Sarier, N., Onder, E., Organic phase change materials and their textile applications: An overview, *Thermochim. Acta*, 540, 7, 2012.
132. Shin, Y., Yoo, D.-I., Son, K., Development of thermoregulating textile materials with microencapsulated phase change materials (PCM). IV. Performance properties and hand of fabrics treated with PCM microcapsules, *J. Appl. Polym. Sci.*, 97, 910, 2005.
133. Sánchez, P., Sánchez-Fernandez, M.V., Romero, A., Rodríguez, J.F., Sánchez-Silva, L., Development of thermo-regulating textiles using paraffin wax microcapsules, *Thermochim. Acta*, 498, 16, 2010.

134. Pause, B., High-performance architectural membranes: Phase-change materials. In: *Fabric structures in architecture*, J.I.D. Llorens (Ed.), pp. 187–199, Elsevier, Amsterdam, 2015.
135. Yu, Z., Gao, Y., Di, X., Luo, H., Cotton modified with silver-nanowires/polydopamine for a wearable thermal management device, *RSC Adv.*, 6, 67771, 2016.
136. Guo, Y., Li, K., Hou, C., Li, Y., Zhang, Q., Wang, H., Fluoroalkylsilane-modified textile-based personal energy management device for multifunctional wearable applications, *ACS Appl. Mater. Interfaces*, 8, 4676, 2016.
137. Stoppa, M., Chiolerio, A., Wearable electronics and smart textiles: A critical review, *Sensors*, 14, 11957, 2014.
138. Weng, W., Chen, P., He, S., Sun, X., Peng, H., Smart electronic textiles, *Angew. Chem. Int. Ed.*, 55, 6140, 2016.
139. Dai, L., Chang, D.W., Baek, J.-B., Lu, W., Carbon nanomaterials for advanced energy conversion and storage, *Small*, 8, 1130, 2012.
140. Hu, L., Pasta, M., La Mantia, F., Cui, L., Jeong, S., Deshazer, H.D. *et al.*, Stretchable, porous, and conductive energy textiles, *Nano Lett.*, 10, 708, 2010.
141. Liu, Z.F., Fang, S., Moura, F.A., Ding, J.N., Jiang, N., Di, J. *et al.*, Hierarchically buckled sheath-core fibers for superelastic electronics, sensors, and muscles, *Science*, 349, 400, 2015.
142. Atwa, Y., Maheshwari, N., Goldthorpe, I.A., Silver nanowire coated threads for electrically conductive textiles, *J. Mater. Chem. C*, 3, 3908, 2015.
143. Zhou, Y., Zhao, Y., Fang, J., Lin, T., Electrochromic/supercapacitive dual functional fibres, *RSC Adv.*, 6, 110164, 2016.

Section 3

PRODUCTION TECHNOLOGIES FOR SMART NANOTEXTILES

9

Production Methods of Nanofibers for Smart Textiles

Rajkishore Nayak

Fashion and Textiles Merchandising, School of Communication and Design, RMIT University Vietnam, Saigon South Campus, Vietnam

Abstract

Over the last two decades, the demand for polymeric nanofibers has increased manifold for various applications in areas of smart textiles. Smart textiles are those textiles that can sense stimuli from the environment, react to them, and adapt to them by integration of functionalities in the textile structure. Nanofibers are increasingly used in smart textiles for sensing, self-cleaning, tissue engineering, communication, wearable computer, and medicine areas. There are several techniques that can be used for the fabrication of nanofibers for smart textile applications, which includes electrospinning, melt blowing, bicomponent spinning, force spinning, and flash spinning. Among these processes, electrospinning is the most popular technique used for producing nanofibers. This chapter discusses about the production methods used to fabricate nanofibers that can be utilized for smart textile applications. The differences among various processes and the recent research works on the fabrication of nanofibers for smart textile applications have also been discussed.

Keywords: Nanofiber, electrospinning, melt spinning, wet spinning, force spinning, bicomponent spinning, flash spinning

9.1 Introduction

Recently, various engineering fields are paying intense attention to nanoscale materials, e.g. nanofibers in the fiber industry for applications in various areas. Nanofibers are fibers with diameters of 100 nm or less [1], notable

Email: rajkishore.nayak@rmit.edu.au

Nazire D. Yilmaz (ed.) Smart Textiles, (303–346) © 2019 Scrivener Publishing LLC

for their characteristic features such as large surface-area-to-volume ratio, extremely small pore dimensions, and superior mechanical properties [2]. Due to these features, nanofibers have a wide range of applications in areas such as high-performance filtration, battery separators, wound dressing, vascular grafts, enzyme immobilization, electrochemical sensing, composite materials, reinforcements, blood vessel engineering, and tissue engineering [3–5].

The existing fiber spinning technologies cannot produce robust fibers with diameters smaller than 2 μm due to limitations in the process. The process widely used for the fabrication of nanofibers is electrospinning because of its simplicity and suitability for a variety of polymers, ceramics, and metals. Other processes include melt blowing, flash spinning, bicomponent spinning, force spinning, phase separation, and drawing [6, 7]. In most of these processes, the fibers are collected as nonwoven random fiber mats known as nanowebs, consisting of fibers having diameters from several nanometers to hundreds of nanometers.

The research on smart textiles is rapidly growing and getting increased attention from the research community [8, 9]. Smart textiles are those textiles that can sense stimuli from the environment, react to them, and adapt to them by integration of functionalities in the textile structure. The stimulus as well as the response can have an electrical, thermal, chemical, magnetic, or other origin. These breakthrough innovations in smart textile field particularly in the last 10 years would achieve significance use in our day-to-day life. Many of the nanofibers such as conductive nanofibers, nanofibers of shape memory polymers, and carbon nanofibers are some of the smart fibers that show promise for use in smart textiles [10, 11]. These fibers can be fabricated by electrospinning or many other techniques as mentioned above.

Chapter 2 focuses on the types of nanofibers for various smart textile applications. This chapter investigates the fabrication techniques used for nanofibers. Among different fabrication techniques, electrospinning has been covered in detail. In addition, the other fabrication techniques such as melt spinning, wet spinning, and melt blowing are also discussed. Depending on the nature of the polymers and the fiber properties needed, one or several of these techniques can be employed for the fabrication of nanofibers for smart textile applications.

The fibers of inherently conducting polymers (ICPs) such as polyacetylene [12, 13], polypyrrole [14, 15], polyaniline [16, 17], and polythiophene [17] are produced by electrochemical or chemical oxidation procedures. The addition of a variety of dopants to the monomer in solution or in vapor phase results in fabrication of ICPs of varying characteristics. Figure 9.1 shows the chemical structures of inherently conducting fibers.

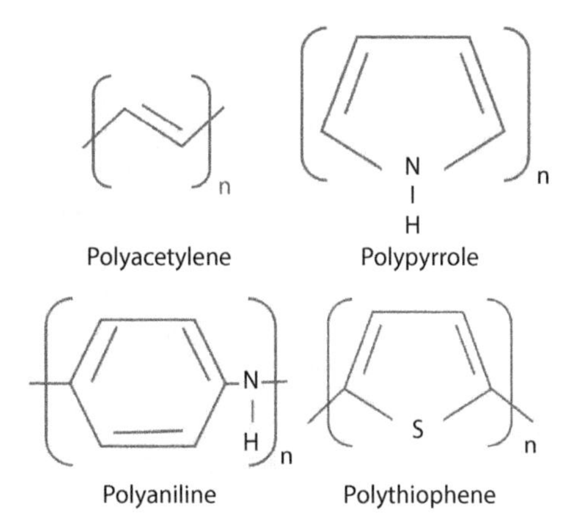

Figure 9.1 Chemical structures of inherently conducting fibers. (The image has been prepared by the author.)

9.2 Electrospinning

The most common method to produce nanofibers is electrospinning. The origin of electrospinning as a viable technique for the production of nanofibers can be traced back to Formhals patent in 1934, on the production of artificial filaments using high electric field [18]. The work was based on the effect of electrostatic force on liquids, i.e. when a suitably electrically charged material is brought near a droplet of liquid held in a fine capillary, it forms a cone shape and small jets may be ejected from the tip of the cone if the charge density is very high.

Electrospinning or electrostatic spinning is a simple and unique process to produce fine fibers by electrostatic forces. It is based on the basic principle of effect of electrostatic force on liquids, i.e. when a suitably electrically charged material is brought near to a droplet of liquid held in a fine capillary, it would form a cone shape and small droplets would be ejected from the tip of the cone if the charge density is very high. Though the principle of electrospinning dates back to the 1800s, the first noticeable work done was by Formhals. It combines the principles of both electrospraying and conventional wet spinning of fibers. Some other techniques that work similarly to electrospinning are electrostatic precipitators and pesticide sprayers.

The schematic of an electrospinning device is shown in Figure 9.2a. Generally, in electrospinning, a liquid droplet is delivered to the tip of a capillary. When an electric field is applied, charges accumulate on the surface of the pendant droplet formed at the tip of the capillary and create an instability that deforms the hemispherical droplet into a conical shape,

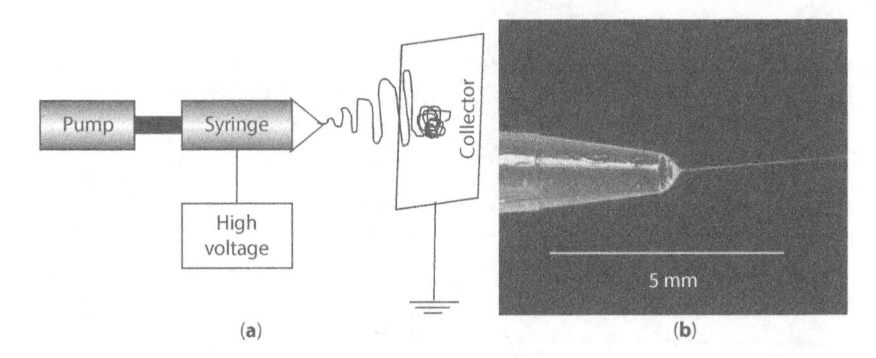

Figure 9.2 Schematic diagram of: (a) electrospinning device and (b) Taylor cone. (The image has been prepared by the author.)

referred to as a Taylor cone [19, 20]. At this stage, there is a competition between the coulombic repulsion of alike charges favoring droplet distortion and surface tension opposing droplet division. When the applied electric field strength is sufficiently high, a liquid jet is continuously ejected from the apex of the cone and travels towards the grounded plate as a barely visible nanoscale fiber (Figure 9.2b). The high charge density on the surface of the fine jet leads to electrical instability, making it whip around rapidly. The jet diameter decreases due to stretching (whipping) and evaporation of the solvent.

9.2.1 Types of Electrospinning

Electrospinning can be classified into two groups: solution electrospinning and melt electrospinning, based on the nature of the polymer used [21]. In the last two decades, research activities in electrospinning have mainly focused on solution electrospinning [22–25], whereas there are few works reported on melt electrospinning [26–28]. However, solution electrospinning of polyolefins, including polyethylene and polypropylene, and polyesters has been limited due to high solvent resistance and high electrical resistivity [22]. The following section discusses on both types of electrospinning.

9.2.1.1 Solution Electrospinning

In solution electrospinning, the polymer is dissolved in a suitable solvent and the resultant solution is used for electrospinning. In this method, high voltage is applied to the polymeric solution and the collector is grounded.

When the charges exceed the threshold, a fluid jet is ejected from the tip of the needle and forms the Taylor cone. This jet travels towards the grounded collector and gets deposited on the collector surface. The nanofibers deposited on the collector are randomly arranged on the surface. The nanofibers can also be aligned in specific direction by adopting various techniques.

Figure 9.2a shows the basic setup for a solution electrospinning equipment. The basic setup used for electrospinning is very simple in construction that consists of three major components, namely, high-voltage power supply, syringe, and collector. Though many researchers use the setup as shown in the figure, some researchers modify to generate a wide variety of fine fibers.

9.2.1.1.1 High-Voltage Power Supply

Normally, direct current (DC) is used as the source of power supply, though there is feasibility of using alternating current (AC). Very high voltage (usually in the range of 1–30 kV) is required for electrospinning. The polarity of the electrospinning system is arbitrary and can be reversed depending upon the polymer type and final product [28]. Free charges are induced to the polymer solution through an immersed electrode. The charged ions of the polymer solution move in response to the external applied electric field towards the collector of opposite polarity.

9.2.1.1.2 Syringe

The syringe or pipette is a very fine capillary tube. It holds the polymer solution or melt into which a metal electrode is inserted. It is mounted horizontally or vertically on an adjustable electrically insulating stand. The spinneret is connected to the syringe at one end. During the spinning process, the syringe pump is used to supply polymer at a constant and controllable rate.

9.2.1.1.3 Collector

The collector or the collecting surface is used to collect the electrospun fibers. Surfaces of different geometry and configurations are used to alter the alignment of the nanofibers. The collector is mounted on an insulating stand so that its potential can be controlled.

9.2.1.1.4 Working of Solution Electrospinning

The polymer solution held in the syringe is electrified by the application of a very high voltage, and the charges are evenly distributed on the surface.

When an electric field is applied to the end of the capillary tube, a charge is induced on the surface of the liquid. Now the polymer drop, which is held by its surface tension at the tip of the spinneret, experiences an electrostatic repulsion force between the surface charges and the columbic force exerted by the external electric field. As the electric field intensity is increased, the repulsion force is increased, which distorts the hemispherical liquid drop into a conical form commonly known as Taylor cone (Figure 9.2b).

When the electric field reaches a threshold value at which the repulsive electrical forces overcome the surface tension, a charged jet of the solution is ejected from the tip of the Taylor cone. As the charged liquid jet is accelerated in the air, it undergoes a chaotic motion or bending instability that causes bending and allows the electrical forces to elongate the jet. In addition, the solvent is evaporated, and its diameter is greatly reduced from hundreds of micrometers to tens of nanometers. As the jet becomes thinner in the electric field, radial charge repulsion results in splitting of the primary jet into multiple filaments, in a process known as splaying.

The final fiber diameter is determined primarily by the number of subsidiary jets formed. The charged jet is attracted by the grounded collector plate and is collected randomly on its surface. The setup used for melt electrospinning is slightly different from the solution electrospinning. The following section describes the melt electrospinning process.

9.2.1.2 *Melt Electrospinning*

In melt electrospinning, the polymer is heated to melt for electrospinning. Melt electrospinning has many advantages over solution electrospinning. The initial work on melt electrospinning was carried out by Larrondo and Manley [29]. Generally, the setup for melt electrospinning consists of a provision for melting the polymer and other parts like solution electrospinning. The polymers for melt electrospinning can be heated by different means such as heating ovens [30], heat guns [28, 31], laser melting devices [32, 33], and electric heating [34]. The fabrication of nanofibers of various polymers such as polylactic acid (PLA) [30], polylactide [32], PP [28, 31, 34], polyethylene terephthalate (PET) [34], (polyethylene glycol)-block-(poly-ε-caprolactone) (PEG_{47}-*b*-$PCL_{9}5$), and (poly-ε-caprolactone) (PCL) [28] by melt electrospinning has been reported by several researchers. The nozzle size; temperatures of the spinneret and the spinning region [30]; polymer molecular weight [34], shear, and extensional viscosities [35]; and the polymer flow rate [34] were found to be the important factors for producing submicron fibers in melt electrospinning. In majority of the cases, the melt electrospun web consists of fibers both in nanometer and micrometer scales. For example, it has been demonstrated that most of the melt-blown fibers had diameters in the range of 10–20 μm with some fibers in the range 250–500 nm [31], 1–30 μm with some random scattered fibers of 247 nm [36] in the web. The diameter of the melt electrospun fiber is large in the initial duration of experiment, which decreases substantially after a few seconds [28]. Melt electrospinning is preferred where the problems of solvent recovery and toxicity are a concern. In melt electrospinning, the dissolution of polymers in organic solvents and the subsequent removal of the solvents are not required. Hence, it is free from the extra cost of solvents and complex and expensive solvent-recycling equipment.

While *in vitro* electrospinning, directly onto cells, is considered, solution electrospinning fails to meet the criteria. This is due to the toxicity of the solvent used in electrospinning, which needs to be fully removed prior to *in vitro* use [25]. Electrospinning directly onto cells is possible via melt electrospinning as the fibers contain no residual solvent and are free from the toxicity [35].

Melt electrospinning can commercially fabricate fibers of polymers with no proper solvent at room temperature such as polyethylene (PE), propylene (PP), and polyethylene terephthalate (PET). In the case of PP, one of the most widely used polymers in commercial spinning, it is difficult to find a suitable solvent at room temperature. Hence, melt electrospinning provides an edge over solution electrospinning. Melt electrospinning also favors the production of multicomponent systems such as blends and composites, as in many cases no common solvent for all the components may exist [30]. Besides, the modelling of melt electrospinning is much easier as it is free from the complications associated with solvent evaporation.

Despite these advantages, there has been limited work on melt electrospinning. This may be due to the limiting constraints associated with the process such as (i) the complex equipment used; (ii) the electric discharge problem associated with the equipment design; and (iii) the intrinsic difficulties associated with the polymer, such as high viscosity and low electrical conductivity [37]. The following section focuses on melt electrospinning (including the components, various configurations, polymers used, and fiber diameters obtained).

The designing of equipment for melt electrospinning is complex compared to that for solution electrospinning. A melt electrospinning system is composed of the following components: high-voltage power supply, heating assembly, syringe pump, temperature controller, and collector (Figure 9.3).

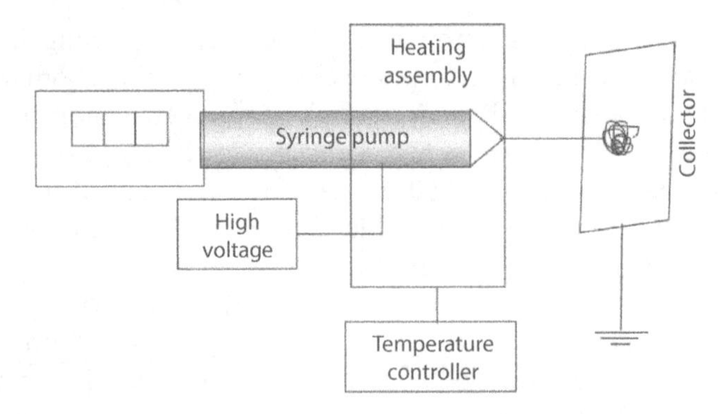

Figure 9.3 Melt electrospinning equipment. (The image has been prepared by the author.)

9.2.1.2.1 High-Voltage Supply

Generally, direct current (DC) is used as the source of power supply, although it is feasible to use alternating current (AC). Very high voltage (usually in the range of 10–30 kV) is required for electrospinning. Lyons [38] claimed that the polarity of the electrospinning system is arbitrary and can be reversed depending on the polymer type and final product. The same production efficiency can be obtained without the damage to the extruder by changing the poles. If the spinneret is grounded and the collector is positively charged, the same electric field strength is created compared to the reverse approach.

9.2.1.2.2 Heating Assembly

The heating assembly is designed to melt the polymer to a suitable viscosity, which can be electrospun easily. In all heating assemblies, the major objective is to melt the polymer by providing heat from various sources such as heating element, heating gun, laser heating, and ultrasound heating [39]. The heating assembly works as a reservoir for the polymer melt. Some pieces of the equipment used by various researchers for melt electrospinning are shown in Table 9.1. A spinneret is connected to the heating assembly at one end to produce nanofibers.

9.2.1.2.3 Syringe Pump

During the melt electrospinning process, a syringe pump is used to pump the polymer melt at a constant and controllable rate through the spinneret.

Table 9.1 Melt electrospinning setup for the fabrication of nanofibers (the table has been arranged by the author).

Process	Features	Advantages	Disadvantages
1. Melt electrospinning (electrical heating [30])	Polymer used: polylactic acid (PLA) Avg. fiber diameter: 800 nm	Simple setup, solvent-free approach	Mostly amorphous fibers, thermal degradation
2. Melt electrospinning (electrical heating [40])	Polymer used: polyethylene glycol-block-poly-ε-caprolactone Fiber diameter: 16 ± 10.7 μm for molten fibers and 560 ± 90 nm for solid fibers	Defect free, continuous and consistent fibers Nontoxic method	Presence of few poor-quality molten fibers
3. Coaxial melt electrospinning [41]	Fabricated PCM-based nanofibers (sheath TiO_2 and core octadecane) Avg. fiber diameter: 150 nm	One step process for encapsulation Production of composite nanofibers Suitable for wide range of materials	Complex setup

(*Continued*)

Table 9.1 Melt electrospinning setup for the fabrication of nanofibers (the table has been arranged by the author). (*Continued*)

Process	Features	Advantages	Disadvantages
4. Melt electrospinning (heating gun [28])	Polymers used: PP and polyethylene glycol (PEG) PEG-*b*-polycaprolactone (PCL) and PCL blend Fiber diameter: 35 ± 1.7 μm (PP-no additive) and 840 ± 190 nm (PP-with viscosity reducing agent); 2 ± 0.3 μm (blend-no additive) and 270 ± 100 nm (blend-with gap method)	Production of blended nanofiber	Coiling and buckling instabilities of the jet near to the collector
5. Melt electrospinning (Laser heating [32])	Polymer used: Polylactide Avg. fiber diameter: ~1 μm	Free from electric discharge problem of the conventional melt electrospinning Suitable for polymers with relatively high melting point Reduced thermal degradation as local and instantaneous heating is possible	Amorphous fibers Unstable fiber formation with higher laser output power

9.2.1.2.4 Collector

The collector or the collecting surface is used to collect the electrospun fibers. Surfaces of different geometry and configurations are used as collectors. The collector is mounted on an insulating stand so that its electric potential can be controlled. In melt electrospinning devices, the high voltage is applied to the collector and the syringe is grounded. The list of polymers used for melt electrospinning has been discussed in Table 9.1.

9.2.2 Use of Electrospinning for Smart Textiles

Electrospinning has been widely used by many researchers to fabricate nanofibers of various polymers to be used as smart textiles [42–44]. Both the electrospinning processes as described above can be used to fabricate fibers with diameters less than 100 nm. The SMPs, polymers of conducting polymers, and polymers for carbon nanofibers can be used in the above two process routes to fabricate nanofibers. Thermoplastic polymers are generally used in melt electrospinning, and the polymers with a suitable solvent can be used in the solution electrospinning.

Several polymers such as polyaniline/polyethylene oxide (PEO) [45, 46], polypyrrole/PEO [46], pure polyaniline [47] and polypyrrole [48, 49], poly(3-hexyl-thiophene)/PEO [50], and polyaniline/PEO/carbon nanotubes [51] have been used in electrospinning to fabricate micro- and nanofibers to be used in smart textiles. Furthermore, electrospinning has also been used for the fabrication of fibers for chemical sensors [52, 53] and field effect transistors [54, 55].

In the case of nonconducting polymers such as PEO, additives are added to increase the electrical conductivity of the solution [47, 48]. Once the solution has been extracted, the fibers will have decreased electrical conductivity. Similarly, reducing the PEO content [47, 48] or incorporation of carbon nanotubes (CNTs) can help to increase the conductivity of the nanofibers [49].

The electrospinning process was used for the first time by Jung *et al.* [56] to fabricate nanofibers (400 nm diameter) of shape memory polyurethane (SMPU). The electrospinning of SMPU resulted in shape memory fibres (SMFs) with good shape recovery and shape retention properties. The incorporation of styryl dye (StD) changed the properties of the nanofibers. For example, the water vapor permeability was significantly changed with the SMPU nanofibers with humidity.

The electrospun nanofibers of PCL-based SMPU with 50–700 nm diameter were reported recently by Zhuo *et al.* [57]. The effect of feed rate, solution concentration, and applied voltage on the fiber diameter and morphology was investigated. In comparison to cotton and wool fabrics, the SMPUs exhibited good shape memory effect (SME) (98% shape recovery and 80% shape fixation) in addition to the excellent water vapor permeability (WVP). The researchers also fabricated PCL-SMPU nanofiber mats with silver nitrate ($AgNO_3$). The nanofiber mats were of higher tensile strength and lower elongation compared to the control PCL-SMPU nanofiber mats.

The electrospinning of electrically conducting polymers has gained much attention recently. Nanofibers with conductivity like metals can be fabricated by electrospinning. Among these polymers, some are inherently conducing polymers (ICPs) and others are the polymers where conductivity can be incorporated during the fiber formation process. The following section discusses the research works on the fabrication of nanofibers of ICPs as well as the polymers where conductivity has been achieved by adding some dopants.

The ICPs such as polypyrrole (PPy) have drawn much attention of the researchers to be applied in the smart textiles such as sensors, actuators, batteries, electronic devices, drug release, and tissue engineering. Due to its aqueous conductivity and low oxidative potential, PPy is one of the polymers most widely used in electrospinning. PPy is used in solution electrospinning process for the fabrication of nanofibers of polymers such as PPy, and polyaniline.

PPy has been electrospun for various applications such as batteries [58], actuators [42], electromagnetic shielding [43], and antistatic applications [43]. The major drawback of PPy for solution electrospinning is its poor solubility in common solvents, which is an impediment in the research on ICPs. In addition to the poor solubility of PPy, other problems associated with electrospinning are low molecular weight and high molecular chain rigidity.

The electrical conductivity of ICPs can be increased by the addition of dopants. For example, electrospinning of PPy was performed in chloroform solution by using additional amount of dodecyl benzene sulfonic acid (DBSA) as dopant, which increased the electrical conductivity and reduced the intermolecular interactions [48]. The functionalization of PPy with sodium salt of di(-2ethyl hexyl) sulfosuccinate [(PPy3) + (DEHS)−]x followed by dissolution in dimethyl formamide (DMF) was performed by Chronakis *et al.* [49].

PPy particles were synthesized in $FeCl_3$ with DBSA dopant by Merlini *et al.* [44], which were subsequently added to polyvinylidene fluoride (PVDF). Nanofibers of 460 nm were fabricated using the electrospinning process. Vapor phase deposition techniques were employed by Han and Shi [59] to coat electrospun nanofibers of sodium 1,4-bis(2-ethylhexyl) sulfosuccinate-ferric chloride with PPy. Similarly, other approaches were adopted by various researchers to increase the electrical conductivity while electrospinning.

Nanofibers of PPy/poly(vinyl pyrrolidone) were fabricated by Tavakkol *et al.* using electrospinning of pyrrole solution followed by oxidation. The low solubility and low viscosity of PPy are detrimental to smooth electrospinning [60]. Hence, poly(vinyl pyrrolidone) (PVP) (9% w/w) was added to increase the viscosity of pyrrole solution, which made the electrospinning possible. This was a one-step method where the nanofibers of pyrrole-PVP were collected in a solution of $FeCl_3 \cdot 6H_2O$ (2% w/v) and dopant in ethanol and oxidized *in situ*. In the second approach, pyrrole-PVP nanofibers were deposited onto aluminum foil and then detached from it. The nanofibrous mats of pyrrole-PVP were immersed in $FeCl_3 \cdot 6H_2O$ (2% w/v) solution to polymerize pyrrole; the dopant consisted of P-toluene sulfonic acid (PTSA) and anthraquinone-2-sulfonic acid sodium salt (AQSA).

Like the SMPs, shape changing polymers (SCPs) are also important in textile applications. The SCPs change their shape on heating or cooling. The limitations associated with SCPs are their high cost and low productivity. The SCPs are only investigated in the laboratory scale now. For example, the liquid crystal elastomers (LCEs) were synthesized from the melt and well-aligned fibers were fabricated. The LCEs showed reversible contraction on heating and elongation on cooling, which are suitable for thermal actuators. Scaling up the production and the characterization of fibers is essential to understand their properties.

The use of carbon nanofibers (CNFs) is also rapidly increasing due to their better electrical conductivity and tensile properties compared to other similar materials. CNFs also exhibit high porosity network and large specific surface area. These properties make the CNFs an ideal candidate for various applications such as filtration, electrode, and hydrogen storage materials. The fabrication of CNFs from renewable cellulosic sources is one of the major research areas. In comparison to PAN, the CNFs are biodegradable and derived from renewable resources.

Electrospinning of cellulose can be carried out by using either the solution electrospinning or the melt electrospinning. For example, Deng *et al.* [61] used solution electrospinning to fabricate nanofibers. Similarly, electrospinning of carbon nanofibers from regenerated cellulose has also been conducted by researchers [62, 63]. Deng *et al.* [61] fabricated cellulose nanofibers from a cellulose acetate solution by electrospinning followed by deacetylation, which were then carbonized at 800–2200°C.

The use of bicomponent nanofiber structures, i.e. core-sheath nanofibers, has been reported by some researches. For example, Wei *et al.* [64] fabricated the core-sheath structure with polyaniline core for conductivity and polystyrene or polycarbonate layer as sheath for insulation. It was found that blending of polyaniline with poly(methyl methacrylate) (PMMA) and poly(ethylene oxide) (PEO) resulted in bead-like structures with the formation of isolated domains of polyaniline. The morphology of the bicomponent structures was dependent on the molecular weight of the polymer and surface tension of the solution. Low molecular weight and polymer incompatibility are the key factors for the formation of core-sheath structure.

Chemical cross-linked liquid crystal polymeric fibers (CLCP) were fabricated by Yoshino *et al.* [65]. The fibers showed photoinduced bending effect in addition to the thermally induced shape changing. Nanofibers with cross-linking and uniform alignment of the nematic directors were also electrospun.

The fibers of SMPs cannot be directly used in textiles due to poor surface properties. The fibers need to be bound together to provide an integrated structure. Hence, the fibers need to be converted to fabric format by various processes such as weaving and knitting. As such, the fibers can also be collected as nonwoven web and directly used. These fibers can also be blended with other fibers as well to achieve improved properties.

Several studies have also been done on the fabrication of fibers of conducting polymer of polyaniline (PANi) and its blends. For example, MacDiarmid [66] fabricated PANi fibers doped with sulfuric acid and other polymers such as PEO, polyacrylonitrile, and polystyrene. Similarly, El-Aufy [67] fabricated nanofibers of poly(3,4-ethylenedioxythiophene) (PEDOT) with polystyrene sulfonate (PSS) and polyacrylo nitrile (PAN) as a carrier.

Unlike other production methods, electrospinning is a versatile method to produce fibers of several polymers, ceramics, and metallic oxides. This process has the limitation of lower productivity compared to some other methods. Several researchers have modified the basic setup for solution electrospinning to increase the productivity [68–73]. The basic principle to obtain higher productivity in solution electrospinning is based on increasing the number of jets by adopting different techniques. These techniques can be summarized as (a) multijets from single needle, (b) multijets from multiple needles, and (c) needleless systems [72].

9.2.3 Multijets from Single Needle

Generally, in single needle electrospinning (SNE), a single jet is initiated from the Taylor cone with the application of electric field. Multijets were observed for the first time by Yamashita *et al.* [74] during electrospinning from a single home-made stainless-steel needle with a grooved tip mounted on a glass syringe. Beadless membranes of polybutadiene (PB) were successfully prepared by growing multijets at the needle tip of a SNE setup. The formation of multijets on a SNE setup was attributed to two possible mechanisms: significant discrepancy in electric field distribution and some degree of solution blockage at the needle tip. In another approach, a curved collector was used by Vaseashta [75] for the formation of multijets from multiple Taylor cones in a SNE system.

Multijets can be obtained from a single needle by splitting the polymer jet into two separate subjets in its path to the collector [69, 76]. Jet splitting (i.e. a sequence of secondary jets emanating from the primary jet) has been observed under certain conditions, where fluid jets interact with the large axial electric fields [77–79]. Although the mechanism of jet splitting is yet to be fully analyzed, the experimental investigation of controlled jet splitting will be a fascinating challenge for increasing the productivity of SNE systems.

9.2.4 Multijets from Multiple Needles

Some work has been done to increase the productivity of electrospinning using multiple needle electrospinning (MNE) systems [80–83]. While designing MNE systems, the needle configuration, number of needles, and needle gauge are important factors to be taken into consideration. Needles can be arranged in linear configuration or two-dimensional configurations such as elliptical, circular, triangular, square, and hexagonal. For example, a linear configuration with four needles has been designed to fabricate nanowebs and to scale up the production [84]. The nanofibers were unevenly deposited on fibrous substrates because of the distorted electric field on the multineedle setup. Similarly, Ding *et al.* [85] fabricated biodegradable nanofibrous mats by MNE setup with four movable syringes and a rotatable grounded tubular collector. Uniform thickness of nanofibrous mats with good dispersibility was achieved. In another attempt, where seven and nine needles were arranged in a linear configuration, it was observed that the behaviors of central jets and border jets were different [86].

Generally, MNE systems require large operating space and careful design of the relative spacing between the needles to avoid strong charge repulsion between the jets. The spacing between the needles depends on nozzle gauge as well as the solution properties to be electrospun. Non-uniform electric field on needle tips at different positions, needle clogging, instability problems (such as dripping or nonworking needles), and uneven fiber deposition are some of the key limitations of MNE systems [87].

9.2.5 Multijets from Needleless Systems

Needleless electrospinning systems are becoming popular as the productivity can be substantially improved by provoking numerous polymeric jets from free liquid surfaces [88]. The basic principle of formation of multijets from a needleless system is as follows: the waves of an electrically conductive liquid self-organize on a mesoscopic scale and finally form jets when the applied electric field intensity is above a critical value [89]. The pioneering work on needleless electrospinning was reported by Yarin and Zussman [72].

The work was based on a combination of normally aligned magnetic and electric fields acting on a two-layer system, where the lower layer was a ferromagnetic suspension and the upper layer was a polymeric solution. Numerous steady spikes of polyethylene oxide (PEO) were generated at the free surface of the magnetic fluid by the application of a magnetic field. With the addition of the polymer layer and application of high voltage, some perturbations were visible at the free surface of the polymer layer. After a threshold voltage, multiple jets were ejected towards the grounded electrode and fibers were collected on a glass slide.

Another needleless electrospinning system for fabricating polyvinyl alcohol (PVA) nanofibers by using a conical metal wire coil as the spinneret has been reported [90]. The needleless approach produced finer nanofibers on a much larger scale compared to conventional electrospinning. Kumar *et al.* [91] developed an electrospinning setup for the formation of multiple jets with controlled fiber repulsion using a plastic filter. Apart from increased throughput, this setup reduced fiber repulsion as compared to a multineedle setup. The fiber repulsion was reduced by controlling emitter voltage and emitter/collector distance. It was found that the plastic filter setup produced fibers with lower average diameter and better uniformity.

Needleless electrospinning comprising a rotating disk and a cylinder nozzle, for the fabrication of PVA nanofibers, was reported by Niu *et al.* [89]. Under identical operating conditions, the fibers produced from the disk nozzle were finer than those from the cylinder nozzle. The disk nozzle needed a relatively low voltage to initiate the fiber formation. The cylinder nozzle showed a higher dependence on the applied voltage and polymer concentration.

9.2.6 Other Potential Approaches in Electrospinning

In addition to the needle and needleless systems discussed above, electrospinning can also be classified into confined feed system (CFS) and unconfined feed system (UFS) based on the way the solution or melt is dispensed. In CFS, the polymer solution or melt is injected at a constant rate, whereas in UFS, it flows unconstrained over the surface of another material. The advantages of CFS are restricted flow rate (needed for maintaining a continuous stable electrospinning), uniform fiber diameter, and better-quality fiber. However, CFS increases the system complexity (as a control system is required for each jet or multijets) and is prone to clogging. CFS includes electrospinning systems that use a syringe pump, whereas UFS includes different systems such as bubble electrospinning [92], electroblowing, and electrospinning by using porous hollow tube, microfluidic manifold, and roller electrospinning. In the following section, various UFSs have been illustrated.

9.2.7 Bubble Electrospinning

Liu and He [92] explored the feasibility of mass production of nanofibers by bubble electrospinning. The device consisted of a high-voltage DC generator, a gas pump, a vertical liquid reservoir having a top opening, a gas tube installed at the bottom center of the reservoir, a thin metal electrode fixed along the center line of the gas tube, and a grounded collector. The gas tube and the electrode were inserted through the bottom of the reservoir and were connected to the gas pump and the DC generator, respectively. One or several bubbles were formed on the free surface of the solution, when the gas pump was turned on slowly. The shape of each bubble changed from spherical to conical (similar to a Taylor cone [93]) as the DC voltage was applied. Multiple jets were ejected from the bubbles when the applied voltage exceeded a threshold value.

The polymer jets in bubble electrospinning also exhibited an instability stage similar to that in the conventional electrospinning. The fibers produced were a mixture of straight, coiled, and helical fibers along with a few beaded and thick fibers. The number of bubbles was affected by the gas pressure, solution viscosity, nozzle diameter, and height between the nozzle tip and liquid surface. Later, Liu and coworkers [94–96] investigated the effect of applied voltage on fiber diameter and morphology in bubble electrospinning. The average fiber diameter increased with the applied voltage, which is quite different from the results in basic electrospinning.

9.2.8 Electroblowing

Electroblowing is an electrospinning process assisted with air blowing. The method comprises preparation of a polymer solution by dissolving the polymer in a solvent, feeding the polymer solution through a spinning nozzle applied with high voltage, injecting compressed air through the lower end of the spinning nozzle, and collecting the fibers in the form of a web on a suitable grounded collector. In electroblowing, two simultaneously applied forces (an electrical force and an air-blowing shear force) interact to fabricate the nanofibers from the polymeric fluid. Nanofibers of both thermoplastic and thermosetting resins can be produced by electroblowing.

Wang *et al.* [97] were one of the first to modify an electrospinning apparatus by the attachment of an air-blowing system to fabricate nanofibers from hyaluronic acid (HA). In the setup, high positive voltage was supplied between the spinneret and the grounded collector plate. The air temperature and blowing rates were achieved by controlling the power output of the heater and the flow rate of air, respectively. The air flow was also used to control the cooling rate of the fluid jet and the rate of solvent evaporation. The factors affecting the fiber morphology and diameter were the air-blowing rate, polymer concentration, solution feeding rate, electric field strength, and type of collector.

Kim *et al.* [98] prepared polyacrilonitrile (PAN) fibers with diameters ranging from several nanometers to hundreds of nanometers by electroblowing a 20 wt% solution of PAN in dimethyl formamide (DMF). The apparatus consisted of a storage tank for polymer solution, a spinning nozzle, an air nozzle adjacent to the lower end of the spinning nozzle for injecting compressed air, a source of high voltage, and a grounded collector. The equipment had a higher productivity over the conventional electrospinning.

Arora *et al.* [99] produced nanowebs of PP with average diameter of 850–940 nm by electroblowing. In the setup, compressed and heated air was supplied from air nozzles positioned around the sides of the spinning nozzle. The air forwarded the newly issued polymeric solution from the nozzle and attenuated to nanofibers, which were collected on a grounded porous collection belt. Other materials used in the process were the addition and condensation polymers such as polyamide (PA), polyester, polyolefins, polyacetal, polyalkylene sulfide, polyarylene oxide, polysulfone, cellulose ether and ester, polyvinylchloride (PVC), polymethylmethacrylate (PMMA), polystyrene (PS), polyvinylidene fluoride, polyvinylidene chloride, and PVA.

9.2.9 Electrospinning by Porous Hollow Tube

Varabhas *et al.* [100] used a polytetrafluroethylene (PTFE) tube with a porous wall for increasing the productivity of the electrospinning process. The tube containing the pores with an average diameter of 20–40 μm was oriented horizontally. Holes of 0.5mm diameter, spaced 1 cm apart from each other, arranged in two rows parallel to the axis of the tube and penetrating 1 mm into the wall, were drilled along the bottom of the tube. To maintain an equal electrical potential in the vicinity of each hole, a wire electrode made out of a square wire mesh having 5-mm spacing between the wires was inserted inside the tube.

The porous tube was suspended on the frame of a PVC pipe with an adjustable distance of 12–15 cm above the grounded aluminum foil collector. The solution of polyvinylpyrrolidone (PVP) in ethanol (15 wt%) was pushed at low air pressure (1–2 kPa) through the tube wall with a voltage of 40–60 kV. During electrospinning, each hole produced one jet, which finally became a continuous long fiber. As the jet traveled a few centimeters, the bending instability became dominant and the jet formed an expanding coil. The production rate obtained by this method was about 3–50 times higher than the SNE system, which was dependent on the number of rows of holes, spacing between the holes, and the collector geometry.

9.2.10 Electrospinning by Microfluidic Manifold

Srivastava *et al.* [101, 102] designed a microfluidic device to fabricate multicomponent nanofibers and to scale up the production of electrospinning. A polydimethylsiloxane (PDMS) based multilayer microfluidic device capable of spinning several hollow fibers in parallel order was used to produce nanofibers. Two layers of microchannels (four spinnerets in each layer) were used to flow PVP solution as sheath material. An array of spinnerets was used for the fabrication of the core consisting of heavy mineral oil or pyrrole. Each of the eight-outlet spinneret was provided constant pressure by two layers of nonintersecting stacked microchannels arranged in a branching tree pattern. The sheath polymer solution was introduced through the microchannels at the bottom layer, and the core material was introduced through the top microchannels. During the electrospinning process, mutually interacting electrified jets underwent bending instabilities and were repulsed from their neighbors due to coulombic repulsion. Uniform nanofiber mats were produced at the rate of 0.1 g/h with the inter nozzle distance of approximately 8 mm.

Srivastava *et al.* [103] also produced bicomponent Janus nanofibers using a PDMS-based multiple outlet microfluidic device capable of spinning eight nanofibers in parallel. The ability of microfluidic device to synthesize multiphase nanofibers with controlled distribution of functionalities (which is otherwise difficult to obtain) provides a useful technology to the next generation of smart materials. Microfluidic manifold method has many advantages over conventional electrospinning such as rapid prototyping, ease of fabrication, and parallel electrospinning within a single, monolithic device. The fabrication of complex networks of microchannels is easily accomplished using PDMS-based micromolding technology.

9.2.11 Roller Electrospinning

Fabrication of nanofibers by electrospinning using rotational setup dates back to the 1980s [104]. Roller electrospinning process was first developed by Jirsak *et al.* [105] in 2005 at the Technical University of Liberec (Czech Republic). The mechanism of formation of Taylor cones on the surface of roller was described by Lukas *et al.* [106]. Cengiz and Jirsak [107] studied the effect of tetraethylammonium bromide (TEAB) salt on the spinnability of polyurethane (PU) nanofibers by roller electrospinning. The roller electrospinning setup consisted of a rotating cylinder to spin nanofibers directly from the polymer solution. An aluminum rotating roller was partially immersed in the PU solution contained in a polypropylene (PP) dish. The rotating roller was applied with high voltage and the collector was grounded. It was found that the salt concentration had an important effect on conductivity, viscosity, fiber diameter, and morphology.

Later, Cengiz and coworkers [108] also investigated the influence of solution properties on the fiber diameter of PVA using the device mentioned above. It was found that the electric conductivity and surface tension of the solution did not affect either the throughput or the fiber diameter significantly. Molecular weight had an important influence on spinnability, whereas solution concentration influenced the throughput and properties of nanofibers (Table 9.2).

Table 9.2 Modifications of basic electrospinning process for the production of nanofibers (the table has been arranged by the author).

Process	Features	Advantages	Disadvantages
1. Needleless electrospinning [72]	Polymer used: PEO (M_w = 600,000), fiber diameter: 200–800 nm	Production rate: 12 times of the conventional electrospinning	Wider fiber diameter distribution
2. Bubble electrospinning [92, 109]	Polymers used: PVP (M_w = 40,000 g/mol) and PEO (M_w = 500,000 g/mol) Solution concentration for PVP = 30 wt% and PEO = 2 wt%	High production rate, high efficiency, free from clogging, ease of operation, simple process and low cost	Solvent recovery problem
3. Electroblowing [97]	Polymer used: hyaluronic acid (HA) 2–3% (w/v) Fiber diameter: 40–120 nm	High production rate and simple process	Solvent recovery problem
4. Cylindrical porous hollow tube [100]	Polymer used: PVP (M_w = 360,000) Fiber diameter: 300–600 nm	High productivity	Complex design of the equipment
5. Microfluidic manifold [102]	Polymer used: PVP Fiber diameter: 85–350 nm	Simple process, flexible control over channel size, rapid prototyping and the ability to spin multiple fibers in parallel through arrays of individual microchannels	Solvent recovery problem
6. Roller electrospinning [107]	Polymer used: PU (M_w = 2000 g/mol) Avg. fiber diameter: 144 nm	Production rate is higher, simple process	Solvent recovery problem, low molecular weight polymers are difficult to be spun

9.3 Other Techniques without Electrostatic Force

Apart from the techniques discussed above, which are mainly based on the application of electrostatic force for the fabrication of nanofibers, several other approaches such as melt blowing, flash spinning, bicomponent spinning, force spinning, phase separation, and drawing are already used for the fabrication of nanofibers. These techniques are highlighted in the flowing section.

9.3.1 Melt Blowing

Melt blowing is a simple, versatile, and one-step process to produce materials in micrometer and smaller scale [110]. The technology of melt blowing was first developed in the 1950s at the Naval Research Laboratory of the United States. In melt-blowing process, a molten polymer is extruded through the orifice of a die. The fibers are formed by the elongation of the polymer streams coming out of the orifice by air drag and are collected on the surface of a suitable collector in the form of web [111]. The average fiber diameter mainly depends on the throughput rate, melt viscosity, melt temperature, air temperature, and air velocity [112]. A brief review of the melt-blowing process and the factors affecting the properties of the web has been reported by various researchers [7, 97, 110]. The schematic of the melt-blowing equipment has been shown in Figure 9.4.

The extruder is a heated barrel responsible for the melting of the polymer to achieve appropriate viscosity. The polymer in the form of powder, chips, pellets, or granules is gravity-fed from the hopper to the extruder at a certain feed rate and heated in different heating zones. The rotating screw(s) carries the polymer towards the die assembly where it is metered by a metering pump. The metering pump delivers the polymer melt uniformly and consistently to the die assembly. The die assembly is responsible for the production of good quality fibers. In this section, the fiber attenuation is achieved by the action of cool air. Subsequently, the fibers are collected as web and forwarded to the winding section.

The difficulty in fabricating nanofibers in melt blowing is due to the inability to design sufficiently small orifice in the die and the high viscosity of the polymeric melt. Nanofibers can be fabricated by special die designs with small orifice, reducing the viscosity of the polymeric melt and suitable modification of the melt-blowing setup. For example, Ellison *et al.* [113] produced melt-blown nanofibers of different polymers by a specially designed single-hole die with small orifice using the processing conditions

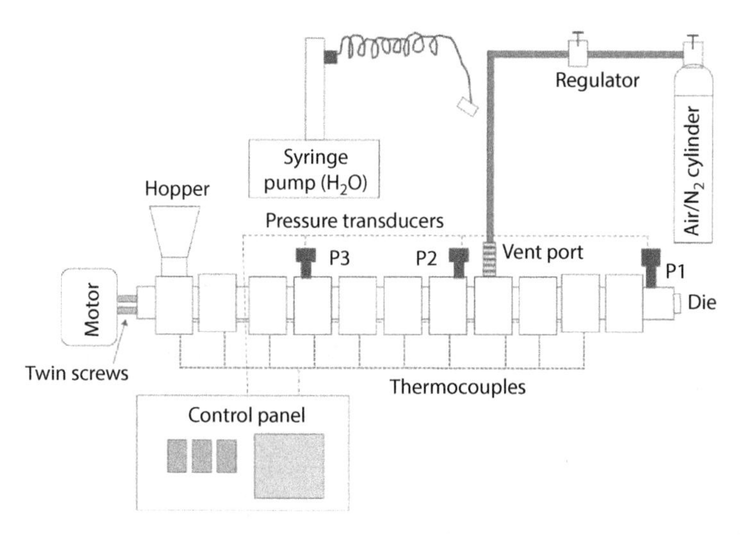

Figure 9.4 Schematic of the melt-blowing equipment. (The image has been prepared by the author.)

used in industry. Similarly, the special die design by Podgórski *et al.* [114] where the polymer nozzles were surrounded by air nozzles produced nanofibers with diameters ranging from 0.74 to 1.41 μm.

Bodaghi and Sinangil [115] fabricated melt-blown nanofibers by changing the rheology of the polymers. The melt-blowing apparatus consisted of two extruders with different barrel diameters to create different shear rates. A spin pack coupled with both extruders directed high-velocity air towards the melt-blown fibers to attenuate and split them into nanoscale. The process was suitable for many melt-spinnable commercial polymers, copolymers, and their blends such as polyesters, polyolefins (PE and PP), PA, nylons, PU, PVC, PVA, and ethylene vinyl acetate. The melt-blown fibers had significantly reduced average diameters and enhanced surface area-to-mass ratios compared to conventional melt-blown fibers. Hills Inc. (West Melbourne) produced melt-blown nanowebs from low viscosity (or high melt flow index 1500–1800) homopolymers with an average diameter of 250 nm [116]. Apart from the low viscosity, smaller diameter orifice with higher L/D ratio (500+) and low flow rate assisted in nanofiber fabrication.

In melt blowing, the sudden cooling of the fiber as it leaves the die can prevent the formation of nanofibers. This can be improved by providing hot air flow in the same direction of the polymer around the die. The hot air stream flowing along the filaments helps in attenuating them to smaller diameter. The viscosity of polymeric melt can be lowered by increasing the temperature, but there is a risk of thermal degradation at high temperature.

9.3.2 Wet Spinning

As discussed in Chapter 2, the fibers fabricated from shape memory polymers (SMPs) can sense external stimuli of different forms such as chemical, thermal, electrical, and magnetic. The fibers of SMPs can also be fabricated by the process of wet spinning [117]. In wet spinning, the polymers with a suitable solvent can be used. The name is derived from the method as the fibers are extruded to a liquid bath after extrusion. The liquid medium can provide larger drag force on the extruded fiber as compared to air. Then the fibers are dried and drawn for fiber orientation and crystallinity.

In the liquid medium, the solvent is extracted from the fiber and only the single phase of the fibers exists. The faster the process of solvent extraction, the more will be the voids in the fiber. The fibers are formed by extrusion and driven into a nonsolvent.

The use of wet spinning for the fabrication of SMFs has been reported by several researchers. For example, Ji *et al.* [117] used poly(ethylene) adipate (PEA) to fabricate smart fibers with shape memory effects. PEA of 600 Mw and 1,4-butanediol were dehydrated at 65°C in vacuum. 4,4'-Diphenylmethane diisocyanate flakes were distilled at 200°C in vacuum. N,N'-Dimethyl formamide (DMF) was also dried using 4 Å molecular sieves. Two-step copolymerization process was used to prepare with varying amounts of hard-segment content of 50%, 55%, and 60%. The polyurethane solution was casted into a rectangular Teflon mold followed by drying at varying temperatures to prepare 0.1-mm SMPU films.

The preparation of polymer solution involved filtering and degassing at 100°C followed by wet spinning using the equipment. The extrusion of SMPU was done with 0.08-mm pinholes in the die (total 30) into a water coagulation bath at a spinning velocity of 6 m/min under nitrogen pressure. The fibers thus coagulated were drafted using a set of rolls at 10 m/min followed by rinsing in a water bath. The rinsed fibers were dried in a hot chamber and collected by winding. The wound packages were heat-set to eliminate residual stress during wet spinning. The heat treatment was done in a hot oven (2 m long, maintained at 120°C). A draft of ≤ 1 was used by using the drawing rolls 1 and 2.

The earlier work on the fabrication of nanofibers was reported by Ji *et al.* in 2006 [117]. The fibers were heat-set and wound on the packages using a winder. It was observed that the hard segments in the SMFs were readily aggregated into hard domains. Furthermore, the SMFs showed higher recovery stress and higher shape recovery. Wet spinning is playing an important role in the formation of SMPs. For example, the SMFs fabricated by weaving can be used in weaving looms or knitting machines to prepare the fabrics with shape memory effect.

Although wet spinning plays an important role in fabricating fibers with shape memory effects, there are certain limitations. The major limitation is the spinning speed, which needs to be improved for large-scale production. Furthermore, the properties of fibers such as tactility, tenacity, and breaking elongation need to be carefully considered while fabricating the SMFs. The other problem is the low tenacity of the fibers fabricated by wet spinning. The other spinning process, i.e. melt spinning, is also used for the fabrication of SMFs, which is discussed in the following section.

9.3.3 Melt Spinning

The melt-spinning devices are used for the thermoplastic polymers. The polymers are melted by suitable heating devices and extruded through specific die with varying number of spinnerets. The molten polymer is controlled by a metered pump, and the hard lumps are filtered before extrusion. As the fibers exit the die hole, they are cooled by moving air and drafted by drawing rolls. A lubricant can be added at this stage to avoid the problem of static charge generation. The fibers are then heat-set to remove the residual stress and structural faults during the spinning and drafting process. This method can be used for the fabrication of SMP fibers as discussed in the following section.

One of the earlier works in 2006, Kaursoin and Agrawal [118] used SMPs commercially available (MM-4510). It was suggested that the mechanical and shape recovery properties of SMPs can be improved by post spinning operations such as heat setting. This can be attributed to the increase in the alignment of the polymeric chain, which helps to increase the crystallinity of the fibers. Varying the amount of draft and setting temperatures can result in different levels of increase in the properties.

It has been established experimentally that SMPs produced by melt spinning possess higher tenacity, shape recovery, and shape fixation properties compared to the SMPs prepared by wet spinning. This can be attributed to higher phase separation and well-developed hard and soft phase in the structure of the SMFs in the melt spinning compared to the wet spinning [119]. Furthermore, melt spinning provides the benefit of introducing additives during the spinning process to modify the fiber properties. For example, multiwalled carbon nanotubes (MWCNs) were introduced to achieve shape recovery and increase the recovery force. Hence, it can be concluded that the SMFs fabricated by dry spinning have superior properties to the fibers obtained by wet spinning.

The other advantage of melt spinning is higher productivity. In wet spinning, as the solvent is extracted from the fibers, the net productivity is reduced due to mass loss. However, as melt spinning is free from this, the productivity is higher due to no mass loss. The drafting of melt-spun fibers is easier compared to the wet-spun fibers, which help to achieve finer fibers. There are some drawbacks associated with melt spinning as well, for example, polymer degradation, which can lead to lowering of the molecular weight and change in the thermal properties.

9.3.4 Template Melt Extrusion

Li *et al.* [120] combined the extrusion technology with the template method for the production of polymeric nanofibers of thermoplastic polymers. In this process, the molten polymer was forced through the pores of an anodic aluminum oxide membrane (AAOM) and then subsequently cooled down to room temperature. A special stainless-steel appliance was designed to support the thin AAOM, to bear the pressure, and to restrict the molten polymer movement along the direction of the pores. The appliance containing the polymer was placed on the hot plate of a compressor (with temperature-controlled functions) followed by the forcing of the polymeric melt. The hot plate was stopped after 2 h of heating, and the pressure was maintained until the system cooled to room temperature (Table 9.3).

Isolated nanofibers of PE were obtained by the removal of the AAOM with sodium hydroxide/ethanol (20 wt%). Finally, the nanofibers were broken down from the bulk feeding film by ultrasonic bath (in ethanol for 5 min) to form isolated fibers. The diameter of the PE fibers ranged from 150 to 400 nm (diameter of AAOM pores = 200 nm), and the length of the fibers corresponds to the length of the pores in AAOM (i.e. 60 μm).

9.3.5 Flash Spinning

In flash-spinning process, a solution of a fiber forming polymer in a liquid spin agent is spun into a zone of lower temperature and substantially lower pressure to generate plexi-filamentary film-fibril strands [121]. A spin agent is required for flash spinning, which: 1) should be a nonsolvent to the polymer below its normal boiling point; 2) can form a solution with

Table 9.3 Processes to produce nanofibers based on polymeric melt (the table has been arranged by the author).

Process	Features	Advantages	Disadvantages
1. Melt-blowing setup [113]	Polymers used: polybuta styrene, PP and PS Avg. fiber diameter: less than 500 nm	Nanofibers feasible at commercial processing condition	Dispersion of spherical particles among the fiber mat, fiber breakup between the die and collector
2. Melt blowing by Podgórski *et al.* [114]	Polymers used: PP Avg. fiber diameter: 210 nm to 37.5 μm	High production rate	High variation of fiber diameter
3. Melt-blowing setup [115]	Polymers used: polyesters, polyolefins (PE and PP), PA, PU, PVC, PVA, and ethylene vinyl acetate Avg. fiber diameter: 940 nm along with some microfibers	Favorable for many polymers High production rate	Complex process
4. Template melt extrusion [120]	Polymers used: PE Avg. fiber diameter: 150 nm to 400 nm Fiber length: 60 μm	Less variation in fiber diameter	Lengthy process Very short length fiber with some breaks

the polymer at high pressure; 3) can form a desired two-phase dispersion with the polymer when the solution pressure is reduced slightly; and 4) should vaporize when the flash is released into a substantially low-pressure zone.

Flash-spinning process was described by Blades and White of Du Pont in 1963, and since then, several patents have been filed. Weinberg *et al.* [122] produced nanofibers of polyolefins using flash spinning with fiber length of 3–10 μm and at a production rate that is at least two orders of magnitude higher than the conventional electrospinning. The nonwoven fibrous webs produced had morphologies (i.e. complex interconnecting networks or webs of large and small polyolefin filaments or fibers like spider webs) significantly different than those produced by other technologies.

Flash spinning is more suitable for difficult-to-dissolve polymers such as polyolefins and high molecular weight polymers. The spinning temperature should be higher than the melting point of the polymer and the boiling point of the solvent to affect solvent evaporation prior to the collection of the polymer. Flash-spinning process does not produce fibrous webs consisting of completely nanofibers.

9.3.6 Bicomponent Spinning

Bicomponent spinning is a two-step process that involves spinning two polymers through the spinning die (which forms the bicomponent fiber with island-in-sea [IIS], side by side, sheath core, citrus, or segmented-pie structure) and removal of one polymer [123]. Although bicomponent fibers of different cross-sectional shapes and geometries with micrometer diameter can be produced with the existing fiber-forming techniques, fabricating smaller diameters especially in nanometers is a real challenge.

The production of webs of IIS structure (Nylon 6 island and PLA sea) by spun bonding process and subsequent removal of sea for the production of micro- and nanofibers have been reported [124]. Hill Inc. produced nanofibers of 300-nm diameter from the IIS structure [125]. Lin *et al.* [126] fabricated side-by-side bicomponent nanofibers of elastomeric (polyurethanes) and thermoplastic (PAN) polymers using a microfluidic device as the spinneret in electrospinning. The silicone microfluidic spinneret consisted of three capillary channels: two for the inlet of polymer solutions and the other for the outlet. They observed self-crimping of PAN after the PU was removed from the bicomponent fiber by dissolving in tetrahydrofuran (THF).

Bicomponent spinning can be used for the fabrication of smaller nanofibers by sacrificing one of the polymer components as well as to create multicomponent nanofibers. Several research studies of bicomponent polymeric nanofibers of sheath-core structure by electrospinning process using a coaxial two-capillary spinneret have been reported [127–129]. The use of melt coaxial electrospinning for the fabrication of core-shell nanofibers having potential for temperature sensors [130] and composites based on phase change materials (PCMs) [41] has been investigated. The segmented-pie structure forms micro- and nanofibers (diameters 500 nm to 2 μm) with noncircular cross section. Recently, a new modified coaxial electrospinning process has been developed to prepare polymer fibers from high-concentration solutions of PVP [131]. This process involved a pure solvent concentrically surrounding polymer fluid in the spinneret and was able to produce fibers with a smooth surface morphology and good structural uniformity.

9.3.7 Other Approaches

In addition to the above-mentioned techniques, several other innovative methods such as template synthesis, self-assembly, phase separation, and drawing have been reported for nanofiber fabrication [3]. In template synthesis, nanofibers of polymers, metals, semiconductors, and ceramics are formed within the numerous cylindrical pores of a nonporous membrane (5–50 mm thickness) by oxidative polymerization accomplished electrochemically or chemically. In electrochemical synthesis, one surface of a membrane is coated with a metal film that works as an anode for the polymer, whereas in chemical synthesis, the membrane is immersed in a solution of the monomer and its oxidizing agent. The template synthesis process has been used to prepare nanofibers of PAN, PCL, polyaniline, CNF, polypyrrole, and poly(3-methylthiophene) [132].

Self-assembly is a manufacturing method where small molecules are used as basic building blocks, which add up to give nanofibers [133]. The small molecules are arranged in concentric manner, which upon extension in a normal plane produces the longitudinal axis of the nanofibers. In self-assembly, the final (desired) structure is 'encoded' in the shape from small blocks, as compared to traditional techniques (such as lithography) where the desired structure must be carved out from a large block of matter. Self-assembly is thus referred to as a 'bottom-up' manufacturing technique, whereas lithography is a 'top-down' technique. The synthesis of molecules for self-assembly often involves a chemical process called convergent synthesis. This process requires standard laboratory equipment and is limited to specific polymers.

In self-assembly, the shape and properties of nanofibers depend on the molecules and the intermolecular forces that bring the molecules together. Nanofibers of various polymeric configurations such as diblock copolymers, triblock copolymers, triblock polymers (of peptide amphiphile and dendrimers), and bolaform (of glucosamide and its deacetylated derivatives) can be assembled by this process.

In phase separation, the gel of a polymer is prepared by storing the homogeneous solution of the polymer at required concentration in a refrigerator set at the gelation temperature [134]. The gel is then immersed in distilled water for solvent exchange followed by the removal from the distilled water, blotting with filter paper and finally transferring to a freeze-drying vessel leading to a nanofiber matrix. The phase-separation process was used for the fabrication of nanofiber matrices of poly-L-lactic acid and blends of poly-L-lactic acid–polycaprolactone [135]. Although the phase-separation process is very simple, it is only limited to the laboratory scale.

In drawing process [136, 137] a millimetric droplet of a solution is allowed to evaporate after it is deposited on a silicon dioxide (SiO_2) surface. The droplet becomes more concentrated at the edge because of evaporation due to capillary flow. A micropipette is dipped into the droplet near the contact line with the surface and then withdrawn at a speed of 100 μm/s, resulting in a nanofiber being pulled out. The pulled fiber is then deposited on another surface by touching it with the end of the micropipette. From each droplet, nanofibers can be drawn for several times. Nanofibers of sodium citrate were formed by dissolving it in chloroauric acid through the drawing process [138]. The drawing process is suitable for viscoelastic materials, which can undergo strong deformations while being cohesive enough to support the stresses developed during pulling. This process is simple but limited to laboratory scale as nanofibers are formed one by one.

Reneker *et al.* [139] fabricated nanofibers by using pressurized gas where an expanding gas jet supplied the mechanical force required to create nanofibers. In this process, a polymeric solution or melt is fed into an annular column having an exit orifice and subjected to the action of a gas jet that pushes the material through the orifice forming the fibers. After the fibers are ejected from the orifice, solidification can occur in many ways such as cooling, chemical reaction, coalescence, or solvent removal. The polymers used in this process include nylon, polyolefins, PA, polyesters, and fluoropolymers. Various factors affecting the fiber diameter are the temperature of the gas jet, flow rate of the gas, and flow rate of the polymeric material.

Recently, nanofibers of wide range of materials were fabricated by a new process called force spinning [140, 141]. In this process the electric field of electrospinning is replaced by centrifugal force. The process involves heating a fiber-forming material in a heated structure and rotating the heated structure (with at least one nozzle) at very high speed to extrude nanofibers of the material. Rotational speed of the heated structure, nozzle configuration, collection system, and temperature are the key factors governing the geometry and morphology of the nanofibers. The limitations of electrospinning process such as very high electric field, low productivity, and high cost of production are eliminated in force spinning. Force spinning also broadens the selection of materials as both conductive and nonconductive materials can be spun into nanofibers. Several solid materials can be melted and directly spun into nanofibers without any chemical preparation. In addition, the process is free from extra process of solvent recovery as no solvent is used.

A new jet-blowing technique was used for the fabrication of micro- and nanofibers of polymers with high melt viscosity such as PTFE and PMMA [142]. In the process, the mixture of a polymer and a pressurized gas was blown through the aperture of a nozzle having two segments with different diameter. Polymeric fibers having a diameter in the range of 10 nm to 50 μm were produced by this method.

Huang [143] produced nanofibers by melt spinning where a polymeric melt of at least one thermoplastic polymer was supplied to the inner surface of a heated rotating distribution disc having a forward fiber discharge edge. The melt is then distributed into a thin film and attenuated by hot gas to produce polymeric nanofibers. The process of nanofiber (diameter less than 100 nm) production by rapid expansion of a supercritical solution into a liquid solvent (RESSLS) was developed by Meziani *et al.* [144]. The RESSLS process is the modification of the traditional rapid expansion of a supercritical solution (RESS) process used for the production of polymeric particles and fibers [145–147].

9.4 Comparisons of Different Processes

Table 9.5 summarizes the relative merits and demerits of various processes employed for nanofiber fabrication. The table also highlights the potential for scale-up, repeatability, and the ease of control of fiber dimensions by these processes.

Table 9.4 Other approaches for nanofiber production (the table has been arranged by the author).

Process	Features	Advantages	Disadvantages
1. Flash spinning [122]	Polymers used: PP Fiber diameter: 500 nm	High production rate	Short fiber length (3 to 10 μm)
2. Drawing [136, 137]	Polymers used: polytrimethylene terephthalate Fiber diameter: 60 nm	Simple and one-step process, longer fibers	Lower productivity, nonuniform fiber size
3. Application of pressurized gas [139]	Polymers used: nylon, polyolefins, polyimides, polyesters, fluoropolymers; and spinnable fluids include molten glassy materials, molten pitch, polymeric melts, polymers that are precursors to ceramics Fiber diameter: less than 300 nm	Versatile process for many spinnable fluids, polymeric melts and solutions for nanofiber production	Complex machines and equipment necessary
4. Force spinning [140, 141]	Polymers used: PEO, PLA, bismuth, PP, PS, acrylonitrile-butadiene-styrene, and polyvinyl pyrrolidone	Simple process, free from high electric field and solvent and high production rates	Sometimes heating to very high temperature is necessary

5. Jet-blowing [142]	Polymers used: PTFE, polyaramide, PMMA, organic polymers, and their blends Fiber diameter: 10 nm to 50 μm	Environmental advantages as it is nontoxic, uses chemically inert gas and no deleterious solvents are used Superior fiber properties Formation of composites	Complex process
6. Melt-spinning [143]	Polymers used: polyolefins, polyesters, PA, vinyl polymers, polystyrene-based polymers, biopolymers, polycarbonates, cellulose esters, acrylics, fluoropolymers, and chlorinated polyethers	Simple and versatile technique suitable for many polymers	Complex setup

Table 9.5 Comparison of various nanofiber fabrication techniques (the table has been arranged by the author).

Manufacturing process	Scope for scaling-up	Repeatability	Control on fiber dimension	Advantages	Disadvantages
Electrospinning (solution)	Yes	Yes	Yes	Long and continuous fibers	Solvent recovery issues, low productivity, jet instability
Electrospinning (melt)	Yes	Yes	Yes	Long and continuous fibers	Thermal degradation of polymers, electric discharge problem
Melt blowing	Yes	Yes	Yes	Long and continuous fibers, high productivity, free from solvent recovery issues	Polymer limitations, thermal degradation of polymers
Template synthesis	No	Yes	Yes	Easy to change diameter by using different templates	Complex process
Drawing	No	Yes	No	Simple process	Discontinuous process
Phase separation	No	Yes	No	Simple equipment required	Only works with selective polymers
Self-assembly	No	Yes	No	Easy to get smaller nanofibers	Complex process
Force spinning	Yes	Yes	Yes	Free from very high voltage, eco-friendly	Requirement of high temperature at times

9.5 Conclusions

Over the last two decades, there has been tremendous growth in the fabrication techniques for nanofibrous materials because of their unique features and many useful applications. Many of the nanofibers are being widely used for the fabrication of smart textiles, which can sense stimuli from the environment, react to them, and adapt to them by integration of functionalities in the textile structure. Nanofibers are increasingly used in smart textiles in numerous applications such as sensing, self-cleaning, tissue engineering, communication, wearable computers, and medicine. There are several techniques such as electrospinning, melt blowing, bicomponent spinning, force spinning, and flash spinning that can be used for the fabrication of nanofibers for smart textile applications.

This chapter has discussed about the production methods used to fabricate nanofibers for various smart textile applications. Among the several processes, electrospinning is the most popular technique used from producing nanofibers. Other techniques are gaining wider attention for the research community in various countries. However, fundamental analysis of these fabrication techniques is needed to develop nanofibers with desired properties on a commercial scale. Some of the techniques are still in their infancy, and much research is required for standardization and commercialization. Although electrospinning is so far the only method with potential for commercial production, the major issue yet to be resolved is how to substantially scale up the production to match the demands from a range of potential markets. In the future, there will be wider applications of nanofibers in smart textiles due to their superior properties and improvements in the production methods for easier fabrication.

References

1. Ramakrishna S., *An introduction to electrospinning and nanofibers*, World Scientific. Pub. Co. Inc., 2005.
2. Huang Z., Zhang Y., Kotaki M., Ramakrishna S., A review on polymer nanofibers by electrospinning and their applications in nanocomposites. *Composites Science and Technology*, 63, 2223–2253, 2003.
3. Nayak R., Padhye R., Kyratzis I.L., Truong Y.B., Arnold L., Recent advances in nanofibre fabrication techniques. *Textile Research Journal*, 82, 129–147, 2012.
4. Feng C., Khulbe K., Matsuura T., Tabe S., Ismail A., Preparation and characterization of electro-spun nanofiber membranes and their possible applications in water treatment. *Separation and Purification Technology*, 102, 118–135, 2013.

5. Huang L., Arena J.T., Manickam S.S., Jiang X., Willis B.G., McCutcheon J.R., Improved mechanical properties and hydrophilicity of electrospun nanofiber membranes for filtration applications by dopamine modification. *Journal of Membrane Science*, 460, 241–249, 2014.
6. Nayak R., Padhye R., Arnold L., Islam S., Production of novel surfaces by electrospinning. *Acta Universitatis Cibiniensis*, 58, 128–138, 2011.
7. Nayak R., Kyratzis I.L., Truong Y.B., Padhye R., Arnold L., Structural and mechanical properties of polypropylene nanofibres fabricated by meltblowing. *The Journal of the Textile Institute*, 106, 629–640, 2014.
8. Senić Ž., Bauk S., Vitorović-Todorović M., Pajić N., Samolov A., Rajić D., Application of TiO_2 nanoparticles for obtaining self-decontaminating smart textiles. *Scientific Technical Review*, 61, 63–72, 2011.
9. Gheibi A., Bagherzadeh R., Merati A.A., Latifi M., Electrical power generation from piezoelectric electrospun nanofibers membranes: Electrospinning parameters optimization and effect of membranes thickness on output electrical voltage. *Journal of Polymer Research*, 21, 571, 2014.
10. Weng W., Chen P., He S., Sun X., Peng H., Smart electronic textiles. *Angewandte Chemie International Edition*, 55, 6140–6169, 2016.
11. Zhang F., Zhang Z., Liu Y., Lu H., Leng J., The quintuple-shape memory effect in electrospun nanofiber membranes. *Smart Materials and Structures*, 22, 085020, 2013.
12. Bi K., Weathers A., Matsushita S., Pettes M.T., Goh M., Akagi K., Shi L., Iodine doping effects on the lattice thermal conductivity of oxidized polyacetylene nanofibers. *Journal of Applied Physics*, 114, 194302, 2013.
13. Das T.K., Prusty S., Review on conducting polymers and their applications. *Polymer-Plastics Technology and Engineering*, 51, 1487–1500, 2012.
14. Ghenaatian H., Mousavi M., Rahmanifar M., High performance hybrid supercapacitor based on two nanostructured conducting polymers: Self-doped polyaniline and polypyrrole nanofibers. *Electrochimica Acta*, 78, 212–222, 2012.
15. Yanilmaz M., Kalaoglu F., Karakas H., Sarac A.S., Preparation and characterization of electrospun polyurethane–polypyrrole nanofibers and films. *Journal of Applied Polymer Science*, 125, 4100–4108, 2012.
16. Kulkarni S.B., Patil U.M., Shackery I., Sohn J.S., Lee S., Park B., Jun S., High-performance supercapacitor electrode based on a polyaniline nanofibers/3D graphene framework as an efficient charge transporter. *Journal of Materials Chemistry A*, 2, 4989–4998, 2014.
17. Gao Z., Yang W., Wang J., Yan H., Yao Y., Ma J., Wang B., Zhang M., Liu L., Electrochemical synthesis of layer-by-layer reduced graphene oxide sheets/polyaniline nanofibers composite and its electrochemical performance. *Electrochimica Acta*, 91, 185–194, 2013.
18. Formalas A., 1975504 UP ed, 1934.
19. Pantano C., Gañán-Calvo A., Barrero A., Zeroth-order, electrohydrostatic solution for electrospraying in cone-jet mode. *Journal of Aerosol Science*, 25, 1065–1077, 1994.

20. Taylor G., Electrically driven jets. *Proceedings of the Royal Society of London A Mathematical and Physical Sciences*, 313, 453–475, 1969.
21. Liu Y., Deng R., Hao M., Yan H., Yang W., Orthogonal design study on factors effecting on fibers diameter of melt electrospinning. *Polymer Engineering & Science*, 50, 2074–2078, 2010.
22. Watanabe K., Kim B.S., Kim I.S., Development of polypropylene nanofiber production system. *Polymer Reviews*, 51, 288–308, 2011.
23. Bhattarai N., Edmondson D., Veiseh O., Matsen F.A., Zhang M., Electrospun chitosan-based nanofibers and their cellular compatibility. *Biomaterials*, 26, 6176–6184, 2005.
24. Choi J.S., Lee S.W., Jeong L., Bae S.-H., Min B.C., Youk J.H., Park W.H., Effect of organosoluble salts on the nanofibrous structure of electrospun poly (3-hydroxybutyrate-co-3-hydroxyvalerate). *International Journal of Biological Macromolecules*, 34, 249–256, 2004.
25. Min B.M., Lee G., Kim S.H., Nam Y.S., Lee T.S., Park W.H., Electrospinning of silk fibroin nanofibers and its effect on the adhesion and spreading of normal human keratinocytes and fibroblasts *in vitro*. *Biomaterials*, 25, 1289–1297, 2004.
26. Dalton P., Joergensen N., Groll J., Moeller M., Patterned melt electrospun substrates for tissue engineering. *Biomedical Materials*, 3, 034109, 2008.
27. Chen C., Wang L., Huang Y., Electrospinning of thermo-regulating ultrafine fibers based on polyethylene glycol/cellulose acetate composite. *Polymer*, 48, 5202–5207, 2007.
28. Dalton P., Grafahrend D., Klinkhammer K., Klee D., Möller M., Electrospinning of polymer melts: Phenomenological observations. *Polymer*, 48, 6823–6833, 2007.
29. Larrondo L., Manley SJ., *Journal of Polymer Science: Polymer Physics Ed 19*, 909, 1981.
30. Zhou H., Green T., Joo Y., The thermal effects on electrospinning of polylactic acid melts. *Polymer*, 47, 7497–7505, 2006.
31. Khurana H., Patra P., Warner S., Nanofibers from melt electrospinning. *Polymer Preprints*, 44, 67, 2003.
32. Ogata N., Yamaguchi S., Shimada N., Lu G., Iwata T., Nakane K., Ogihara T., Poly(lactide) nanofibers produced by a melt electrospinning system with a laser melting device. *Journal of Applied Polymer Science*, 104, 1640–1645, 2007.
33. Ogata N., Shimada N., Yamaguchi S., Nakane K., Ogihara T., Melt electrospinning of poly (ethylene terephthalate) and polyalirate. *Journal of Applied Polymer Science*, 105, 1127–1132, 2007.
34. Lyons J., Li C., Ko F., Melt-electrospinning part I: Processing parameters and geometric properties. *Polymer*, 45, 7597–7603, 2004.
35. Dalton P., Klinkhammer K., Salber J., Klee D., Möller M., Direct *in vitro* electrospinning with polymer melts. *Biomacromolecules*, 7, 686–690, 2006.
36. Lyons J., Pastore C., Ko F., Developments in melt-electrospinning of thermoplastic polymers. *Polymer Preprints*, 44, 122–123, 2003.

37. Nayak R., Padhye R., Kyratzis I.L., Truong Y.B., Arnold L., Effect of viscosity and electrical conductivity on the morphology and fibre diameter in melt electrospinning of polypropylene. *Textile Research Journal*, 83, 606–617, 2013.
38. Lyons J.M., *Melt-electrospinning of thermoplastic polymers: An experimental and theoretical analysis*, Drexel University, 2004.
39. Nayak R., Kyratzis L., Truong Y., Padhye R., Arnold L., Peeters G., O'Shea M., Fabrication of submicron fibres by meltblowing and melt electrospinning. In *ICNFA 2011: 2nd International Conference on Nanotechnology: Fundamentals and Applications*. ASET Inc, 338-331–338-337, 2011.
40. Dalton P., Lleixà Calvet J., Mourran A., Klee D., Möller M., Melt electrospinning of poly (ethylene glycol block caprolactone). *Biotechnology Journal*, 1, 998–1006, 2006.
41. McCann J., Marquez M., Xia Y., Melt coaxial electrospinning: A versatile method for the encapsulation of solid materials and fabrication of phase change nanofibers. *Nano Letters*, 6, 2868–2872, 2006.
42. Jager E.W., Smela E., Inganäs O., Microfabricating conjugated polymer actuators. *Science*, 290, 1540–1545, 2000.
43. Kaynak A., Unsworth J., Clout R., Mohan A.S., Beard G.E., A study of microwave transmission, reflection, absorption, and shielding effectiveness of conducting polypyrrole films. *Journal of Applied Polymer Science*, 54, 269–278, 1994.
44. Merlini C., Barra G., Araujo T.M., Pegoretti A., Electrically pressure sensitive poly (vinylidene fluoride)/polypyrrole electrospun mats. *RSC Advances*, 4, 15749–15758, 2014.
45. Norris I.D., Shaker M.M., Ko F.K., MacDiarmid A.G., Electrostatic fabrication of ultrafine conducting fibers: Polyaniline/polyethylene oxide blends. *Synthetic Metals*, 114, 109–114, 2000.
46. MacDiarmid A., Jones W., Norris I., Gao J., Johnson A., Pinto N., Hone J., Han B., Ko F., Okuzaki H., Electrostatically-generated nanofibers of electronic polymers. *Synthetic Metals*, 119, 27–30, 2001.
47. Cardenas J., De França M., De Vasconcelos E., De Azevedo W., da Silva E., Jr.., Growth of sub-micron fibres of pure polyaniline using the electrospinning technique. *Journal of Physics D: Applied Physics*, 40, 1068, 2007.
48. Kang T.S., Lee S.W., Joo J., Lee J.Y., Electrically conducting polypyrrole fibers spun by electrospinning. *Synthetic Metals*, 153, 61–64, 2005.
49. Chronakis I.S., Grapenson S., Jakob A., Conductive polypyrrole nanofibers via electrospinning: Electrical and morphological properties. *Polymer*, 47, 1597–1603, 2006.
50. Laforgue A., Robitaille L., Fabrication of poly-3-hexylthiophene/polyethylene oxide nanofibers using electrospinning. *Synthetic Metals*, 158, 577–584, 2008.
51. Shin M.K., Kim Y.J., Kim S.I., Kim S.-K., Lee H., Spinks G.M., Kim S.J., Enhanced conductivity of aligned PANi/PEO/MWNT nanofibers by electrospinning. *Sensors and Actuators B: Chemical*, 134, 122–126, 2008.

52. Liu H., Kameoka J., Czaplewski D.A., Craighead H., Polymeric nanowire chemical sensor. *Nano Letters*, 4, 671–675, 2004.
53. Huang J., Virji S., Weiller B.H., Kaner R.B., Polyaniline nanofibers: facile synthesis and chemical sensors. *Journal of the American Chemical Society*, 125, 314–315, 2003.
54. Pinto N., Johnson A., MacDiarmid A., Mueller C., Theofylaktos N., Robinson D., Miranda F., Electrospun polyaniline/polyethylene oxide nanofiber field-effect transistor. *Applied Physics Letters*, 83, 4244–4246, 2003.
55. Liu H., Reccius C.H., Craighead H., Single electrospun regioregular poly (3-hexylthiophene) nanofiber field-effect transistor. *Applied Physics Letters*, 87, 253106, 2005.
56. Jung Y., Kim J., Chun B., Chung Y., Cho J., Shape memory polyurethane nanofibers. *Quality Textiles for Quality Life Vol. 1*, 4, 43–46, 2004.
57. Zhuo H., Hu J., Chen S., Yeung L., Preparation of polyurethane nanofibers by electrospinning. *Journal of Applied Polymer Science*, 109, 406–411, 2008.
58. Penner R.M., Martin C.R., Electrochemical investigations of electronically conductive polymers. 2. Evaluation of charge-transport rates in polypyrrole using an alternating current impedance method. *The Journal of Physical Chemistry*, 93, 984–989, 1989.
59. Han G., Shi G., Novel route to pure and composite fibers of polypyrrole. *Journal of Applied Polymer Science*, 103, 1490–1494, 2007.
60 Tavakkol E., Tavanai H., Abdolmaleki A., Morshed M., Production of conductive electrospun polypyrrole/poly (vinyl pyrrolidone) nanofibers. *Synthetic Metals*, 1, 231, 95–106, 2017.
61. Deng L., Young R.J., Kinloch I.A., Zhu Y., Eichhorn S.J., Carbon nanofibres produced from electrospun cellulose nanofibres. *Carbon*, 58, 66–75, 2013.
62. Lu P., Hsieh Y.-L., Multiwalled carbon nanotube (MWCNT) reinforced cellulose fibers by electrospinning. *ACS Applied Materials & Interfaces*, 2, 2413–2420, 2010.
63. Ma Z., Kotaki M., Ramakrishna S., Electrospun cellulose nanofiber as affinity membrane. *Journal of Membrane Science*, 265, 115–123, 2005.
64. Wei M., Lee J., Kang B., Mead J., Preparation of core-sheath nanofibers from conducting polymer blends. *Macromolecular Rapid Communications*, 26, 1127–1132, 2005.
65. Yoshino T., Mamiya J.-I., Kinoshita M., Ikeda T., Yu Y., Preparation and characterization of crosslinked azobenzene liquid-crystalline polymer fibers. *Molecular Crystals and Liquid Crystals*, 478, 233/[989]–243/[999], 2007.
66. MacDiarmid A.G., "Synthetic metals": A novel role for organic polymers (Nobel lecture). *Angewandte Chemie International Edition*, 40, 2581–2590, 2001.
67. El-Aufy A., Doctor of Philosophy Thesis. Drexel University, 2004.
68. Vaseashta A., Controlled formation of multiple Taylor cones in electrospinning process. *Applied Physics Letters*, 90, 093115, 2009.
69. Paruchuri S., Brenner M., Splitting of a liquid jet. *Physical Review Letters*, 98, 134502, 2007.

70. Tomaszewski W., Szadkowski M., Investigation of electrospinning with the use of a multi-jet electrospinning head. *Fibres and Textiles in Eastern Europe*, 13, 22, 2005.
71. Bocanegra R., Galán D., Márquez M., Loscertales I., Barrero A., Multiple electrosprays emitted from an array of holes. *Journal of Aerosol Science*, 36, 1387–1399, 2005.
72. Yarin A., Zussman E., Upward needleless electrospinning of multiple nanofibers. *Polymer*, 45, 2977–2980, 2004.
73. Fujihara K., Kotaki M., Ramakrishna S., Guided bone regeneration membrane made of polycaprolactone/calcium carbonate composite nano-fibers. *Biomaterials*, 26, 4139–4147, 2005.
74. Yamashita Y., Ko F., Tanaka A., Miyake H., Characteristics of elastomeric nanofiber membranes produced by electrospinning. *Journal of Textile Engineering*, 53, 137–142, 2007.
75. Vaseashta A., *Applied Physics Letters*, 90, 093115, 2007.
76. Burger C., Hsiao B.S., Chu B., Nanofibrous materials and their applications. *Materials Research*, 36, 333, 2006.
77. Huebner A., Chu H., Instability and breakup of charged liquid jets. *Journal of Fluid Mechanics*, 49, 361–372, 1971.
78. Yarin A., Kataphinan W., Reneker D., Branching in electrospinning of nanofibers. *Journal of Applied Physics*, 98, 064501, 2005.
79. Koombhongse S., Liu W., Reneker D.H., Flat polymer ribbons and other shapes by electrospinning. *Journal of Polymer Science Part B: Polymer Physics*, 39, 2598–2606, 2001.
80. Teo W., Gopal R., Ramaseshan R., Fujihara K., Ramakrishna S., A dynamic liquid support system for continuous electrospun yarn fabrication. *Polymer*, 48, 3400–3405, 2007.
81. Regele J., Papac M., Rickard M., Dunn-Rankin D., Effects of capillary spacing on EHD spraying from an array of cone jets. *Journal of Aerosol Science*, 33, 1471–1479, 2002.
82. Lee J.R., Jee S.Y., Kim H.J., Hong Y.T., S S.K., KR Patent WO2006123879, 2006.
83. Kim G.H., Cho Y.S., Kim W.D., Stability analysis for multi-jets electrospinning process modified with a cylindrical electrode. *European Polymer Journal*, 42, 2031–2038, 2006.
84. Bowman J., Taylor M., Sharma V., Lynch A., Chadha S., *Multispinneret methodologies for high throughput electrospun nanofiber*. Warrendale, Pa. Materials Research Society, 15–20, 2003.
85. Ding B., Kimura E., Sato T., Fujita S., Shiratori S., Fabrication of blend biodegradable nanofibrous nonwoven mats via multi-jet electrospinning. *Polymer*, 45, 1895–1902, 2004.
86. Theron S., Yarin A., Zussman E., Kroll E., Multiple jets in electrospinning: Experiment and modeling. *Polymer*, 46, 2889–2899, 2005.
87. Yang Y., Jia Z., Li Q., Hou L., Liu J., Wang L., Guan Z., Zahn M., A shield ring enhanced equilateral hexagon distributed multi-needle electrospinning

spinneret. *Dielectrics and Electrical Insulation, IEEE Transactions on*, 17, 1592–1601, 2010.
88. Jirsak O., Sanetrnik F., Lukas D., Kotek V., Martinova L., Chaloupek J., Method of nanofibres production from a polymer solution using electrostatic spinning and a device for carrying out the method. Google Patents, 2009.
89. Niu H., Lin T., Wang X., Needleless electrospinning. I. A comparison of cylinder and disk nozzles. *Journal of Applied Polymer Science*, 114, 3524–3530, 2009.
90. Wang X., Niu H., Lin T., Needleless electrospinning of nanofibers with a conical wire coil. *Polymer Engineering and Science*, 49, 1582–1586, 2009.
91. Kumar A., Wei M., Barry C., Chen J., Mead J., Controlling fiber repulsion in multijet electrospinning for higher throughput. *Macromolecular Materials and Engineering*, 295, 701–708, 2010.
92. Liu Y., He J., Bubble electrospinning for mass production of nanofibers. *International Journal of Nonlinear Sciences and Numerical Simulation*, 8, 393, 2007.
93. Li D., Xia Y., Electrospinning of nanofibers: Reinventing the wheel? *Advanced Materials*, 16, 1151–1170, 2004.
94. Liu Y., He J.H., Yu J.Y., Bubble-electrospinning: a novel method for making nanofibers. In. IOP Publishing, 012001, 2008.
95. Liu Y., Ren Z.F., He J.H., Bubble electrospinning method for preparation of aligned nanofibre mat. *Materials Science and Technology*, 26, 1309–1312, 2010.
96. Liu Y., He J.H., Xu L., Yu J.Y., The principle of bubble electrospinning and its experimental verification. *Journal of Polymer Engineering*, 28, 55–65, 2008.
97. Wang X., Um I., Fang D., Okamoto A., Hsiao B., Chu B., Formation of water-resistant hyaluronic acid nanofibers by blowing-assisted electro-spinning and non-toxic post treatments. *Polymer*, 46, 4853–4867, 2005.
98. Kim Y.M., Ahn K.R., Sung Y.B., Jang R.S., Manufacturing device and the method of preparing for the nanofibres via electro-blown spinning process US Patent 7,618,579, 2009.
99. Arora P., Chen G., Frisk S., Graham D., Marin R., Suh H., Nanowebs. Google Patents, 2009.
100. Varabhas J., Chase G., Reneker D., Electrospun nanofibers from a porous hollow tube. *Polymer*, 49, 4226–4229, 2008.
101. Srivastava Y., Marquez M., Thorsen T., Multijet electrospinning of conducting nanofibers from microfluidic manifolds. *Journal of Applied Polymer Science*, 106, 3171–3178, 2007.
102. Srivastava Y., Loscertales I., Marquez M., Thorsen T., Electrospinning of hollow and core/sheath nanofibers using a microfluidic manifold. *Microfluidics and Nanofluidics*, 4, 245–250, 2008.
103. Srivastava Y., Marquez M., Thorsen T., Microfluidic electrospinning of biphasic nanofibers with Janus morphology. *Biomicrofluidics*, 3, 012801, 2009.

104. Simm W., Gösling C., Bonart R., Falkai B., Fibre fleece of electrostatically spun fibres and methods of making the same, USP 4143196. USA: Bayer Aktiengesellschaft, Leverkusen, Fed. Rep. of Germany, 1979.
105. Jirsak O., Sanetrnik F., Lukas D., Kotek V., Martinova L., Chaloupek J., WO2005-024101. *Patent Czech Republic*, vol. 4, 2005.
106. Lukas D., Sarkar A., Pokorny P., Self organisation of jets in electrospinning from free liquid surface-a generalized approach. *Journal of Applied Physics*, 103, 084309, 2008.
107. Cengiz F., Jirsak O., The effect of salt on the roller electrospinning of polyurethane nanofibers. *Fibers and Polymers*, 10, 177–184, 2009.
108. Cengiz F., Dao TA., Jirsak O., Influence of solution properties on the roller electrospinning of poly (vinyl alcohol). *Polymer Engineering & Science*, 50, 936–943, 2010.
109. Liu G., Qiao L., Guo A., Diblock copolymer nanofibers. *Macromolecules*, 29, 5508–5510, 1996.
110. Nayak R., *Fabrication and characterisation of polypropylene nanofibres by melt electrospinning and meltblowing*, RMIT University, 2012.
111. Nayak R., *Polypropylene nanofibers: Melt electrospinning versus meltblowing*, Springer, 2017.
112. Nayak R., Kyratzis I.L., Truong Y.B., Padhye R., Arnold L., Peeters G., O'Shea M., Nichols L., Fabrication and characterisation of polypropylene nanofibres by meltblowing process using different fluids. *Journal of Materials Science*, 48, 273–281, 2013.
113. Ellison C., Phatak A., Giles D., Macosko C., Bates F., Melt blown nanofibers: Fiber diameter distributions and onset of fiber breakup. *Polymer*, 48, 3306–3316, 2007.
114. Podgórski A., Balazy A., Gradon L., Application of nanofibers to improve the filtration efficiency of the most penetrating aerosol particles in fibrous filters. *Chemical Engineering Science*, 61, 6804–6815, 2006.
115. Bodaghi H., Sinangil M., Meltblown nonwoven webs including nanofibres and apparatus and method for forming such meltblown nonwoven webs. Patent U ed. USA, 2009.
116. Willkie A., Haggard J., Nanofiber Meltblown Nonwovens a New Low. *International Fibre Journal*, 22, 48–49, 2007.
117. Ji F., Zhu Y., Hu J., Liu Y., Yeung L-Y., Ye G., Smart polymer fibers with shape memory effect. *Smart Materials and Structures*, 15, 1547, 2006.
118. Kaursoin J., Agrawal A.K., Melt spun thermoresponsive shape memory fibers based on polyurethanes: Effect of drawing and heat-setting on fiber morphology and properties. *Journal of Applied Polymer Science*, 103, 2172–2182, 2007.
119. Meng Q., Hu J., Zhu Y., Lu J., Liu Y., Morphology, phase separation, thermal and mechanical property differences of shape memory fibres prepared by different spinning methods. *Smart Materials and Structures*, 16, 1192, 2007.

120. Li H., Ke Y., Hu Y., Polymer nanofibers prepared by template melt extrusion. *Journal of Applied Polymer Science*, 99, 1018–1023, 2006.
121. Xia L., Xi P., Cheng B., A comparative study of UHMWPE fibers prepared by flash-spinning and gel-spinning. *Materials Letters*, 147, 79–81, 2015.
122. Weinberg M.G., Dee G.T., Harding T.W., Flash spun web containing submicron filaments and process for forming same, US Patent 0135020 A1, 2006.
123. Naeimirad M., Zadhoush A., Kotek R., Esmaeely Neisiany R., Nouri Khorasani S., Ramakrishna S., Recent advances in core/shell bicomponent fibers and nanofibers: A review. *Journal of Applied Polymer Science*, 135, 46265, 2018.
124. Fedorova N., Pourdeyhimi B., High strength nylon micro and nanofiber based nonwovens via spunbonding. *Journal of Applied Polymer Science*, 104, 3434–3442, 2007.
125. Alcantar N., Aydil E., Israelachvili J., Polyethylene glycol-coated biocompatible surfaces. *Journal of Biomedical Materials Research Part A*, 51, 343–351, 2000.
126. Lin T., Wang H., Wang X., Self crimping bicomponent nanofibers electrospun from polyacrylonitrile and elastomeric polyurethane. *Advanced Materials*, 17, 2699–2703, 2005.
127. Sun Z., Zussman E., Yarin A.L., Wendorff J.H., Greiner A., Compound core–shell polymer nanofibers by co electrospinning. *Advanced Materials*, 15, 1929–1932, 2003.
128. Zhang Y., Huang Z.M., Xu X., Lim C.T., Ramakrishna S., Preparation of core-shell structured PCL-r-gelatin bi-component nanofibers by coaxial electrospinning. *Chem Mater.*, 16, 3406–3409, 2004.
129. Li D., Xia Y., Direct fabrication of composite and ceramic hollow nanofibers by electrospinning. *Nano Letters*, 4, 933–938, 2004.
130. Li F., Zhao Y., Wang S., Han D., Jiang L., Song Y., Thermochromic core–shell nanofibers fabricated by melt coaxial electrospinning. *Journal of Applied Polymer Science*, 112, 269–274, 2009.
131. Yu D.G., Branford-White C., White K., A modified coaxial electrospinning for preparing fibers from a high concentration polymer solution. *Express Polymer Letters*, 5, 732–741, 2011.
132. Song L.-T., Wu Z.-Y., Liang H.-W., Zhou F., Yu Z.-Y., Xu L., Pan Z., Yu S.-H., Macroscopic-scale synthesis of nitrogen-doped carbon nanofiber aerogels by template-directed hydrothermal carbonization of nitrogen-containing carbohydrates. *Nano Energy*, 19, 117–127, 2016.
133. Hartgerink J.D., Beniash E., Stupp S.I., Self-assembly and mineralization of peptide-amphiphile nanofibers. *Science*, 294, 1684, 2001.
134. Ma P.X., Zhang R., Synthetic nano-scale fibrous extracellular matrix. *Journal of Biomedical Materials Research*, 46, 60–72, 1999.
135. Mo X., Yang F., Teoh S., Seeram R., Studies on nanofiber formation from PLLA and PCL blends through phase separation. *2002 POLY Biennial: Polymeric Nanomaterials, California, USA*, 2002.

136. Xing X., Wang Y., Li B., Nanofibers drawing and nanodevices assembly in poly (trimethylene terephthalate). *Optics Express*, 16, 10815–10822, 2008.
137. Tong L., Mazur E., Glass nanofibers for micro-and nano-scale photonic devices. *Journal of Non-Crystalline Solids*, 354, 1240–1244, 2008.
138. Ondarcuhu T., Joachim C., Drawing a single nanofibre over hundreds of microns. *EPL (Europhysics Letters)*, 42, 215, 1998.
139. Reneker D.H., Chun I., Ertley D., Process and apparatus for the production of nanofibres, US Patent 6382526 B1, 2002.
140. Lozano K., Sarkar K., Methods and apparatuses for making superfine fibers. Google Patents, 2009.
141. Lozano K., Sarkar K., Superfine fiber creating spinneret and uses thereof. Google Patents, 2009.
142. Sen A., Bedding J., Gu B., Process for forming polymeric micro and nanofibers. Google Patents, 2004.
143. Huang T., Production of Nanofibres by Melt Spinning, US Patent 0242171 A1, 2008.
144. Meziani M., Pathak P., Wang W., Desai T., Patil A., Sun Y., Polymeric nanofibers from rapid expansion of supercritical solution. *Industrial & Engineering Chemistry Research*, 44, 4594–4598, 2005.
145. Matson D., Fulton J., Petersen R., Smith R., Rapid expansion of supercritical fluid solutions: solute formation of powders, thin films, and fibers. *Industrial & Engineering Chemistry Research*, 26, 2298–2306, 1987.
146. Matson D., Petersen R., Smith R., Production of powders and films by the rapid expansion of supercritical solutions. *Journal of Materials Science*, 22, 1919–1928, 1987.
147. Petersen R., Matson D., Smith R., The formation of polymer fibers from the rapid expansion of supercritical fluid solutions. *Polymer Engineering & Science*, 27, 1693–1697, 1987.

10

Characterization Methods of Nanotechnology-Based Smart Textiles

Mamatha M. Pillai[1], R. Senthilkumar[2], R. Selvakumar[1] and Amitava Bhattacharyya[2*]

[1]Tissue Engineering Laboratory, PSG Institute of Advanced Studies, Coimbatore, India
[2]Nanoscience and Technology, Department of Electronics and Communication Engineering, PSG College of Technology, Coimbatore, India

Abstract
Research and development towards nanotechnology-based smart textiles have become a new trend in the field of wearable nanotechnology. Nanoparticles or nanomaterials are incorporated into textiles to enhance the efficacy for different smart applications in biomedical (drug delivery, wound healing, health monitoring, scaffolds and implants, etc.), defence (composites, camouflage, UV resistant, stain removal, water proof, etc.), and other high-performance sectors. The main challenge in this field is the precise control of size, shape and distribution of nanomaterials anchored with the textile substrate. Hence, suitable and extensive characterization techniques are essential for successful implementation of smart textiles. This chapter highlights different characterization techniques used to study the performance enhancements in nanotechnology-based smart textiles. The basic principle and applications of various characterization techniques such as particle size analysis, electron microscopy, X-ray diffraction, spectroscopy, etc. have been discussed in detail referring to their applications in such smart textile products.

Keywords: Smart textiles, nanotechnology, characterization, spectroscopy, electron microscopy, X-ray, particle size analyzer

**Corresponding author*: amitbha1912@gmail.com

Nazire D. Yilmaz (ed.) Smart Textiles, (347–378) © 2019 Scrivener Publishing LLC

10.1 Introduction

Nanotechnology is an interdisciplinary area having huge impact virtually in all fields of science and technology due to its potential vast range of applications. Chemical, thermal, mechanical and electrical properties of nanomaterials can be tailored for various applications in different fields. Nanomaterials and nanotechnology-based developments are attracting a great deal of attention from the textile researchers and industrialists. With the advent of nanoscience and nanotechnology, conventional and technical textile industries are experiencing change at a tremendous growth rate. Many chemical companies have started investing on nanotechnology-based textile processing chemicals due to the vast application possibilities of nanomaterials in textile field. Almost all textile products have significant impact of nanotechnology in terms of materials, processing or testing technologies. Nanofibers, nanocomposite fibers, nanocoated and nanofinished fabrics are commercialized not only in high-performance advanced applications but also in conventional textiles for their new functionality and improved performance. The use of nanoparticles in conventional textile processing techniques like dyeing, coating and finishing can impart new functionality as well as enhance the product performance manifold. The advanced processing techniques like sol–gel, nanoemulsion, layer-by-layer deposition, plasma polymerization, etc. can introduce multi-functionality with excellent durability and weather resistance functionalities to fabrics.

Chemical, physical and physiochemical modifications are the heart of a textile product irrespective of its application area. Fabrics with some exceptional functional properties like antibacterial, mosquito repellent, oil/water repellent, fire retardant, etc. are defined as high-performance or functional textiles. They should possess optimized material properties like color fastness, tear and rubbing strength, heat and cold resistance, etc. keeping the aesthetics of fabric. The nanomaterials and nanotechnology are gradually replacing the old conventional processing methods and materials. Nanotechnology may be defined as a technology dealing with dimensions of roughly 1 to 100 nm (1 nm = 10^{-9} m). A particle possessing a well-defined grain size of less than 100 nm in any one of its dimensions is called as a nanoparticle. Nanoparticles when dispersed and well separated in a medium lead to tremendous increase in the surface area, and most of their atoms are exposed to the surface. Thus, it is possible to observe size-dependant exceptional physical properties in nanoparticles of a single material [1]. The scope of nanotechnology is huge in the textile sector in terms of fiber or yarn manufacturing as well as fabric processing. Nanoparticles have higher surface area compared to larger particles.

They have better optical clarity and do not alter the color or brightness of the fabrics. Nanomaterials allow good breathability while maintaining the good hand and feel of the original material due to the limited use of the material to achieve the desired output. Nano-processed garments have such a thin layer on that it is difficult to detect with the naked eye; they stay bright and fresh looking and are more durable. Besides the conventional improvements in the functional properties, an emerging class of textiles, namely smart and intelligent textiles, is developing rapidly. They can sense external stimuli, react to them and help the fabric structure to adapt to the changes. The wearable electronics or e-textiles possess microelectronic sensors; these are a combination of intelligent, functional and fashionable textiles. The other responsive behaviors or the smart functionalities available in the market include odor release or odor prevention on demand, individually adjustable heat insulation, microcapsules and phase change materials, protection from environmental impacts like UV radiation, etc. So far, nanotechnology has found its application in various areas including biomedical and bio-based materials, electronics, human safety and environmental protection equipment, high-precision sensors, high-performance nanocomposites as well as consumer products [1, 2].

Perhaps, the textile and fiber industries were the first to have implemented nanotechnological advances successfully for consumer usage. Nanotechnology can also be used in textiles for fashion to bring novel, creative and thrilling properties. The applications and scope of nanotechnology in textiles are enormous ranging from fiber or yarn manufacturing to the development of fabric finishes. The application potentials are sky-high, some of which can be given as encapsulated textiles that release scent, vitamin or drugs with time, self-cleaning fabrics, garments with displays, bulletproof lightweight fabrics and color-change textiles. Till now, the nanotechnological research relevant to the conventional textile industry has mostly focused on using nanoparticles and creating nanostructures during fabric finishing processes to develop various functional properties on the textiles such as water repellent, soil resistant, anti-bacterial and anti-static, flame retardant, dyeable, etc. [3, 4]. Nanoparticles present very high surface areas; hence, they are able to introduce better functional properties at low loading amounts compared to the larger-sized particles. The color and brightness of the nano-processed fabric remain almost unaffected, and the structure becomes more compact, which allows good ventilation, enhanced breathability with sustained good hand feel. The quality of nano-processed products is far better than their conventional counterparts, as the garments stay bright, fresh looking, toxic free and more durable.

The nanofibers, nanocomposite fibers and garment-embedded printed electronics for various sensing and actuating purposes are the areas where nanotechnology is extensively utilized to develop high-performance smart and intelligent textiles [5]. The challenges against creating comfortable wearable computing devices are addressed by using nanotechnology. Carbon nanotube and graphene show prospect for fabrication of such devices [6, 7]. Functionalized carbon nanotubes attached on silicon surfaces in different patterns lead to fabricating wearable devices using silane coupling agents [7]. Such initiatives are the foundation of wearable nanotechnology-based electronic devices in the textile and garment industries. Studies in smart and intelligent textiles focus on wearable computing devices integrated in conventional textile structures. Fabric-based sensors to monitor gesture, posture or respiration have been investigated. These smart textiles can measure and monitor the physiological conditions of the wearer [8]. One such product was launched by a US Company named Sensoria Inc. The product is termed as Smart socks [9]. This can monitor the foot pressures at different location, temperature, speed, etc. and display those parameters in mobile applications (Apps). During intense activities, phase change and shape memory materials are used in some intelligent textiles for the sportsmen so that the body temperature does not increase beyond a certain limit [10].

As the properties of materials change dramatically when their sizes are reduced to the nanometer range, it is important to measure them accurately. However, measuring these nano dimensions is not an easy task. Characterization of these nano-sized materials or nano-finished products is challenging. In order to design a new nanomaterial for a specific application, it has to be analyzed and characterized via suitable techniques. Research activities in this area lead to discovery and invention of sophisticated nano characterization techniques. There are different techniques available such as spectroscopy techniques (Raman spectroscopy, Fourier transform infrared spectroscopy, UV–Vis spectroscopy, etc.), microscopic techniques (scanning or transmission electron microscopy, scanning probe microscopy), characterization methods using X-rays (X-ray diffraction, X-ray photoelectron spectroscopy) and other techniques like biocompatibility tests, surface area and porosity analyses, thermal characterization, electrical conductivity, barrier property measurements, etc. [3]. These methods help in determining the chemical composition, crystal, phase and defect structures, crystal or grain sizes, etc. These techniques are extensively used for characterization of nanoparticles and are explained in this chapter. The major characterization techniques used for nanotechnology-based smart textiles covering areas like nano-finishing, nano-coating, nanocomposites and nanofibers are discussed in the following sections with examples from suitable application areas.

10.2 Nanomaterial Characterization Using Spectroscopy

Common spectroscopy techniques include Raman, FTIR (Fourier transform infrared) and UV–Vis spectrums. These play a significant role in characterizing the nanoparticles for smart textile applications. While Raman is inelastic scattering due to Raman active bonds, FTIR spectroscopy suggests the emission or absorption due to the motion of asymmetric bonds. These two techniques are complementary to each other. UV–Vis spectroscopy is a fundamental and efficient technique to understand the size and shape of nanoparticles in colloidal suspensions [11].

10.2.1 Raman Spectroscopy

Raman spectroscopy is a scattering technique. Raman effect is based on the frequency of scattered radiation, which is different from the frequency of monochromatic incident radiation. Raman spectroscopy is used to determine molecular motions, especially the vibrational ones [11].

10.2.1.1 Principle

In Raman spectroscopy, monochromatic laser beam interacts with the molecules of a sample and originates a scattered light (Figure 10.1). The scattered light frequency differs from the incident light (inelastic scattering) and is used to construct a Raman spectrum. Raman spectra arise due to inelastic collision between the incident monochromatic laser beam and the molecules

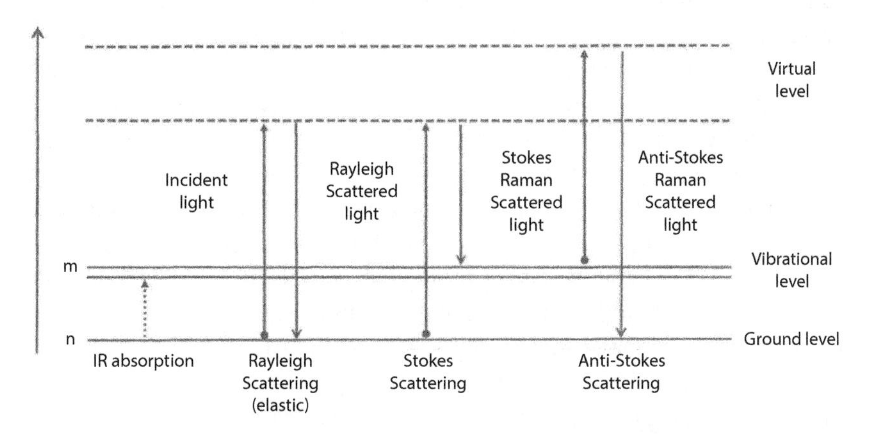

Figure 10.1 Scheme of IR absorption and different Raman scattering. Note: The image has been prepared by the authors.

of the sample. The monochromatic laser beam strikes and interacts with sample molecules, and it scatters in all directions. The scattering of the radiation gives information about the molecular structure. The frequency of scattered radiation and incident radiation is equal in Rayleigh scattering. A small fraction of the scattered radiation frequency and the frequency of the incident radiation is different from each other and constitutes Raman scattering. When the incident radiation frequency is higher than the scattered radiation frequency, Stokes lines appear in the Raman spectrum. But when the incident radiation frequency is lower than the scattered radiation frequency, anti-Stokes lines appear in the Raman spectrum. Scattered radiation is frequently measured at a right angle to the incident radiation. Due to the distribution of thermal energy, a small portion of the energy is transferred to the incident photon, and the molecules in the scattered photon have the higher-energy state. In order for a molecule to exhibit Raman effect, deformation change of the electron cloud or polarization change with respect to vibration is required. This will determine the Raman scattering intensity [11].

10.2.1.2 Applications

Raman spectrum gives information about the chemical bonds of the molecules; thus, it gives molecular fingerprint, which is in the range of 500–2000 cm^{-1}. It is also used to characterize the crystallographic orientation of the specimens [11].

Figure 10.2a demonstrates Raman shift of cotton fibers [12]. In this study, Raman shifts of the cotton fibers, with Au/TiO_2 and TiO_2, were studied and found that these are Raman-active materials. Carbon nanomaterials such as graphene, carbon nanotubes, carbon nanofibers or nanodiamonds exhibit different Raman spectra. The intensity and position of Raman bands give confirmatory evidence for the formation structures of these nanoparticles as well as their qualities following the synthesis process. In polymer nanocomposites, the strain on these nanoparticles can be quantified by using their peak intensities. Applying this technique, these can be accurately measured on a nanocomposite fiber.

Bhattacharyya *et al.* [13] studied the impact of a nanofiller (graphene) on the mechanical properties of an ultra-high molecular-weight polyethylene (UHMWPE) material for biomedical and defence applications. The synthesized UHMWPE films with and without the nanofillers were characterized using Raman spectroscopy (Figure 10.2b). They have discussed two different routes to produce the reduced graphene oxide-based nanocomposite films (pre-reduced and *in situ*), and from the intense D band Raman spectra, it was observed that *in situ* reduction of graphene oxide resulted in better exfoliation of graphene in the polymer matrix.

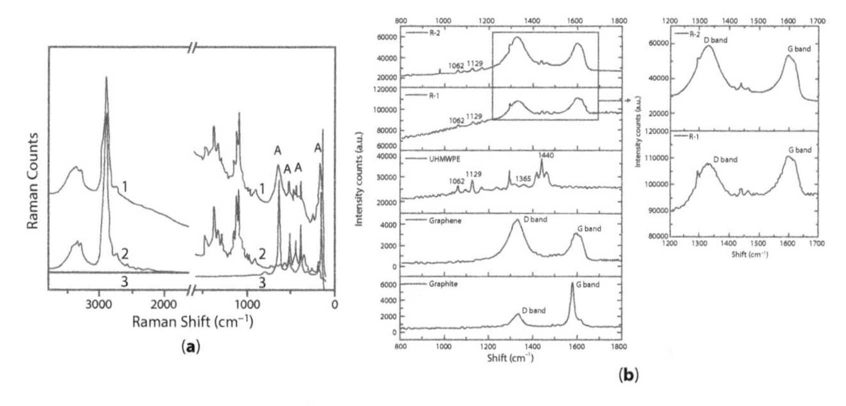

Figure 10.2 Raman spectra for (a) (1) Au/TiO_2-covered cotton; (2) pure cotton fiber and (3) anatase TiO_2. Label A indicates anatase phase TiO_2 on cotton fiber. (Reprinted from reference [12], with permission of Elsevier.); (b) graphite, synthesized graphene, UHMWPE film and graphene/UHMWPE nanocomposites films. (Reprinted from reference [13], with permission of Express Polymer Letters.)

10.2.2 Fourier Transform Infrared Spectroscopy

Fourier transform infrared spectroscopy (FTIR) is a technique that is used to obtain an infrared spectrum of the absorption or emission of a solid, liquid or gas. An FTIR spectrometer simultaneously collects high-spectral-resolution data over a wide spectral range. The FTIR measures infrared spectrum by the Fourier transform of an interferogram [14].

10.2.2.1 Principle

Figure 10.3 describes the scheme of the basic working principle. The light source is usually a silicon carbide, heated above 1200 K, which emits 5000–400 cm^{-1} mid IR radiation. Other sources like tungsten-halogen or mercury lamps were used for near-IR or far-IR regions, respectively. Light passes through the aperture, and the light beam is turned into a parallel beam by the collimator mirror and enters the beam splitter. A germanium film, deposited on a potassium bromide substrate via evaporation, comprises the beam splitter; it splits the single beam into two, reflecting one to the fixed mirror and transmitting the other to the moving mirror. Both mirrors reflect their beams back to the beam splitter; part of each returning beam is reflected and transmitted. The transmitted light from the fixed mirror and the reflected light from the moving mirror recombine and interfere with each other as they travel towards the collecting mirror. The interference is either constructive or destructive. This interferogram is converted to a

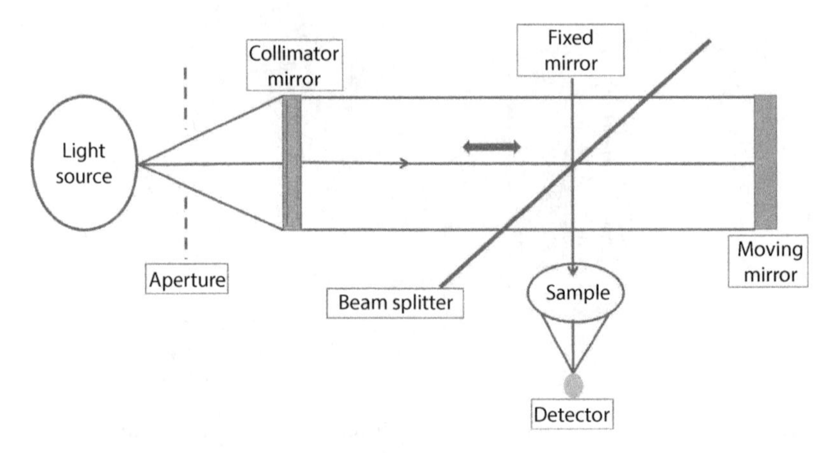

Figure 10.3 Scheme of FTIR working principle. Note: The image has been prepared by the authors.

spectrum by the Fourier transformation technique. The spectrum obtained presents molecular vibrations in the range of 200–4000 cm^{-1}, which gives details of functional groups present in the specimens [14].

The attenuated total reflection (ATR) crystal attachment to the FTIR helps direct measurement of samples without need of tedious sample preparation. The ATR crystal has a higher refractive index compared to the sample under test. The sample is mounted in such a way that one of its surfaces is attached to the crystal surface. When a beam of IR passes through the ATR crystal with an incidence angle larger than the critical angle, the total internal reflection occurs and an evanescent wave forms at the reflection surface. In this process, this internal surface of the crystal is in contact with the sample, and the evanescent wave extends into the sample and penetrates up to 2 μm from the crystal surface depending on the nature of the sample. The beam is then collected by a detector after it exits from the crystal. A very small quantity of liquid sample is required in this study, only to cover the surface of the crystal. Solid sample should be in direct contact with the crystal, so it is clamped against the crystal ensuring no trapped air in between the surceases of the sample and crystal [14].

10.2.2.2 Applications

FTIR (Fourier transform infrared spectroscopy) is a technique particularly used for identifying organic chemicals in a whole range of applications, although it can also characterize some inorganics. Examples can be given as paints, adhesives, resins, polymers, coatings and drugs.

Pillai *et al.* [15] studied the functional groups present in raw egg shell membrane (RESM) and autoclaved egg shell membrane (AESM). The spectrum absorbance was measured in the range of 600–4000 cm^{-1}. Different functional groups corresponding to the samples at different wavelengths are visible in Figure 10.4.

Presence of biomolecules in polycaprolactone nanofibers has been identified by Gopinathan *et al.* [16]. In this work, successful incorporation of two biomolecules (biotin and galactose) was identified by observing their characteristic peaks at 3400–3250 and 3600–3200 cm^{-1}, respectively. In some cases, functional molecules have been introduced after the textile structure development. Mostly, this relies on surface adsorption of molecules to impart the desired functionality. Though X-ray photoelectron spectroscopy is widely used to understand the surface groups, ATR-FTIR can also give significant information in this aspect. As nanocoatings and nanocomposite coatings

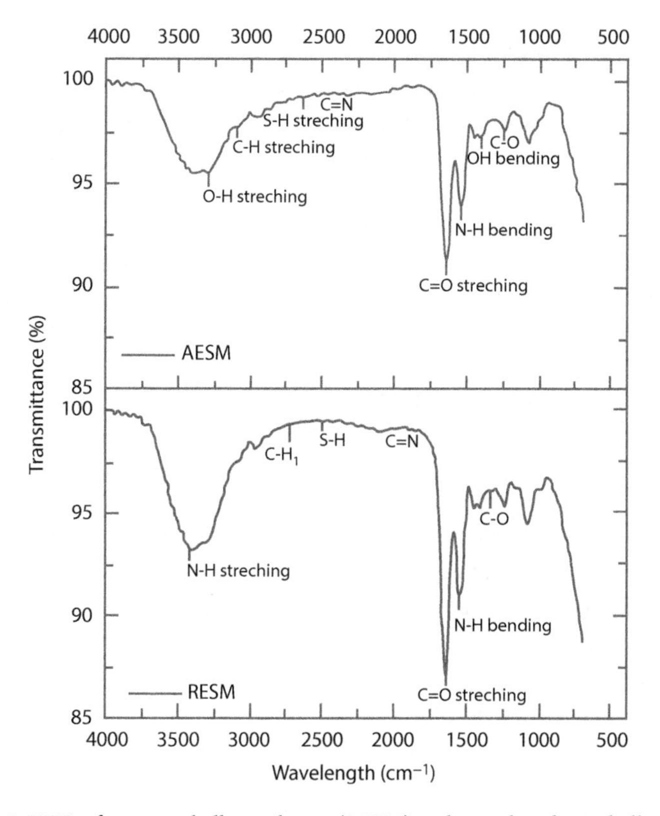

Figure 10.4 FTIR of raw egg shell membrane (RESM) and autoclaved egg shell membrane (AESM). (Reprinted from reference [15], with permission of The Royal Society of Chemistry.)

usually possess submicron thicknesses, the functional groups present in surface coatings of a fiber, film or fabric can be analyzed using this technique.

10.2.3 Ultraviolet UV–Vis Spectroscopy

Ultraviolet (UV)–visible spectroscopy refers to absorption in the UV spectral region. In this region, the electromagnetic spectrum, atoms and molecules undergo electronic transitions. UV–Vis spectroscopy is the most commonly used characterization technique for evaluating nanoparticle formation after the synthesis process. Most of the nanoparticulate structures show characteristic peaks in the UV–Vis region [17].

10.2.3.1 Principle

UV–Vis spectrophotometer measures the transmittance or absorbance of sample over the spectral region. Transmittance is the intensity ratio of a particular wavelength of a light after and before passing through the sample. It can also be used to measure reflectance. A beam of light passes through a sample or reflected from the sample surface. Samples are mostly liquids, either in solution or dispersion forms, though the gas and solid samples can also be measured for their absorbency. The sample container or cuvette used must allow radiation to pass over the spectral region of interest. Generally high-quality fused silica or quartz glass is used for this purpose. The work system determines that the wavelength of the sample absorbed a large amount of the ultraviolet or visible light by scanning for the largest gap between two beams (Figure 10.5). The measured absorption can be a single wavelength or extended spectral range. The spectrophotometer consists of a light source, sample and reference beams, diffraction grating in a monochromator to separate different wavelengths and a detector. Both single-beam and double-beam systems are in use. In a single-beam system, all of the light passes through the sample cell, and the intensity ratio after and before sample placing was plotted as a function of wavelength. In a double-beam system, a beam splitter splits the light and one of the resultant beams is used as the reference while the other passes through the sample. The UV spectrum for a compound is obtained by exposing a sample to an ultraviolet light from a xenon lamp light source. The UV radiation interacts with the matter that causes electronic transitions (promotion of electrons from the ground state to a high energy state). The ultraviolet region falls in the range between 190 and 380 nm; the visible region falls in between 380 and 750 nm. The detector detects a single-wavelength light at one time. The monochromator step through each wavelength and the intensity is detected

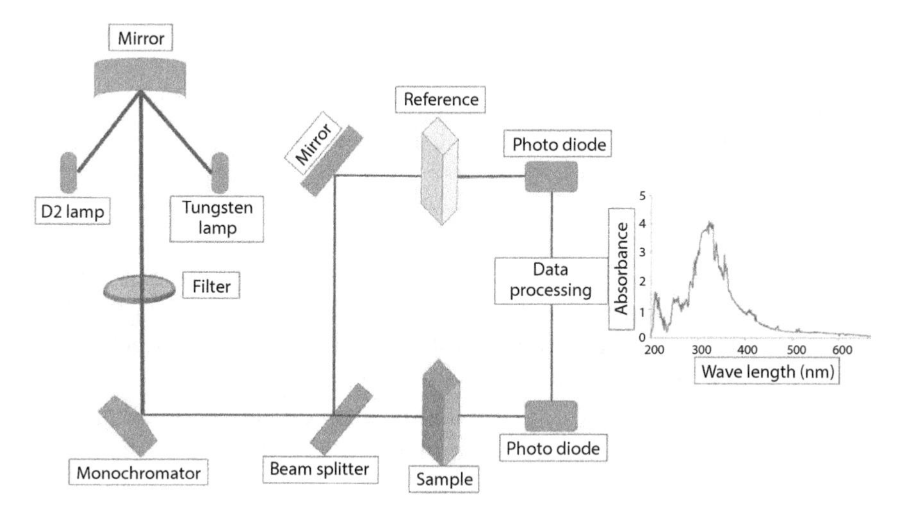

Figure 10.5 Scheme for UV spectroscopy working principle. Note: The image has been prepared by the authors.

as a function of the wavelength. A number of detectors can be used to collect different wavelength data simultaneously [17].

UV–Vis spectrometer uses a fixed path length to determine the concentration. The absorbency is directly proportional to the concentration of the absorbing material in the solution (Beer–Lambert law) except at very high concentration levels or inhomogeneous solutions. Hence, quantitative analysis can easily be performed using this technique. For more accuracy, often calibration curves are plotted for such analyses. Thus, UV–Vis is also used as a detector for HPLC. The Beer–Lambert law is applicable for most of the compounds but is not a universal relationship, e.g. it fails for large complex organic molecules [17].

10.2.3.2 Applications

UV–Vis spectroscopy is in extensive use in analytical chemistry for the quantitative analysis of transition metal ions, conjugated organic compounds, etc. Solutions of transition metal ions absorb visible light due to d-orbital electrons. As the color is strongly affected by other anions or ligands, the change in UV–Vis peak confirms their presence. Organic compounds with high degree of conjugation also absorb UV or visible light. All organic solvents are not suitable for use in UV spectroscopy as they have significant UV absorption. The thickness and optical properties of thin films are also measured using UV–Vis spectroscopy in the semiconductor industry [17].

UV–Vis spectroscopy is a valuable tool to study the synthesized nanoparticles. Optical properties of nanoparticles are strongly dependent on their size, shape, concentration, agglomeration state and dispersion media. Gold nanoparticles (AuNP) have distinct physical and optical properties depending upon their surface properties, size and shape [18]. AuNP possess unique optical features and localized surface plasmon resonance (LSPR). LSPR is the oscillation of electrons on the conduction of the AuNP with respect to specific incident light. LSPR is measured using UV spectroscopy and is in the range of 500–600 nm. Similarly, silver nanoparticles also exhibit surface plasma resonance at the visible region of the spectrum, which is used to identify the presence of silver nanoparticles by using UV [19]. Pillai *et al.* [20] have synthesized lignin capped copper-based fluorescent nanocolorants for live cell staining and fabric dyeing. The UV absorption spectra confirm that the synthesized nanoparticles are not copper or copper oxide nanoparticles, though a small amount of copper is present in that hybrid nanoparticle structure.

10.3 Nanomaterial Characterization Using Microscopy

In the field of nanotechnology, microscopy plays an important role in characterization of structure and properties of materials. Optical microscopes are generally used for observing micron-level structures with reasonable resolution. Further magnification cannot be achieved through optical microscopes due to aberrations and limits in the wavelength of light. Advanced imaging techniques such as scanning electron microscopy, transmission electron microscopy, atomic force microscopy, scanning tunneling microscopy, etc. make it possible to study materials with sizes of submicron, nano or even atomic level. Though the principles of all the techniques are different, all of them produce highly magnified images of the surface or the bulk of a sample using a raster scanning technique. In this technique, an electron beam or probe sweeps across the sample area in a rectangular pattern, the response is captured in each point and the image is reconstructed on the display screen [21]. Basic principles and applications of these imaging techniques used in nanotechnology research are described below:

10.3.1 Scanning Electron Microscopy

A scanning electron microscope (SEM) focuses an electron beam over a surface to create an image. The energy beam of electrons interacts with the sample, producing various signals that can be used to obtain information about the surface topography and composition [21].

10.3.1.1 Principle

Scanning electron microscopes operate at high vacuums. The basic principle can be described as follows. A beam of electrons is generated by a suitable source, typically a tungsten filament or a field emission gun (Figure 10.6). The high energy electron beam is accelerated through a high voltage field and passes through a system of apertures and electromagnetic lenses to produce a thin beam of electrons. Then the beam scans the surface of the specimen. The primary electrons strike the sample and it produces secondary electrons. These secondary electrons are collected by a positively charged electron detector, which, in turn, gives a three-dimensional image of the sample. The two most common electron images are the secondary electron image and the backscattered electron image. The secondary electron image is mainly used to image fracture surfaces and gives high-resolution images. SEM can produce high-resolution images for dimensions less than 1–5 nm. The backscattered electrons are the scattered electrons of the beam due to the elastic interaction with the sample atom, and used for composition analyses. The focused high-energy electron beam interaction with the specimen surface also produces characteristic X-rays, which are used for EDX analysis [21].

10.3.1.2 Sample Preparation

All samples are prepared in proper dimensions to be mounted in a specimen holder known as the "stub." For SEM imaging, samples must be conducting, because non-conducting materials tend to charge while electron

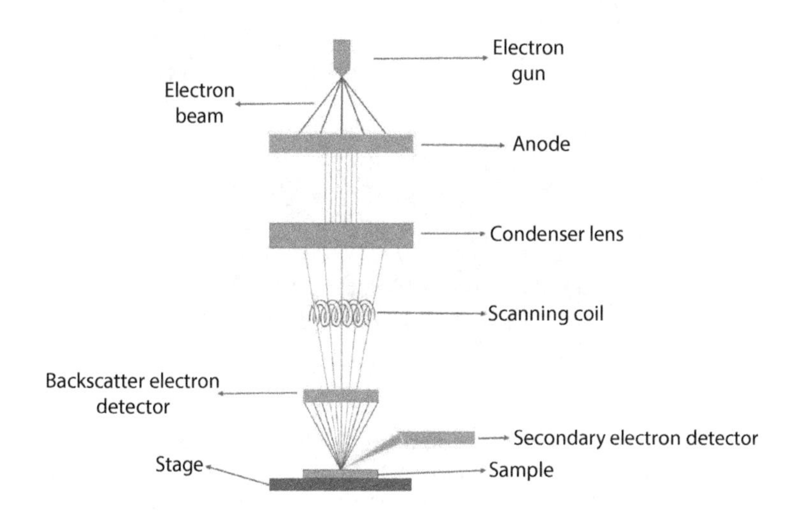

Figure 10.6 SEM setup. Note: The image has been prepared by the authors.

beams scan the sample. Hence, metal samples require easier sample preparation methods as they are already conductive. For these samples, appropriate size and cleaning are only required. In case of non-conducting samples, thin coating of conducting metals is given via vacuum sputtering or by vacuum evaporation. The metals commonly used for sputtering are gold, platinum, tungsten, chromium, etc. Sputtering or coating of conducting materials to surface of samples prevents less artifacts by reducing accumulation of electric charge on the sample. For backscattered imaging, the samples can be embedded in a resin followed by mirror finish polishing. For fracture studies, the surface of the samples must be in suitable size and the surface must be cleaned to remove organic residues [21].

10.3.1.3 Applications

SEM is mainly used to study the surface structures of nanomaterials such as nanofibers, nanoparticles, nanocomposites, etc. The EDS attachment of SEM also gives information on the chemical composition of the materials [21].

The surface characteristics of polycaprolactone (PCL) nanofibers are visible in Figure 10.7a. From this, the nanofiber diameter range and the arrangement

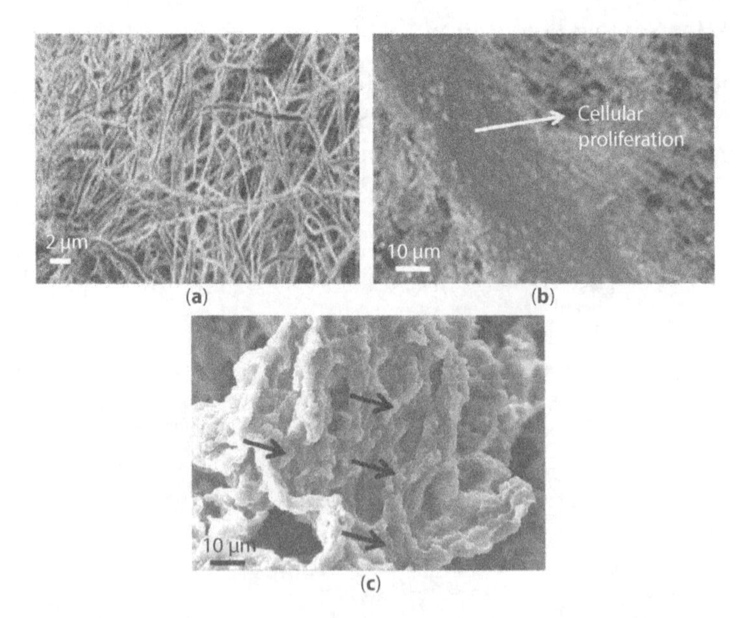

Figure 10.7 SEM image of (a) polycaprolactone nanofibers, (b) cellular proliferation on functionalized polycaprolactone nanofibers ((a) and (b) reprinted from reference [16], with permission of The Royal Society of Chemistry.) and (c) cells attached onto the 3D porous silk–polyvinyl alcohol scaffold. Note: Image (c) has been prepared by the authors.

of nanofibers can be analyzed. SEM can also be used for observing cellular attachment and proliferation on polymeric scaffolds [16]. Figure 10.7b shows the cellular proliferation on biomolecule-functionalized nanofibrous scaffolds. Observation of cell proliferation and growth can also be analyzed by SEM [22]. Cells grown on a 3D porous silk-polyvinyl alcohol (PVA) scaffold are shown in Figure 10.7c. The fracture surface of textile structured composites is usually studied using SEM. In case of multi-scale polymer nanocomposites, the fracture surface morphologies at different regions indicate the interaction and interfacial behavior of each constituent [23]. Other than this, SEM is widely used to analyze uniformity of surface coatings for different textile applications.

10.3.2 Energy Dispersive X-Ray Analysis

An analytical technique used for elemental and chemical characterization of samples is energy dispersive X-ray spectroscopy (EDS). This is based on the interaction between high energy electrons and sample particles resulting in electromagnetic X-ray radiation.

10.3.2.1 Principle

Each element has a fingerprint atomic structure that allows emission of X-rays that can be used for detection of the material composition. Initially, to stimulate the emission of X-rays from the specimen, high energy electron beams (10–20 keV) are passed through the sample. This results in transformation of electrons on the surface of the samples from the ground state to the excited state and finally ejecting from its shell. An electron from higher energy orbits comes to fill this electron hole, and this difference between the higher and lower energy shells releases as a characteristic X-ray radiation from the point of the sample. This can be measured by an energy dispersive spectrometer and thus allows detecting elemental composition of the samples. EDS can be used as an attachment for both SEM and TEM. The interaction is within 2 μm and thus it gives some reference of the bulk of the material. Furthermore, other than getting a spectrum at a specific location, the electron beam can be moved across the material to map the elemental composition of the specimen throughout the scanned area. However, such mapping usually takes a long time to be acquired due to the low X-ray intensity [24].

10.3.2.2 Applications

The surface properties of nanofibers, nanocoatings, nanocomposites and fracture surfaces can be analyzed for textile samples. The EDS analysis of a raw egg shell membrane has given the elemental composition, which is mainly made up of sulphur and calcium. In another study, alumina and titanium nanocomposite-functionalized (thermoplastic polyurethane) TPU nanofibers were developed for efficient removal of fluoride from waste water. EDS analysis of TPU nanofibers confirms the presence of alumina and titanium dioxide, which shows effective functionalization [25]. Using EDS analysis software, EDS mapping is another option from which details of dispersion uniformity and quantitative data on the required metal ions can be availed.

10.3.3 Transmission Electron Microscopy (TEM)

Transmission electron microscope is an instrument used to analyze the structure and chemical nature of nanomaterials. Micro-analytical information, diffraction and imaging produced by this instrument can be combined together and will give details of nanomaterials such as particle or grain size, distribution and morphology.

10.3.3.1 Principle

The basic working principle of TEM can be explained as follows: an electron beam that is generated by an electron gun is passed through a sample and is scattered by the sample. These scattered electrons are focused by optic lenses to form an image (Figure 10.8). The mode of imaging can be controlled by an aperture. Depending upon the imaging mode, there are bright-field (scattered electrons are allowed to pass through) and dark-field images (selected scattered beams form an image). The electrons scattered from the sample are collected by a cathode-ray tube to form the image. The magnifications of images are approximately in the range between 10× and 5,000,000× with a resolution of a few nanometers. TEM can also be used for chemical analysis. The crystalline nature of samples is analyzed by selected area diffraction (SAD) and convergent beam electron diffraction (CBED) methods. Using auxiliary detectors, TEM can also provide elemental analysis such as EDS [26].

10.3.3.2 Sample Preparation

Sample preparation has prime importance for getting a high-resolution image in TEM. The sample thickness must be in the range of 40–150 nm. There are different methods for the preparation of samples such as mechanical milling, chemical itching and ion itching. In mechanical milling, surface polishing of samples is performed by a diamond or boron nitride compound to remove any scratches from the sample surface to avoid contrast

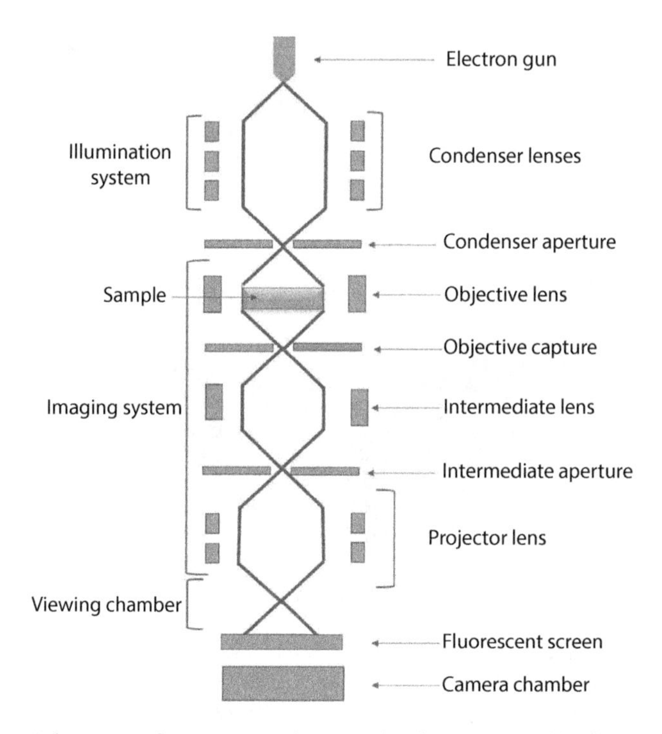

Figure 10.8 Schematics of TEM. Note: The image has been prepared by the authors.

fluctuation caused by varying surface thickness. In case of chemical itching, the sample thickness is reduced by use of an itching agent like an acid. Ion itching uses sputtering process to polish the surface. In this, an electric field of specific kilovolts is passed through the sample in presence of an inert gas like argon, and the formed plasma is directed to the sample surface for fine polishing. In terms of biological and organic samples, they usually need to be fixed using standard protocols, and the samples cannot be tested at higher kilovolts as they degrade under high energy beam. Thus, high-resolution imaging with these sensitive samples is often not possible [26].

10.3.3.3 Applications

TEM is widely used in different fields such as biomedicine and materials science. An important factor for performing TEM is the sample size and preparation. Samples must be thin and must have the ability to withstand the high vacuum and high-energy focused beam inside the chamber. TEM images reveal the size, dispersion and suspension of nanoparticles (Figure 10.9a). They also give information about the orientation and the extent of intercalation and exfoliation of nanomaterials in a nanocomposite.

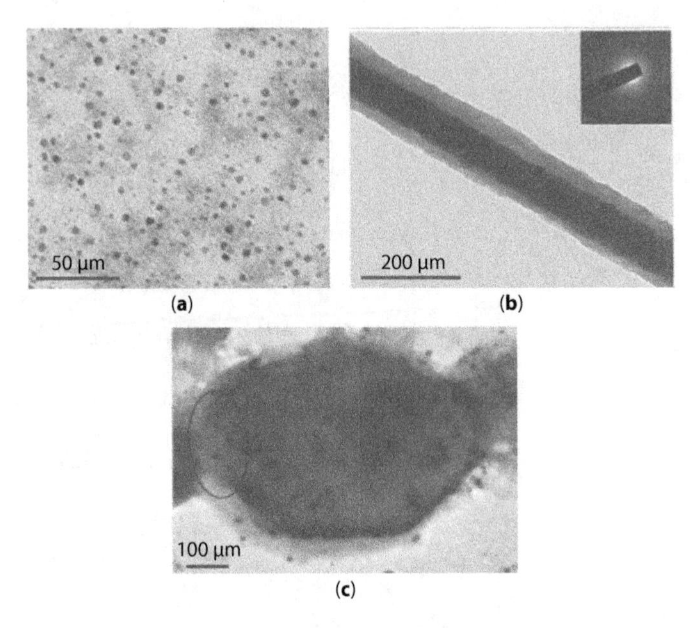

Figure 10.9 TEM image of (a) lignin-based green nanocolorant (Reprinted from reference [20], with permission of Elsevier.), (b) PAN-TPU polymeric core-shell nanofiber (Reprinted from reference [27], with permission of The Royal Society of Chemistry.) [27] and (c) antibacterial efficiency of silver nanoparticles (Reprinted from reference [28], with permission of Taylor & Francis.).

Fathima *et al.* [27] performed co-axial electrospinning for core-shell nanofiber synthesis from hydrophobic thermoplastic polyurethane (TPU) and hydrophilic polyacrylonitrile (PAN). TEM images gave the confirmation of core-shell formation with ~82-nm core and ~156 nm as the average diameter (Figure 10.9b). The silver nanoparticles' interaction with the bacteria cell was studied using TEM. The attack of silver nanoparticles on the cell wall was imaged in this study, which suggests the mechanism of the antibacterial effect of the silver nanoparticles [28] (Figure 10.9c).

10.3.4 Scanning Probe Microscopy (SPM)

SPM is the technique that uses extremely sharp cantilever tips to scan the specimens. Atomic force microscopy is an SPM technique used for surface texture and roughness analysis of nanomaterials. Multi-mode SPM has several modes of operation such as height imaging, lateral force imaging, phase imaging, spreading resistance imaging, magnetic imaging, scanning tunneling microscopy and nanolithography [29].

10.3.4.1 Principle

The atomic force microscope (AFM) measures the force between the sample and the cantilever tip. For height, lateral force or phase imaging, there is no involvement of electric current or magnetic field; hence, samples do not need to be conductive or magnetic in nature. However, for conducting AFM or spreading resistance imaging, the conductivity at a particular point on the sample surface is characterized by giving a voltage bias between the tip and the sample surface to study the I–V characteristics. So, the sample should be conducting or semi-conducting in this case. Further, for scanning tunneling microscopy, samples should be highly conducting and the surface roughness should be in an atomic level, which enables the jump of tunneling electrons for atomic resolution imaging. For magnetic AFM, the tip is coated with a ferromagnetic material to assess the magnetic force field on the surface of the magnetic sample [29].

Three operation modes are present: (1) the contact mode where the sample and the cantilever tip have a small distance and the cantilever repulsive force is set by a piezoelectric positioning element. (2) In the non-contact mode, the force is the weak one like van der Walls. (3) In the tapping mode, the cantilever intermittently oscillates at the surface of the specimens. This oscillation can be generally 50,000–500,000 cycles/s. AFM imaging can achieve a resolution of <1 nm. The AFM can scan both hard and soft materials. A scheme is given to illustrate the principle for the operation of AFM (Figure 10.10). AFM imaging process is the reconstruction of the sample surface in three dimensions. This is advantageous over SEM or TEM, as it

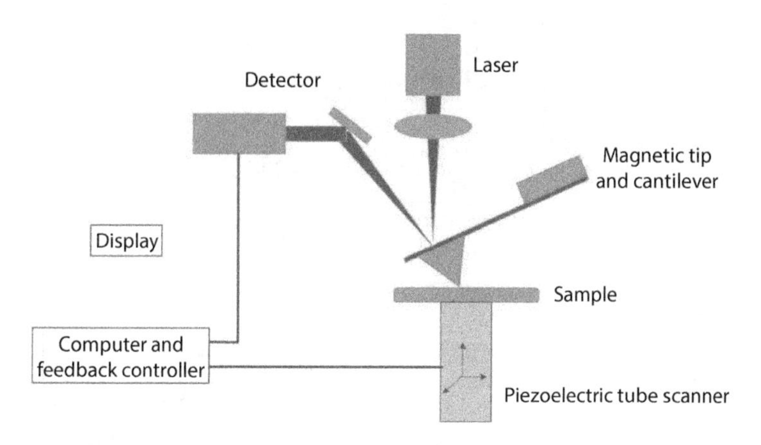

Figure 10.10 Schematics of AFM. Note: The image has been prepared by the authors.

is widely used to analyze the surface roughness and layer thickness of the samples [29].

10.3.4.2 *Applications*

AFM is widely used in textile applications for the analysis of nanomaterial surface properties and the structural characteristics of nanofibers. AFM can be used as a nondestructive characterization technique for textile applications. From the AFM image structure, orientation, size and surface roughness of nanofibers can be analyzed (Figure 10.11a) [30]. Figure 10.11b shows the nanocomposite coated surface morphology under AFM. In height imaging, there is no distinct difference, while in phase imaging, the hard nanoparticles are clearly visible in the soft polymer matrix [31]. In Figure 10.10c, the strong magnetic field lines are observed for magnetic

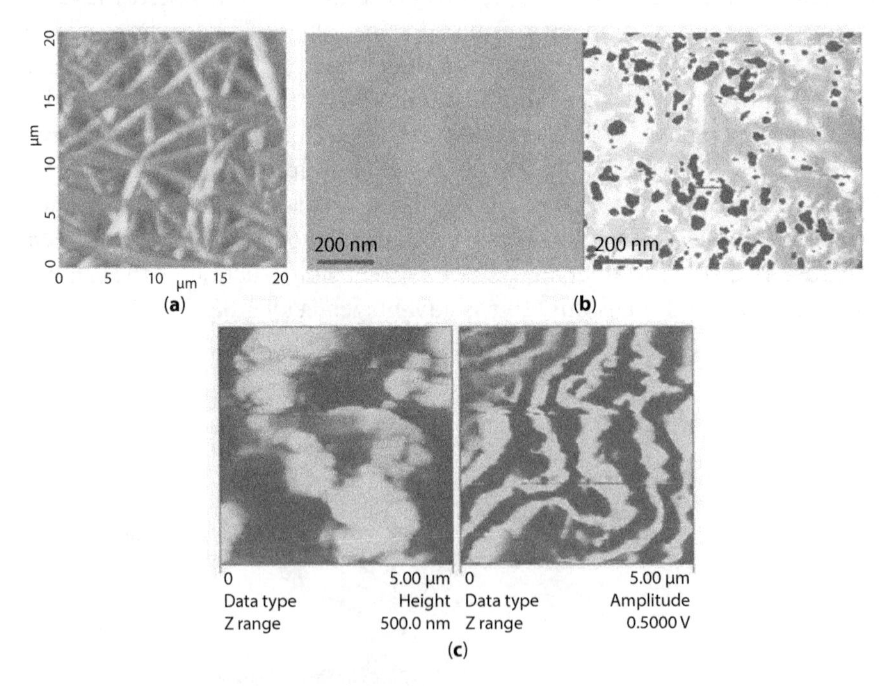

Figure 10.11 AFM image of (a) silk fibroin nanofibers (Reprinted from reference [30], with permission of Springer.), (b) AFM height and phase image of nanocomposite coating of nylon fabric showing the distribution of nanoparticles in the polymer matrix (Reprinted from reference [31], with permission of John Wiley and Sons.) and (c) AFM magnetic imaging of iron-coated nanographite particles height and magnetic force field showing strong magnetic field line at a lift of 1.5 μm (Reprinted from reference [32], with permission of Lap Lambert.).

nanoparticles even at a distance of 1.5 μm (lift) between the tip and the sample surface [32].

10.4 Characterization Using X-Ray

X-rays are electromagnetic radiations with wavelengths between 10 and 0.01 nm. These typical electromagnetic waves present photon energy levels in the range of 100 eV to 100 keV. X-rays can deeply penetrate into solid surfaces, and the wavelength is comparable to the dimensions of atoms or molecules; hence, it is used for analyzing the structure of materials [33].

10.4.1 X-Ray Diffraction

When X-rays interact with a material (crystal phase), a diffraction pattern is produced, which is considered as the fingerprint of that material. This method can be used for analyzing crystal structure, orientation and atomic spacing.

10.4.1.1 Principle

When an X-ray beam hits a crystal lattice, the electrons will start to oscillate with the same frequency of the incident rays, and scattering occurs. This scattering can be destructive or constructive. If scattering of waves interferes itself and eliminates due to self-interaction, it is known as destructive scattering. This occurs when the scattered waves meet with waves from other atomic planes in the opposite phase. Similarly when the diffracted rays of the same direction are in phase, they combine to form enhanced waves, which are known as constructive scattering (Figure 10.12). Under this condition, the reflections combine to form new enhanced wave fronts that mutually reinforce each other (constructive interference). This diffraction patterns are in accordance with Bragg's law. As each crystalline material has a characteristic atomic structure, it will diffract X-rays in a unique characteristic

$$\text{Bragg's law: } n\,\lambda = 2d\,\sin\theta,$$

where λ = wavelength of X-ray beam, 2θ = angle of diffraction, d = distance between each set of atomic planes of crystal lattice and n = order of diffraction.

The wavelength of an X-ray is limited by the anode material. Copper is mostly used and its L shell Kα radiation with 1.54 Å is utilized. In such systems, electrons are emitted from a cathode and accelerated (~50 kV) with copper plate that resulted in emission of bremsstrahlung and some strong

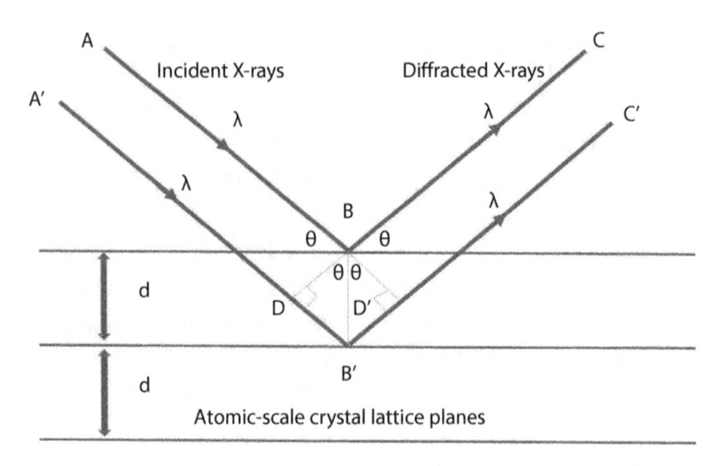

Figure 10.12 Geometry of X-ray diffraction. Note: The image has been prepared by the authors.

spectral lines corresponding to the excitation of inner-shell electrons of the metal (Kα and Kβ lines). The Kβ line is often suppressed with a thin (~10 μm) nickel filter [33].

10.4.1.2 Applications

X-ray diffraction is used to analyze the phase of a material (crystalline/amorphous), crystallite structure, crystallite orientation and size of crystallites. Further, with use of suitable accessories, this technique can be utilized to analyze the modulus of the crystalline phase and to measure the lattice strain, and the structure of polymers can be evaluated by molecular modeling. For polymer nanocomposites, the nanoparticle dispersion can be assessed using X-ray diffraction technique. The layered nanoparticles like nanoclay or graphite have their characteristic peaks indicating the distance between the layers. When intercalated, the distance between the layers is increased resulting in shift of the peak towards lower diffraction angles. With exfoliation in the polymer matrix, the peak becomes broader and eventually disappears indicating extensive layer separation [34].

10.4.2 X-Ray Photoelectron Spectroscopy (XPS)

XPS is used to identify the empirical formula and electronic state of surface elements of the material. XPS is also known as electronic spectroscopy of chemical analysis (ESCA). This technique uses an X-ray source of 1.5 kV to ionize surface electrons of samples [34].

10.4.2.1 Principle

Using XPS, elements of above three atomic numbers can be detected; that means hydrogen and helium cannot be investigated for this analysis. This quantitative spectroscopy analyzes the surface chemistry of samples up to 10-nm depth. To analyze beyond this level, argon gas sputtering can be used to remove surface layers of the sample and for in-depth analysis. The process involves X-ray ionization of the sample surface under ultrahigh vacuum leading to the emission of surface electrons from the core orbitals. Mostly the top 10 nm of surface elements emits the electrons. The measurement of number of electrons emitted from the surface of electrons and kinetic energy gives the photoelectron spectra. The number of electrons emitted from the surface of materials gives the elemental composition. Binding energy of electrons can be calculated from the kinetic energy data. From the binding energy, oxidation state of the material surface elements can be analyzed [34].

10.4.2.2 Applications

XPS can be used to find out the empirical formula depending upon the chemical formula that gives the elemental composition of samples. For textile applications, it can be used to measure the uniformity of surface coatings and its elemental composition after nano-level coating or itching processes. It can be used to detect the purity of sample surfaces. If there is little surface contamination, XPS can determine the empirical formula of the sample with the chemical state of one or more of the elements. This technique can detect nanocoating thickness of different materials up to 10 nm. It can also investigate the coating or finishing uniformity on textile surfaces by analyzing the elemental composition. Carbon, oxygen and nitrogen have distinctive peaks at 281.91, 528.91 and 398.91 eV, respectively. In layer-by-layer deposition of poly(styrene sulfonate) (PSS) and poly (allylamine hydrochloride) (PAH) on a woven cotton fabric, the distinctive peaks of N and S have been recorded at 398.91 and 164.91 eV, respectively, to confirm the layer deposition [34].

10.5 Particle Size and Zeta Potential Analysis

Different techniques are available for particle size analysis such as dynamic light scattering method (DLS), sieve analysis, electro resistance counting method, laser diffraction method, sedimentation method, acoustic method, etc. Among all these techniques, DLS is a widely used technique for analyzing the particle size and zeta potential of the particles. Various physiochemical parameters such as the structure, size, shape, biomolecular conformation, aggregation state and magnitude of electrical charge potential of surface atoms (zeta potential) can be analyzed using DLS [34].

10.5.1 Principle

DLS is also known as quasi-elastic light scattering (QELS) or photon correlation spectroscopy (PCS). DLS is used for the measurement of submicron particles in the range of >1 nm. Zeta potential is the degree of repulsion between the charged particles in a suspension. A high zeta potential confers to the stability of molecules or charged particles. Low zeta potential occurs when attraction of particles exceeds repulsion and results in lower stability of particles and flocculation. In a suspension, particles or emulsions have natural Brownian motion. This Brownian motion is induced further by the bombardment of the solvent and starts to move due to its thermal energy. When a laser source is used to illuminate these moving particles, the scattered light intensity will be proportional to the size of the particles. Using Stokes–Einstein relationship, the particle size is calculated with the help of the velocity of Brownian movement [34].

10.5.2 Applications

The size of nanoparticles is very important in textile applications considering their uniform deposition on the fabric surface. For textile processing applications like finishing and coating, dispersion in water medium is always used. Thus, a stable dispersion of the nanoparticle is essential for such applications. Nanoparticles have a tendency to agglomerate, where size, distribution and charge of nanoparticles are important parameters. Particle size analysis and zeta potential are widely performed for the synthesized nanoparticles to examine stability. The nano-colorants developed by Pillai *et al.* [20] were in the range of 11–151 nm, which suggests the size-dependant coloration. In another study by Pillai *et al.* [15], the impact of egg shell membrane samples (raw and autoclaved) on cellular proliferation was analyzed. The solid surface zeta potential was used to analyze the surface charge of both membranes, and it was found that the raw egg shell membrane showed a cationic nature, whereas the autoclaved egg shell membrane showed a neutral surface charge, which was more suitable for cellular proliferation.

10.6 Biological Characterizations

Biological characterization is an important parameter to analyze the biocompatibility of developed nanomaterials. Biocompatibility tests can be performed via *in vitro* and *in vivo* means. *In vitro* biocompatibility tests are performed by using cytotoxicity assays such as MTT or MTS. MTT (3-(4,5-dimethylthiazol-2-yl)-2,5-diphenyltetrazolium bromide) is a tetrazolium salt, which is widely used to analyze cytotoxicity. MTT binds itself to the mitochondrial dehydrogenase enzyme of live cells and reduces it to purple formazan crystals. These crystals are dissolved by acid/DMSO treatment, and its optical density is measured at 570 nm. MTS (3-(4,5-dimethylthiazol-2-yl)-5-(3-carboxymethoxyphenyl)-2-(4-sulfophenyl)-2Htetrazolium) is also a tetrazolium compound that binds to the NAD(P)H-dependent dehydrogenase enzymes in metabolically active cells. The purple formazan crystals formed after the reduction of enzyme are measured colorimetrically at 490 nm [15]. To perform cytotoxicity analysis, desired cells (human skin cells, mesenchymal stem cells, etc.) are cultured *in vitro* under standard conditions. The nanomaterials are treated with cells and monitored for cell death depending upon the nature of the material.

In vivo characterization of nanomaterials can be conducted in suitable animal models (rats, guinea pigs, rabbits, etc.). Nanomaterials can be tested for skin irritancy or allergic testing; or they can be implanted subcutaneously to analyze the immunological response. Suitable markers can be used to analyze the host body response such as macrophage, natural killer cells, T lymphocytes, etc. Using these specific markers with the help of immunohistochemistry or immunofluorescence index, the rate of immunological response can be analyzed. After implantation, blood/urine biochemical parameters can also be performed for any abnormal parameters in the blood, as well as the kidney or liver functions [35].

10.7 Other Characterization Techniques

Surface area and porosity analyzers are used to measure the specific surface area, pore size, pore distribution, total pore volume and isotherm of specimens. The total porosity is defined as the fraction of void volume over the total volume. Gas adsorption analysis is a commonly used method for surface area determination. The Brunauer, Emmett and Teller (BET) method is widely used for surface area and pore size measurements of porous and powder samples. Nitrogen gas is used as the probe molecule for the analysis of surface and pore sizes. The test is performed under liquid nitrogen conditions. The surface area of a specimen is measured from the cross-sectional area of the probe, i.e. nitrogen molecule (16.2 Å^2) and monolayer capacity of the sample. BET is performed at a relative pressure of 0.3–77 K [36].

Wang *et al.* [36] evaluated the pore size and surface area of zirconia hydroxy/oxy clusters-based MOFs (metal organic frameworks) via experimental and simulation studies (Monte Carlo simulation methods). The experimental studies revealed an ultrahigh surface area of 5646 m^2 g^{-1}. In contrast, the BET analysis showed an overestimation of the surface area (6550 m^2 g^{-1}). This was attributed to the significant pore filling contamination that increased the monolayer loading by ~16%.

The thermal characterization methods used are thermogravimetric analysis (TGA), differential thermal analysis (DTA) and differential scanning calorimetry (DSC) analysis. To study the thermal behavior of polymers, nanoparticles and nanocomposites, each specimen is analyzed by any of these methods under standard experimental conditions. TGA refers to the change in the mass of particles as a function of time and temperature. TGA gives the thermal stability of the specimens at different temperatures. Differential thermal analysis (DTA) gives information about sample kinetics and thermodynamics. DTA measures the thermal transition of specimens induced by cooling or heating the material in a controlled manner. DSC refers to the melting temperature of samples with respect to time and temperature [34].

Electrical conductivity of polymers can be measured by using several techniques like surface conductivity measurement, current–voltage study, impedance analysis and electrochemical impedance spectroscopy (EIS). EIS can be measured by applying AC voltage of low amplitude to the electrode. This results in the generation of a current with respect to the applied voltage. Depending upon the measurement of phase angle and amplitude, materials' conducting behavior can be analyzed. The impedance is measured based on Ohm's law. Poly-ε-caprolactone (PCL)-based nanocomposite films prepared by Devamani *et al.* [37] using carbon nanofibers (CNF), liquid exfoliated graphite and nanographite showed moderately high electrical conductivity. The CNF-based nanocomposite films have conductivity levels as high as 20 S/m, which supports the osteoblast, nerve and meniscal tissue regeneration providing better cell-to-cell communication [38–40]. Further, they exhibit frequency-independent impedance behavior confirming the formation of a continuous nanofibrous network in the dielectric polymer matrix.

The permeability of gases through polymeric surfaces is of great importance in different fields such as textiles, pharmaceutics, packaging, etc. When dispersed in polymer matrices, the layered nanoparticles, such as nanoclay, nanographite, and graphene structures, create a tortuous path for the gases, which increases their path length for diffusion. Thus, they improve the barrier property of the polymers significantly hindering the escape of gases, water or solvent vapors. This can be tested using a pressurized chamber filled with the gases of interest [41].

The modification of surface wettability of textile substrates to hydrophobic and hydrophilic natures is desirable for most of the applications. Surface wettability is widely analyzed by contact angle measurement. Contact angle is the angle formed at the interface of solid–liquid phases. Contact angle is measured by a contact angle goniometer based on Young's equation. Less than 90° is considered as hydrophilic, while more than 150° is defined as super-hydrophobic [42].

A cotton fabric treated with layer-by-layer deposition of nanosilica and fluorocarbon emulsions showed an excellent hydrophobic behavior as observed in Figure 10.13a. The AFM image reveals that the nanosilica

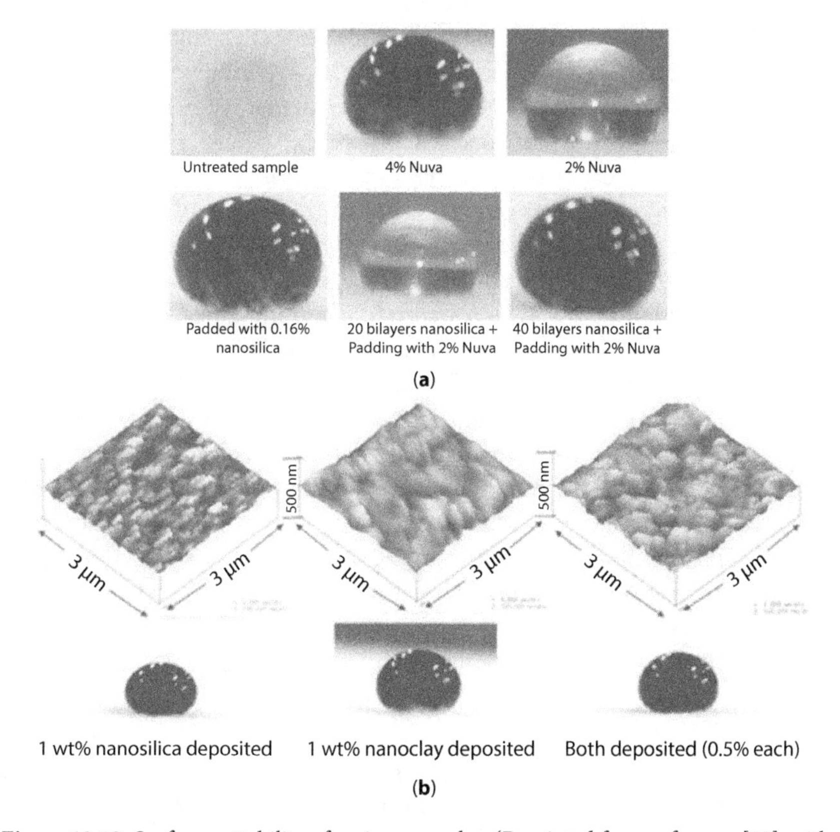

Figure 10.13 Surface wettability of various samples. (Reprinted from reference [39], with permission of Springer Nature.)

creates uniform surface roughness, which helps to achieve this superhydrophobic nature. In Figure 10.13b, nanoparticle-coated cotton fibers were produced to mimic the lotus leaf effect. Nanosilica-deposited cotton fibers showed excellent water-repellent property compared to others due to uniform roughened surface and small dimension [42].

10.8 Conclusions

With the increasing interest on E-textiles or wearable electronics, nanotechnology-based smart textiles are researched not only in high performance, biomedical or defence applications but also to improve the lifestyle of consumers. It became a new trend in the fashion world. Though nanoparticles-incorporated textiles enhance the efficacy for different smart applications, the main challenge in this field is the precise control of size, shape and distribution of nanomaterials anchored with the textile substrate. The advanced characterization techniques such as particle size analysis, electron microscopy, X-ray diffraction, spectroscopy, etc. described in this chapter are essential to study the performance enhancements in nanotechnology-based smart textiles.

The growing interest in using the nanomaterials for different applications brings additional challenges to textile industry. The use of nanomaterials in conventional and advanced textiles is suffering several issues related to nanotoxicity and disposal. Approval and certification for industrial-scale production and marketing are a concern for the companies. In this context, systematic characterization of these materials using sophisticated techniques confirms the material performance and safe use depending upon applications. Before introducing a nanotechnology-based smart textile product into the market, thorough characterization of the product is mandatory to ensure its performance repeatability and reproducibility. Thus, it can be concluded that nanotechnology approach for the development of smart textiles will become efficient, easier and safe with the proper utilization of these characterization techniques in the near future.

References

1. Zhao, Q. Q., Boxman, A., Chowdhry, U., Nanotechnology in the chemical industry opportunities and challenges. *J. Nanopart. Res.*, 5, 567, 2003.
2. Guceri, S., Gogotsi, S., Yury, G., Kuznetsov, V. (Eds.), *Nanoengineered Nanofibrous Materials*, Kluwer Academic Publishers, 245, 2004.
3. Joshi, M., Bhattacharyya, A., Nanotechnology – a new route to high-performance functional textiles. *Textile Progress*, 43, 155, 2011.

4. Mukhopadhyay, S., *Polymeric Nanocomposites and Their Applications in Textiles*, Textile Review, 2007.
5. http://www.physorg.com/news136633890.html (January, 2018).
6. Huang, C. T., Shen, C. L., Tang, C. F., Chang, S. H., A wearable yarn-based piezo-resistive sensor. *Sens. Actuators. A Phys.*, 141, 396, 2008.
7. Flavel, B. S., Yu, J., Shapter, J. G., Quinton, J. S., Patterned attachment of carbon nanotubes to silane modified silicon. *Carbon*, 45, 13, 2551, 2007.
8. Huang, C. T., Shen, C. L., Tang, C. F., Chang, S. H., A wearable yarn-based piezo-resistive sensor. *Sens. Actuators. A Phys.*, 141, 396, 2008.
9. Sensoria socks, http://www.sensoriafitness.com/smartsocks.
10. Mondal, S., Phase change materials for smart textiles–An overview. *Appl. Therm. Eng.*, 28, 1536, 2008.
11. Colthup, N., *Introduction to Infrared and Raman Spectroscopy*, Elsevier, 2012.
12. Uddin, M. J., Cesano, F., Scarano, D., Bonino, F., Agostini, G., Spoto, G., Zecchina, A., Cotton textile fibres coated by Au/TiO_2 films: Synthesis, characterization and self-cleaning properties. *J. Photochem. Photobiol. A Chem.*, 199, 64, 2008.
13. Bhattacharyya, A., Chen, S., Zhu, M., Graphene reinforced ultra high molecular weight polyethylene with improved tensile strength and creep resistance properties. *Express. Polym. Lett.*, 8, 2014.
14. Griffiths, P. R., De Haseth, J. A., *Fourier Transform Infrared Spectrometry* (Vol. 171), John Wiley & Sons, 2007.
15. Pillai, M. M., Akshaya, T. R., Elakkiya, V., Gopinathan, J., Sahanand, K. S., Rai, B. D., Bhattacharyya, A., Selvakumar, R., Egg shell membrane – a potential natural scaffold for human meniscal tissue engineering: An *in vitro* study. *RSC Adv.*, 5, 76019, 2015.
16. Gopinathan, J., Mano, S., Elakkiya, V., Pillai, M. M., Sahanand, K. S., Rai, B. D., Selvakumar, R., Bhattacharyya, A., Biomolecule incorporated poly-ϵ-caprolactone nanofibrous scaffolds for enhanced human meniscal cell attachment and proliferation. *RSC Adv.*, 5, 73552, 2015.
17. Perkampus, H. H., Grinter, H. C., *UV–Vis Spectroscopy and Its Applications*, Springer-Verlag, Berlin, 1992.
18. Kelly, K. L., Coronado, E., Zhao, L. L., Schatz, G. C., The optical properties of metal nanoparticles: The influence of size, shape, and dielectric environment. *J. Phys. Chem. B*, 107, 668–677, 2003.
19. Tao, A., Sinsermsuksakul, P., Yang, P., Tunable plasmonic lattices of silver nanocrystals. *Nat. Nanotechnol.*, 2, 435, 2007.
20. Pillai, M. M., Karpagam, K. R., Begam, R., Selvakumar, R., Bhattacharyya, A., Green synthesis of lignin based fluorescent nanocolorants for live cell imaging. *Materials Letters*, 212, 78, 2018.
21. Reichelt, R., Scanning electron microscopy. *Science of Microscopy*, Springer, New York, NY, 2007.
22. Pillai, M. M., Venugopal, E., Lakshmipriya, H., Gopinathan, J., Selvakumar, R., Bhattacharyya, A., A Novel method to develop three dimensional (3D)

silk-PVA microenvironments for bone tissue engineering–an *in vitro* study. *Biomed. Phys. Eng. Express.*, 4, 027006, 2017.
23. Bhattacharyya, A., Rana, S., Parveen, S., Fangueiro, R., Alagirusamy, R., Joshi, M., Mechanical and thermal transmission properties of carbon nanofiber-dispersed carbon/phenolic multiscale composites. *J. Appl. Polym. Sci.*, 129, 2383, 2013.
24. Bertin, E. P., *Principles and Practice of X-ray Spectrometric Analysis*, Springer Science & Business Media, 2012.
25. Suriyaraj, S. P., Bhattacharyya, A., Selvakumar, R., Hybrid Al_2O_3/bio-TiO_2 nanocomposite impregnated thermoplastic polyurethane (TPU) nanofibrous membrane for fluoride removal from aqueous solutions. *RSC Adv.*, 5, 26905, 2015.
26. Williams, D. B., Carter, C. B., The transmission electron microscope. In *Transmission Electron Microscopy*, Springer, Boston, MA, 1996.
27. Fathima, S., Gopinathan, J., Indumathi, B., Thomas, S., Bhattacharyya, A., Morphology and hydroscopic properties of acrylic/thermoplastic polyurethane core-shell electrospun micro/nano fibrous mats with tunable porosity. *RSC Adv.*, 6, 54286, 2016.
28. Selvakumar, R., Aravindh, S., Ashok, A. M., Balachandran, Y. L., A facile synthesis of silver nanoparticle with SERS and antimicrobial activity using *Bacillus subtilis* exopolysaccharides. *J. Exp. Nanosci.*, 9, 1075, 2014.
29. Bottomley, L. A., Scanning probe microscopy. *Anal. Chem.*, 70, 425–476, 1998.
30. Pillai, M. M., Gopinathan, J., Indumathi, B., Manjoosha, Y. R., Sahanand, K. S., Rai, B. D., Selvakumar, R., Bhattacharyya, A., Silk–PVA hybrid nanofibrous scaffolds for enhanced primary human meniscal cell proliferation. *J. Membr. Biol.*, 249, 2016.
31. Bhattacharyya, A., Joshi, M., Functional properties of microwave-absorbent nanocomposite coatings based on thermoplastic polyurethane-based and hybrid carbon-based nanofillers. *Polym. Adv. Technol.*, 23, 975, 2012.
32. Bhattacharyya, A., *Hybrid Nanographite Based Polymeric Nanocomposites*, Lap Lambert Academic Publishing, Germany, 2012.
33. Wang, G. C., Lu, T. M., X-ray diffraction. In *RHEED Transmission Mode and Pole Figures*, Springer, New York, 2014.
34. Joshi, M., Bhattacharyya, A., Ali, S. W., *Characterization Techniques for Nanotechnology Applications in Textiles*, 2008.
35. Kalambur, V. S., Han, B., Hammer, B. E., Shield, T. W., Bischof, J. C., *In vitro* characterization of movement, heating and visualization of magnetic nanoparticles for biomedical applications. *Nanotechnology*, 16, 1221, 2005.
36. Wang, T. C., Bury, W., Gómez-Gualdrón, D. A., Vermeulen, N. A., Mondloch, J. E., Deria, P., Stoddart, J. F., Ultrahigh surface area zirconium MOFs and insights into the applicability of the BET theory. *J. ACS*, 137, 3585, 2015.
37. Devamani, R. H. P., Deepa, N., Gayathri, J., Morphology and thermal studies of calcium carbonate nanoparticles. *Int. J. Innov. Res. Sci. Eng. Technol.*, 3, 87, 2016.

38. Gopinathan, J., Quigley, A. F., Bhattacharyya, A., Padhye, R., Kapsa, R. M., Nayak, R., Houshyar, S., Preparation, characterisation, and *in vitro* evaluation of electrically conducting poly (ϵ-caprolactone)-based nanocomposite scaffolds using PC12 cells. *J. Biomed. Mater. Res. A*, 104, 853, 2016.
39. Gopinathan, J., Pillai, M. M., Elakkiya, V., Selvakumar, R., Bhattacharyya, A., Carbon nanofillers incorporated electrically conducting poly ϵ-caprolactone nanocomposite films and their biocompatibility studies using MG-63 cell line. *Polym. Bulletin*, 73, 1037, 2016.
40. Gopinathan, J., Pillai, M. M., Sahanand, K. S., Rai, B. D., Selvakumar, R., Bhattacharyya, A., Synergistic effect of electrical conductivity and biomolecules on human meniscal cell attachment, growth and proliferation in poly-ϵ-caprolactone nanocomposite scaffolds. *Biomed. Mat.*, 12, 065001, 2017.
41. Dumont, M. J., Reyna-Valencia, A., Emond, J. P., Bousmina, M., Barrier properties of polypropylene/organoclay nanocomposites. *J. Appl. Polym. Sci.*, 103, 618, 2007.
42. Joshi, M., Bhattacharyya, A., Agarwal, N., Parmar, S., Nanostructured coatings for super hydrophobic textiles. *Bull. Mater. Sci.*, 35, 933, 2012.

Index

Also of Interest

Check out these published and forthcoming related titles from Scrivener Publishing

Smart Textiles
Wearable Nanotechnology
Edited by Nazire D. Yilmaz
Published 2019. ISBN 978-1-119-46022-0

Nanomaterials in the Wet Processing of Textiles
Edited by Shahid-ul-Islam and B.S. Butola
Published 2018. ISBN 978-1-119-45980-4

Advanced Textile Engineering Materials
Edited by Shahid-ul-Islam and B.S. Butola
Published 2018. ISBN 978-1-119-48785-2

Handbook of Renewable Materials for Coloration and Finishing
Edited by Mohd Yusuf
Published 2018. ISBN 978-1-119-40775-1

Textile Finishing: Recent Developments and Future Trends
Edited by K.L. Mittal and Thomas Bahners
Published 2017. ISBN 978-1-119-42676-9

Chemicals and Methods for Conservation and Restoration
Paintings, Textiles, Fossils, Wood, Stones, Metals, and Glass
By Johannes Karl Fink
Published 2017. ISBN 978-1-119-41824-5

Plant-Based Natural Products
Derivatives and Applications
Edited by Shahid-ul-Islam
Published 2017. ISBN 978-1-119-42383-6

CPSIA information can be obtained
at www.ICGtesting.com
Printed in the USA
BVHW071324061218
534780BV00011B/73/P

9 781119 460220